Microbial Applications Vol.1

Vipin Chandra Kalia • Prasun Kumar
Editors

# Microbial Applications Vol.1

## Bioremediation and Bioenergy

Springer

*Editors*
Vipin Chandra Kalia
CSIR-Institute of Genomics & Integrative Biology
Delhi, India

Prasun Kumar
CSIR-Institute of Genomics & Integrative Biology
Delhi, India

ISBN 978-3-319-52665-2 ISBN 978-3-319-52666-9 (eBook)
DOI 10.1007/978-3-319-52666-9

Library of Congress Control Number: 2017934507



Printed on acid-free paper

This Springer imprint is published by Springer Nature
The registered company is Springer International Publishing AG
The registered company address is: Gewerbestrasse 11, 6330 Cham, Switzerland

*Dedicated to my Parents*

# Preface

Plants, humans, and microbes show a strong interaction among themselves. Plants depend upon microbes for their growth and development and for acquiring nutrients. Plants in turn serve as food for human and animals. Microbes in the rhizosphere produce secondary metabolites to protect plants against pathogens and tolerate stress. Human skin and gut harbor a wide range of microbes, which are responsible for their well-being. This interaction of living beings is gaining renewed interest and value. Microbial activities are quite unique and interesting and are finding wider applications ranging from bioremediation, bioenergy, and biomedicine to agriculture and industry. During the past century, there has been a transition from chemical processes to biological methods, largely because the latter are eco-friendly. Now, the emphasis has shifted from only eco-friendly bioprocesses to economical as well. The Green Technologies are the new trend to save the planet Earth through sustainable processes. Scientific progress can be gauged primarily through basic scientific research. As most scientific works are supported through public funds, there is a constant demand to put these findings to applications for human welfare. Students—the rising stars and our scientists in the making—are curious to learn the basic sciences and how these can be translated into products. This book has been brought out to cater to curious young minds and provide economic benefits to the society. In this book, the learned scientific community has put their best efforts to share their expertise, which they have gained through their immense experience targeted toward understanding microbial world. This book is a true reflection of the sincerity of the scientific community, who promptly agreed to contribute their creation for the young minds, who are likely to benefit and take this world a step further into the future. I am truly humbled by the help rendered by the contributing authors. I am indebted to all of them. I am running short of words to adequately acknowledge the worthiness of their efforts. My true inspiration to write this piece of work stems from the faith in me and the constant support of late Mrs. Kanta Kalia and Mr. R.B. Kalia (parents), Amita (wife), Sunita and Sangeeta (sisters), Ravi,

Vinod, and Satyendra (brothers), Daksh and Bhrigu (sons), my teachers especially Dr. A.P. Joshi, and my friends—Rup, Hemant, Yogendra, Rakesh, Atya, Jyoti, Malabika, Neeru, and Ritusree. I must also acknowledge the support of my student friends—Sanjay, Mamtesh, Subhasree, Shikha, Jyotsana, and Ravi.

Delhi, India Vipin Chandra Kalia

# Contents

## Part I Bioremediation


**Phycoremediation: An Eco-friendly Approach to Solve Water Pollution Problems** . . . . . 3
Bala Kiran, Kratika Pathak, Ritunesh Kumar, and Devendra Deshmukh

**Microbial Conversion of Waste and Surplus Materials into High-Value Added Products: The Case of Biosurfactants** . . . . . 29
Constantina Kourmentza, Filomena Freitas, Vitor Alves, and Maria A.M. Reis

**Oil Biodegradation** . . . . . 79
Bhagwan Rekadwad and Chandrahasya Khobragade

**Bacterial Decolourization, Degradation and Detoxification of Azo Dyes: An Eco-friendly Approach** . . . . . 91
Shweta Agrawal, Devayani Tipre, Bhavesh Patel, and Shailesh Dave

**Biological Significance of Degradation of Polyhydroxyalkanoates** . . . . . 125
Subhasree Ray and Vipin Chandra Kalia

**Microbial Biofouling: A Possible Solution to Treat Harmful Microorganisms in Ship Ballast Water** . . . . . 141
Bhagwan Rekadwad and Chandrahasya Khobragade


## Part II Bioenergy


**Xylanases: From Paper to Fuel** . . . . . 153
Gopalakrishnan Menon and Sumitra Datta

**Basic Principles of Microbial Fuel Cell: Technical Challenges and Economic Feasibility** . . . . . 165
Soumya Pandit, Kuppam Chandrasekhar, Ramesh Kakarla, Abudukeremu Kadier, and Velpuri Jeevitha

**Biohydrogen Production: Integrated Approaches to Improve the Process Efficiency** . . . . . 189
Chandrasekhar Kuppam, Soumya Pandit, Abudukeremu Kadier, Chakradhar Dasagrandhi, and Jeevitha Velpuri

**Integrated Biorefinery Approach for the Valorization of Olive Mill Waste Streams Towards Sustainable Biofuels and Bio-Based Products** . . . . . 211
Constantina Kourmentza, Eleni Koutra, Nikolaos Venetsaneas, and Michael Kornaros

**Algae—The Potential Future Fuel: Challenges and Prospects** . . . . . 239
Ramesh Kakarla, Chandrasekhar Kuppam, Soumya Pandit, Abudukeremu Kadier, and Jeevitha Velpuri

**Part III Future Perspectives**

**New Generation DNA Sequencing (NGS): Mining for Genes and the Potential of Extremophiles** . . . . . 255
Bhagwan Rekadwad and Juan M. Gonzalez

**Impact of Next-Generation Sequencing Technology in Plant–Microbe Interaction Study** . . . . . 269
Archana Kumari, Samson Sumer, Bharati Jalan, Pyniarlang Lyngdoh Nongbri, and Mostaque Ahmed Laskar

**An Overview of Next-Generation Sequencing (NGS) Technologies to Study the Molecular Diversity of Genome** . . . . . 295
C. Mohinudeen, Manoharan Melvin Joe, Abitha Benson, and Sa Tongmin

**Marine Polyextremophiles and Their Biotechnological Applications** . . . 319
Bhagwan Rekadwad and Chandrahasya Khobragade

# About the Editors

**Vipin Chandra Kalia** is presently working as Chief Scientist at Microbial Biotechnology and Genomics, CSIR-Institute of Genomics and Integrative Biology, Delhi. He is a Professor of Academy of Scientific and Innovative Research (AcSIR), Delhi. He obtained his M.Sc. and Ph.D. degrees in Genetics from the Indian Agricultural Research Institute, New Delhi. He has been elected as (1) fellow of the Association of Microbiologists of India (FAMI), (2) fellow of the National Academy of Sciences (FNASc), and (3) fellow of the National Academy of agricultural Sciences (FNAAS). His main areas of research are microbial biodiversity, genomics, and evolution, bioenergy, biopolymers, antimicrobials, quorum sensing, and quorum quenching. He has published 101 papers in scientific journals such as (1) Nature Biotechnology, (2) Biotechnology Advances, (3) Trends in Biotechnology, (4) Critical Reviews in Microbiology, (5) Bioresource Technology, (6) International Journal of Hydrogen Energy, (7) PLoS ONE, (8) BMC Genomics, (9) Gene, and (10) Annual Review of Microbiology. His works have been cited 3558 times with a h index of 31 and an i10 index of 66. He has edited 3 books: Quorum sensing versus quorum quenching: A battle with no end in sight (2015, Springer India) and Microbial Factories Vol 1 and 2 (2015). He is presently the editor in chief of the Indian Journal of Microbiology and Editor of (1) PLoS ONE, (2) Journal of Microbiology & Biotechnology (Korea), (3) Applied Biochemistry & Biotechnology (USA), (4) International Scholarly Research Notices (Energy), (5) Dataset Papers in Science (Microbiology), and (6) Journal of Molecular and Genetic Medicine. He is a life member of the following scientific societies: (1) Society of Biological Chemists of India; (2) Society for Plant Biochemistry and Biotechnology, India; (3) Association of Microbiologists of India; (4) Indian Science Congress Association; (5) BioEnergy Society of India; and (6) the Biotech Research Society of India (BRSI). He is also a member of the American Society for Microbiology.

**Prasun Kumar** is a Ph.D. in Biotechnology from Savitribai Phule Pune University, Maharashtra, INDIA. He has worked as a Research Fellow at Microbial Biotechnology and Genomics, CSIR-Institute of Genomics and Integrative Biology, Delhi. His main areas of research are Microbial biodiversity, Bioenergy, and Biopolymers. He has published 19 papers in Scientific journals such as (1) Biotechnology Advances, (2) Bioresource Technology, (3) International Journal of Hydrogen Energy, and (4) Microbial Ecology. He has authored four book chapters. His works have been cited 490 times with an h index of 14 and an i10 index of 17. He is a life member of the following Scientific societies: (1) Association of Microbiologists of India and (2) Indian Science Congress Association. He is serving as Editor to (1) International Journal of Clinical & Medical Allergy (IJCMA) [USA] and (2) Insight Medical Publishing Journals [UK].

# Part I
# Bioremediation

# Phycoremediation: An Eco-friendly Approach to Solve Water Pollution Problems

**Bala Kiran, Kratika Pathak, Ritunesh Kumar, and Devendra Deshmukh**

**Abstract** Increasing anthropogenic and industrial activities along with improper wastewater management strategies are the major causes of elevated levels of pollution in water bodies, a serious environmental challenge globally. The release of water loaded with various chemicals, mainly phosphates and nitrates, causes eutrophication in water bodies. In the present scenario, algae seem to be the possible solution with various advantages and eco-friendly nature. Algae are photosynthetic microorganisms inhabiting marine, brackish, and freshwater environments. They can serve multiple roles: remediation of wastewater simultaneously producing biomass for biofuel generation with parallel sequestration of carbon dioxide. Additionally, treatment of wastewater by means of algae is an environmentally safe process without any secondary pollution. In this chapter, a broad variety of techniques available for algal wastewater treatment (phycoremediation) are discussed such as photobioreactors, open raceway ponds, High Rate Algal Ponds (HRAPs), algal mats, hyper-concentrated culture system, and dialysis culture. These options could prove to be efficient methods for wastewater treatment supporting green technology.

**Keywords** Algae • Wastewate • Phycoremediation • Biofuels

B. Kiran (✉)
Department of Mechanical Engineering, Indian Institute of Technology, Indore 452020, MP, INDIA

Centre for Biosciences & Biomedical Engineering, Indian Institute of Technology, Indore 452020, MP, INDIA
e-mail: kiranb@iiti.ac.in

K. Pathak • R. Kumar • D. Deshmukh
Department of Mechanical Engineering, Indian Institute of Technology, Indore 452020, MP, INDIA



V.C. Kalia, P. Kumar (eds.), *Microbial Applications Vol.1*,
DOI 10.1007/978-3-319-52666-9_1

# 1 Introduction

Water distribution shows that major portion in Earth's atmosphere and crust is saline seawater accounting for 97% of total water resources, primarily contributed by oceans, marginal sea, saline groundwater, and saline closed lakes, while freshwater accounts for only 2.5% of the total. With rise in human population, there is an increase in undesirable anthropogenic activities affecting our freshwater resources. Additionally, the receiving water bodies also do not have adequate water quantity for dilution and self-cleansing mechanism. Water quality data collected by Central Pollution Control Board, India, from rivers, lakes, ponds, tanks, and different groundwater locations suggest that organic and bacterial contamination levels have continued to be alarming since 2005 in India. Chemical and physical wastewater treatment methods, presently in use by municipal corporations, have proved to be incapable of treating sewage flowing into water bodies resulting in increased oxygen demand and microbial flora. All these adverse impacts finally led to increased risks of water-borne diseases. Wastewater treatment using algae seems to be a possible solution of existing wastewater issues. Biodiversity of algae is colossal and expected to be about 2,00,000–8,00,000 species, out of which only about 50,000 are described (Renuka et al. 2015). Variety and propensity of algae to acclimatize severe and inhospitable habitats has facilitated the scientific community in screening and identification of promising strains/species/genera for the development of hopeful algal wastewater remediation options (Fouilland 2012). Algae are able to nurture in nutrient-loaded water environment and can efficiently remove N and P, dropping the content (Kiran et al. 2014a). Algae have also been exploited for their elevated lipid yield, speedy biomass production, and capability to sequester waste carbon dioxide. Furthermore, algae do not compete with existing food commodities and do not utilize water resource which limits the financial and environmental profits of crop feedstocks for biofuel production. Thus, utilization of algae for wastewater treatment seems to be a desirable biological method and produced biomass could be utilized for biofuel production.

# 2 Wastewater and Its Types

General composition of wastewater is 99.93% water and 0.07% residual waste. This residual waste majorly comprises solids with organic and inert in nature. Wastewater coming out of kitchen, bathroom, washing clothes, etc., is known as grey water and from toilets is called as black water. Different factors responsible for varying characteristics of domestic wastewater are daily per capita utilization of water, water quality, and state and scope of sewerage system, and most important is habits/living standard of people. Characteristics of municipal sewage, containing both industrial and domestic wastewater, generally vary from place to place contingent to the type of industry and establishment. Depending upon their origin, wastewater can be classified as follows:

**Sanitary sewage** includes the exhausted water from households and institutions, loaded with ablution water, laundry wastes, kitchen wastes, body wastes, and additional waste products, and is classified as domestic sewage (combination of both black and grey water).

**Commercial sewage** is liquid carrying wastes from stores, small-scale industries, and factories and is usually included in the category of domestic sewage if having almost similar characteristics.

**Industrial wastewater,** composition of industrial wastewater depends on the type of pollutants discharged from specific industry. These mainly comprise COD, BOD, heavy metals, chlorinated organic compounds, mineral oil, phenols, sulfates, mineral acids, salts, etc. Broad classification of industrial effluents includes (a) inorganic industrial wastewater, i.e., wastewater from steel or iron industry majorly constituting settable metal particles which could be easily removed through sedimentation/gravimetric method, wastewater recovered from Blast Furnace usually comprising dissolved $CO_2$, cyanides, and compounds of alkali earth metal, and wastewater from rolling mills composed of mineral oil as a major portion, and (b) organic industrial wastewater, i.e., wastewater from organic industries comprising high carbon compounds, metallic ions, charcoal, organic compounds such as steroids, fermentation waste, etc. Some of the common industries responsible for such waste products are pharmaceutical, fertilizer industry, food and beverages industries, leather and textile industries, etc.

**Surface runoff sewage** is also recognized as storm or overland flow. It includes part of rainfall which runs quickly above the ground surface mixing with a defined channel. This water contains absorbed gases and particulates from atmosphere but majorly constitutes dissolved and leached material from vegetation and fertilized soil (now barren due to anthropogenic activities) and washed debris from urban areas.

# 3 Algae

## *3.1 Algal Composition*

Algae are aquatic, photosynthetic organisms, generally categorized as microalgae or macroalgae. Carbon, nitrogen, and phosphorus along with micronutrients such as cobalt, iron, and zinc are the main components in algae, playing a major role in cellular metabolism. These are mainly composed of lipids (7–23%), proteins (6–52%), and carbohydrates (5–23%), but proportion is mainly species specific (Brown et al. 1997). Their proportion can further be manipulated through varying nutrient composition and other environmental factors such as temperature, pH, light intensity, etc. For example, manipulations have been tried in these eukaryotes,

*Chlamydomonas*, *Chlorella*, *Haematococcus*, and *Dunaliella*, for medical and cosmetic products.

## 3.2 Algal Diversity in Wastewater

Wastewater is usually loaded with contaminants such as heavy metals, nutrients, pesticides, etc. Occurrence of nitrogen (nitrate nitrogen, nitrite nitrogen, and dissolved ammonia) and phosphorus in wastewater leads to eutrophication (Yang et al. 2008; Liu et al. 2010). Algae flourishing in such environment correspond to primary part of microbial diversity, which also helps in self-purification (Sen et al. 2013). Many algal species have been observed and studied in wastewater environment as described in detail:

A case study held at El-Umum drain, west of Alexandria, reported 152 phytoplankton taxa from sewage water including *Dinophyceae* (9), *Euglenophyceae* (17), *Cyanophyceae* (20), *Chlorophyceae* (46), and *Bacillariophyceae* (60), with the dominance of *Bacillariophyta* (39.4%) species (Hussein and Gharib 2012), whereas 58% Cyanophycean species were majorly reported in open channels contaminated with sewage along with Chlorophyta (25%) and Bacillariophyta (17%) (Renuka et al. 2013). Batch reactor studies on municipal wastewater treatment containing dairy waste reported the dominance of microalgae belonging to Cyanophyta, Chlorophyta, and Euglenophyta groups (Bernal et al. 2008). Studies on wastewater treatment plant at Shimoga, Karnataka, reported 71 species from *Cyanophyceae*, *Desmidiaceae*, *Chlorophyceae*, *Euglenophyceae*, and *Bacillariophyceae* where *Chlorella* and *Scenedesmus* (Chlorophyta) were prevailing all over the year.

Warm, stable, and nutrient-enriched water present in wastewater treatment plant positively affects the growth kinetics of Blue Green Algae (BGA) (Vasconcelos and Pereira 2001; Badr et al. 2010). Therefore, not only the microalgal species but cyanobacteria also constitute vital portion of phytoplankton diversity in wastewater treatment plant. These species are predominant in various treatment sections of wastewater treatment plant such as batch reactors and stabilizing ponds (categorized as facultative and maturation ponds). Report from a wastewater stabilization pond suggests cyanobacterial species as dominant form including *Leptolyngbya*, *Synechococcus*, *Limnothrix*, *Merismopedia*, and *Nostoc*, representing >90% of total variety in waste stabilization pond. Studies by Vasconcelos and Pereira (2001) suggested that species in stabilizing ponds of Portugal comprised of *Microcystis aeruginosa*, *Planktothrix mougeotii*, *Phormidium autumnale*, *Pseudoanabaena mucicola*, *Synechocystis* sp., and *Limnothrix* sp. with 15.2–99.8% cyanobacterial community. Similarly, reports by Badr et al. (2010) confirmed the dominance of cyanobacterial species in both ponds of WWTP (facultative and maturation) at El-Sadat city, Egypt, constituting 2–97.8% of phytoplankton density.

It has been recognized that algae symbolize an important biotic constituent of aquatic ecosystems, representing a broad variety of taxonomic diversity which is further dependent on the concentration of various contaminants present in wastewater. Apart from WWTP, information is also accessible on allocation pattern and variety of algal species existing in industrial effluents (Vijayakumar et al. 2007). Chinnasamy et al. (2010) evaluated treated and untreated carpet mill effluent for algal diversity and detected almost equal diversity of Cyanophycean and Chlorophycean species during spring season with dominance of *Lyngbya* sp. (36.7%) and *Chlorococcaceae* sp. (12.79%) in treated wastewater and *Uroglena* sp. (46.49%) and *Stigeoclonium* sp. (38.84%) in untreated wastewater. In paper mill and pharmaceutical effluents, a total of 25 cyanobacterial species were reported by Dubey et al. (2011) with common presence of *Oscillatoria princeps*, *O. curviceps*, *Microcystis aeruginosa*, *P. corium*, and *Phormidium ambiguum*. Vijayakumar et al. (2007) detected the presence of various Blue Green Algae as 93% in sugar mill effluent, 91% in dye effluent, and 76 and 50 % in paper mill and pharmaceutical effluents, respectively, with the dominance of genus *Oscillatoria*, followed by *Microcystis, Phormidium, Synechococcus, and Lyngbya.* In pulp and paper wastewater treatment systems of New Zealand, Brazil, the USA, and Canada, cyanobacterial species from Oscillatoriales and Chroococcales were found to be present with dominating species such as *Geitlerinema*, *Phormidium*, and *Chroococcus*. Even with the accessibility to a big number of reports, it is very difficult to simplify factors contributing to general qualitative and quantitative aspects of algal diversity in varying range of environmental habitats, conditions, and wastewaters.

### *3.3 Advantages of Algae for Wastewater Treatment*

Algae have shown a key function in sustainable wastewater treatment and main points include:

(a) The photosynthetic property of algae provides oxygenic condition ($O_2$) required by bacteria for mineralization of organic matter enhancing the nutrient removal efficiency and elimination of pathogens from wastewater. Studies suggest that these organisms have significant function in the treatment of wastewater in maturation ponds or in aerobic or facultative ponds.
(b) Algae-based treatments are well supported by sunlight, reducing energy input in the process of biomass production.
(c) Lesser energy utilization, collectively with natural assimilation of $CO_2$ from atmosphere and nitrate and phosphate from wastewater for growth, removes a major component of greenhouse gas and nutrients from wastewater, lowering down the cost inputs for sewage treatment. Nitrogen uptake could be further increased with preconditioning through starvation. Algae can be proficiently used to eliminate considerable quantity of nutrients due to their large

requirements of N and P for synthesis of proteins, phospholipids, and nucleic acids. Rise in pH level during photosynthesis further supports for nutrient removal through ammonia stripping or ammonia precipitation.

(d) Production of precious algal biomass to be utilized for production of biofuels or biofertilizer (Mulbry et al. 2006).

(e) Hyper-concentrated algal cultures have shown the decreased demand for land and space removing N and P quickly, i.e., less than 1 h. There are evidences for acclimatization of algae in harsh environments, but in actual environmental conditions, growth and species composition is effected due to certain parameters.

(f) Low pathogenicity, i.e., immunosuppressive mechanism by algal species, has been reported to be very low.

## 4 Wastewater Treatment: Phycoremediation

The major requirement for high algal growth is sufficient nutrient supply, i.e., nitrates and phosphates. Secondary effluents from WWTPs contain loads of these nutrients ($NH_4$, $NO_3^-$, and $PO_3^{4-}$) causing eutrophication in natural water bodies. Hence, wastewater must be treated before liberation into water bodies. There exist a number of methods for removal of nutrients (Fig. 1) but very cost intensive producing large quantities of sludge. Algae have been proposed as a substitute biological remediation option for the exclusion of such nutrients from wastewater (Tables 1 and 2). Wastewater treatment through algae can be achieved through open pond system, photobioreactors, algal mats, hyper-concentrated culture, dialysis culture, and HRAPs (High Rate Algal Ponds).

### *4.1 Open Raceway Ponds*

These are shallow ponds with paddlewheels for proper mixing and circulation of wastewater with direct exposure of culture medium to atmosphere resulting in liquid evaporation with simultaneous temperature regulation. These are classically used at commercial scale for algal cultivation, viz. *Dunaliella salina, Arthrospira platensis, Anabaena* sp., *Pleurochrysis carterae, Phaeodactylum tricornutum, Nannochloropsis* sp., *Chlorella* sp., etc. Open raceway pond systems are also suggested as cost-effective algal cultivation method for growing algae in domestic wastewater for nutrient removal with simultaneous average biomass production up to 0.5 g/L. However, the biomass concentration remains low because of poor mixing condition, bacterial dominance, and low PAR values. Cost-effective, easy and minimal cost operation, efficient utilization/treatment of domestic municipal wastewater, and industrial flue gas for algal biomass production are major advantages related to open raceway ponds. To avoid contamination, closed raceway

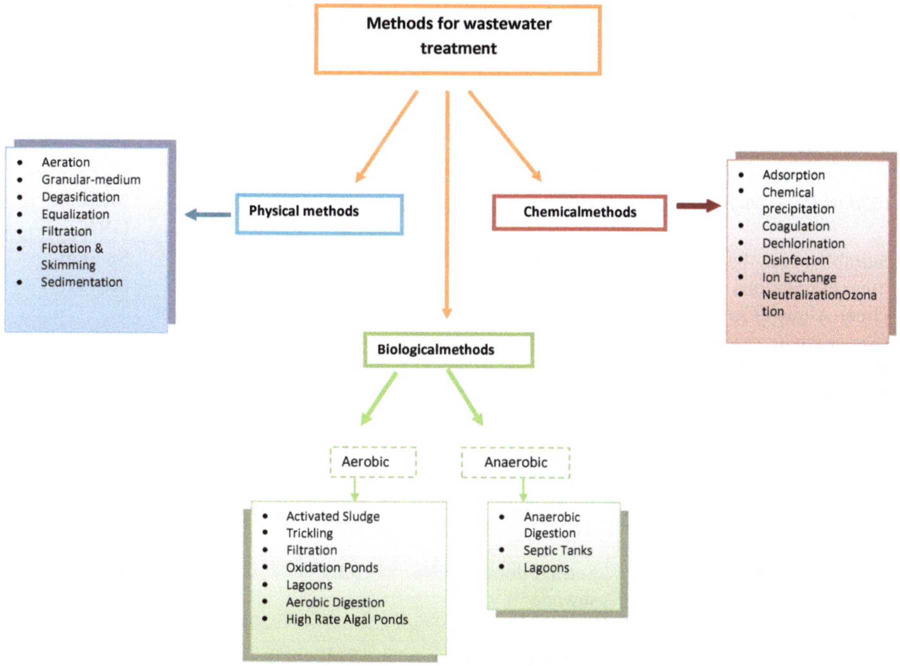


**Fig. 1** Available methods for wastewater treatment

ponds have been proposed which allows maximum sunlight utilization without exposure to contaminants. Further, some of important parameters need to be monitored to determine physiological position of algae in raceway ponds which includes conductivity, cell density, temperature, pH, salinity, light intensity, evaporation rates, dissolved carbon dioxide, TDS (total dissolved solids), DO (dissolved oxygen), and N and P levels.

## *4.2 Photobioreactors*

These have been proved to be highly efficient systems to attain high growth, synchronous monocultures of microalgae for extended periods with better capture of radiant energy and minimal contamination rate with better utilization of cultivation area (Kiran et al. 2014b). Various designs and configurations of photobioreactors have been proposed, classified as tubular (energy consumption up to 2000–3000 W/$m^3$) and flat-plate bioreactors (energy consumption up to 55 W/$m^3$). Davis et al. tested simple reactors in the mid-1990s followed by modern systems in the late 1990s by Pirt et al. Tubular PBRs are the common designs employed for algal cultivation as they provide continuous system with flexibility towards tubes arrangement (vertically or horizontally) for maximal solar capture.

**Table 1** Wastewater treatment using algae in real environment

| Wastewater treatment | Type of wastewater | N removal (%) | P removal (%) | References |
|---|---|---|---|---|
| Symbiotic | | | | |
| Algal–bacterial symbiosis (*Chlorella* + *Nitzchia*) | Settled domestic sewage | 92.0 | 74.0 | Mcgriff and Mckinney (1972) |
| Settable algae–bacterial culture (filamentous Blue–Green Algae + *Flavobacteria, Gammaproteobacteria, Bacteroidia* and *Beta-Proteobacteria*) | Pretreated municipal wastewater | 88.3 | 64.8 | Su et al. (2011) |
| Blue Green Algae | | | | |
| Cyanobacteria | Secondary treated effluent + swine wastewater | 95.0 | 62.0 | Pouliot et al. (1989) |
| Microalgae | | | | |
| *Chlorella pyrenoidosa* | Settled domestic sewage | 93.9 | 80.0 | Tam and Wong (1989) |
| *Chlorella vulgaris* | Diluted pig slurry (suspended solids upto 0.2%) | 54.0–98.0 | 42.0–89.0 | Fallowfield and Garrett (1985) |
| *Chlorella pyrenoidosa* | Domestic sewage and industrial wastewater from pig farm and palm oil mill | 60.0–70.0 | 50.0–60.0 | Aziz and Ng (1992) |
| *Chlorella* sp. | • Primary settling | 68.4 | 83.2 | Wang et al. (2009) |
| | • Wastewater after primary settling | 68.5 | 90.6 | |
| | • Wastewater after activated sludge tank | 50.8 | 4.69 | |
| | • Centrate | 82.8 | 85.6 | |
| *Chlorella vulgaris* | Untreated urban wastewater | 60.1 | 80.3 | Ruiz-Marin et al. (2010) |
| *Chlorella* sp. | Municipal wastewater (raw centrate) | 89.1 | 80.9 | Li et al. (2011) |
| *Chlorella* sp. 227 | Pretreated wastewater (0.2 μm filter) | 92.0 | 86.0 | Cho et al. (2011) |
| *Scenedesmus obliquus* | Untreated urban wastewater | 100 | 83.3 | Ruiz-Marin et al. (2010) |
| *Chlorella minutissima* | Primary treated wastewater | 91.49 | 87.63 | Sharma and Khan (2013) |
| *Scenedesmus* sp. ZTY1 | Primary and secondary treated wastewater | 90.0 | 97.0 | Zang et al. (2014) |
| *Chlorella* sp. IM-01 | Municipal wastewater after settling | 97.81 | 89.39 | Kiran et al. (2014a) |

(continued)

**Table 1** (continued)

| Wastewater treatment | Type of wastewater | N removal (%) | P removal (%) | References |
|---|---|---|---|---|
| *Scenedesmus obliquus* | Wastewater | 92.0–94.0 | 61.0–99.0 | Jebali et al. (2015) |
| *G. sulphurariaia* | Primary effluent | 99.42 | 97.8 | Selvaratnam et al. (2015) |
| Mixed algae culture (*Microspora* sp., Diatoms, *Lyngbya* sp., *Cladophora* sp., *Spriogyra* sp. and *Phixoclonium* sp.) | Municipal wastewater | 97.0 | 93.0 | Ahmad et al. (2012) |
| Algal bioflim wastewater treatment | | | | |
| Algal bioflim | Tertiary treatment of wastewater | NA | 97.0 | Sukacova et al. (2015) |

Further small-scale operations have been reported to use solar-powered photobioreactors in algae-based systems significantly reducing organic matter and nutrients with minimal energy cost, but Net Efficiency Ratio (NER) for tubular reactor is reported to be less than 1. NER is the ratio of total energy produced in terms of biomass and its products over the energy used including construction cost, raw materials, and total energy input in construction, operation, and maintenance. These photobioreactors have been established on pilot level for cultivation of a wide range of algal sp. including *Porphyridium* sp., *Chlorella* sp., *Spirulina* sp., *Dunaliella* sp., *Haematococcus* sp., *Tetraselmis* sp., and *Phaeodactylum* sp.

## 4.3 Algal Mats

Algal mat systems were developed by Adey (1982), which utilizes macroalgae rather than microalgae. This is another option for removal of nutrients from wastewaters. In this, turf forming algal species, viz. *Cladophora*, *Enteromorpha*, *Ceramium*, *Sphacelaria*, *Ectocarpus*, *Oscillatoria*, *Polysiphonia*, and *Herposiphonia*, are cultivated on a mesh or net with nutrient-enriched wastewater passing over them. Algal biomass is regularly removed mechanically from mats. Even though this has demonstrated effective control in nutrient levels of aquarium water, labor-intensive work requirement, large surface area, and high manpower are the major drawbacks of algal mat system. This system also requires artificial light source for the maintenance of sufficient nutrient removal rate, thus increasing the cost of wastewater treatment. Supplementary aquatic plant-based options have also been projected for removal of nutrients like Phragmites, Water hyacinth, and Typha but found less competent than phycoremediation.

**Table 2** Algal wastewater treatment in simulated environment

| Algae | Treatment | % N removal | % P removal | References |
|---|---|---|---|---|
| *C. vulgaris* | Simulated settled domestic wastewater | 100 | 95.0 | Tam and Wong (2000) |
| *Chlorella zofingiensis* | Artificial wastewater using acetic acid as pH regulator | 45.2–73.5 | 92.2–100 | Zhu et al. (2014) |
| *Cladophora* sp. | Varying N/P ratio | 90.0–> 99.0 | 56.3–> 99.0 | Liu and Vyverman (2015) |
| *Klebsormidium* sp. | Varying N/P ratio | 76.4–> 99.0 | 26.0–> 99.0 | |
| *Pseudanabaena* sp. | Varying N/P ratio | 85.4–> 99.0 | 34.5–> 99.0 | |
| *Scenedesmus obliquus* | Artificial wastewater with 30% daily replacements of culture volume | 50.0 | >50.0 | Voltolina et al. (1998) |
| | Artificial wastewater with 50% daily replacements of culture volume | 66.0 | >50.0 | |
| *Scenedesmus* sp. LX1 | Nitrogen/phosphorus ratio of 5:1–12:1 | 83.0–99.0 | 99.0 | Xin et al. (2010) |
| *C. vulgaris* co-immobilized with *A. Brasilense* | Artificial wastewater (batch mode) | 99.0 | 70.0 | De-Bashana et al. (2002) |
| | Artificial wastewater (semi continuous mode) | 100 | 83.0 | |
| | Artificial wastewater (continuous mode) | 91.0 | – | |
| Immobilized *C. vulgaris* | Artificial wastewater (batch mode) | 86.0 | – | De-Bashana et al. (2002) |
| | artificial wastewater (semi continuous mode) | 85.0 | 33.0 | |
| | artificial wastewater (continuous mode) | 59.0 | – | |
| Algal bioflim | Simulated wastewater (batch mode) | 86.6 | 98.2 | Wei et al. (2008) |
| | Simulated wastewater (continuous mode) | 83.9 | 95.4 | |
| Aerobic granules (Algae–bacteria *consortium*) | Synthetic wastewater (sequence batch reactor) | 52.9–58.3 | 90.0 | Huang et al. (2015) |

## *4.4 Hyper-Concentrated Cultures*

Hyper-concentrated cultures have algal density greater than 1.5 g/L, which helps in faster nutrient removal as compared to normal cultures. To obtain such conditions, algal cells are initially flocculated by using flocculant like chitosan (Morales et al.

1985). Studies on *Oscillatoria* sp. and *Scenedesmus obliquus* suggested that cell concentrations of ~1.9 g dry weight can be obtained by using sewage sludge as nutritive source (Hashimoto and Furukawa 1989). Further nitrogen and phosphorous removal was greatly accelerated in these hyper-concentrated cultures and was independent of light limitation due to self-shading (Lavoie and De la Noüe 1985). Small pond requirement for generation of such hyper-concentrated cultures would permit reduced residence time, which is a potential advantage. These have been studied at small scale till now. Economic and engineering feasibility of these systems at large scale still needs further exploration.

## *4.5 Dialysis Cultures*

In this type of system, algal biomass is separated from nutrient media through a semipermeable dialysis barricade and compounds with low molecular weight can diffuse through this barrier with concentration grade (Marsot et al. 1991). The main advantage is maintenance of high cell density of algal cultures for prolonged periods using high membrane surface area/culture volume ratio, showing proficient nutrient consumption (Marsot et al. 1991).

## *4.6 High Rate Algal Ponds (HRAPs)*

This is a modified form of traditional stabilization pond, and algal diversity in HRAPs depends on the composition of wastewater, environmental conditions, as well as the stirring method. Some of the common species in HRAPs are *Chlorella*, *Scenedesmus*, *Cyclotella*, *Micractinium*, *Chlamydomonas*, *Euglena*, and so on. HRAPs are raceway-type ponds with 0.2–1 m depth. Paddle wheel is generally used to mix solid/liquid with an average horizontal velocity of about 0.15–0.3 m/s. Depending on the soil setting and local regulations, pond bottom may be lined or unlined. Carbon dioxide can be added through a gas sparging pump resulting into turbulent flow within the pond. With adjustable $O_2/CO_2$ concentrations in the pond and moderate evaporation rate, temperature can be balanced resulting in the removal of organic matter, phosphorus, and nitrogen. WWT in HRAPs is usually a division of Advanced Pond System (APS), which is normally comprised of advanced facultative ponds incorporating anaerobic digestion pits and High Rate Algal Ponds, followed by settling section and maturation ponds (Fig. 2). Requirement of land area is almost 50 times more for Advanced Pond Systems than activated sludge systems. However, this does not include area needed for disposal of waste-activated sludge. Capital costs requisite for construction of Advanced Pond System are $<50\%$ and operational costs are $<20\%$ to those of activated sludge systems or any other chemical treatment method (Craggs 2005).

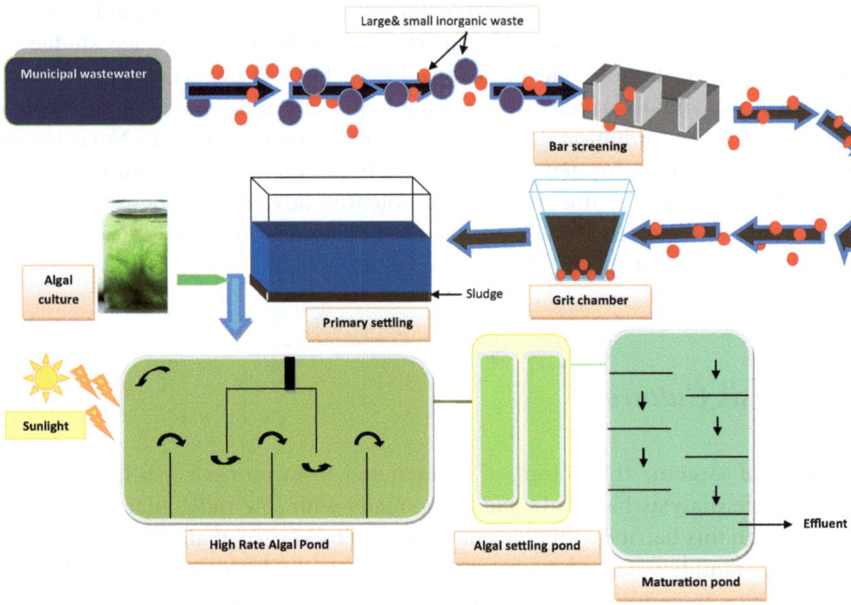


**Fig. 2** Scheme for wastewater Treatment through HRAPs

Algal biomass production and wastewater treatment in HRAPs offers an attractive option. Minor issues of HRAPs during construction and operation can be accommodated for eco-friendly wastewater treatment process resulting in development of algal biofuel feedstock free from environmental burden. Additionally, these systems utilize already existing nutrients in wastewater which otherwise get wasted. Final product from this treatment process is algal biomass, renewable source of energy, and biofertilizer production. Efficiency and working of HRAPs is dependent on several factors as described here:

### 4.6.1 Carbon Dioxide

$CO_2$ addition to HRAPs increases accessibility of carbon for growth and also helps in mitigation of pH inhibition by simply maintaining pH <8.0 (Craggs 2005; Kong et al. 2010). This addition of carbon dioxide may reduce the removal of nutrients through physical/chemical processes like ammonia volatilization and phosphate precipitation. This decreased efficiency in treatment can be compensated by increased production of algal biomass and coupled nutrient assimilation capability.

### 4.6.2 Monoculture Maintenance

Algal species control in HRAP is dependent on environmental, operational, and biological parameters such as temperature, pH, light, nutrient concentration and composition, hydraulic retention time, gazers and parasites, pre-adaptation of algal species and seeding, etc. The main aim is to develop a dynamic condition where algal species could show mutualistic approach with wastewater simultaneously producing biomass. Isolation and identification of already existing species are still supposed to flourish and perform better. Some of the ideal attributes of wastewater treating HRAPs are:

(a) Fast growth kinetics; high biomass productivity of algae in wastewater environment supplemented with ammoniacal-N and phosphate could help in reducing the retention time of effluents in HRAPs leading to efficient treatment and generation of increased levels of important algal components (viz. biofuel, biogas, etc.).
(b) Robust; tolerance towards seasonal variations in outdoor environment as well as sustainability to toxic substances present in wastewater for longer periods could further help.
(c) Easy to harvest; algal species which could form aggregates are much desirable candidates for HRAPs as they can be easily harvested reducing the cost and energy consumption.

### 4.6.3 Control of Parasites

High Rate Algal Ponds are vulnerable to grazing by zooplankton and protozoa reducing biomass concentration to low levels within short duration (Van Harmelen and Oonk 2006; Benemann 2008). Rotifers and cladocerans are reported to decrease algal biomass by 90% in just 2 days at densities >105/L (Oswald 1980). Cauchie et al. (1995) found 99% decrease in algal chlorophylla due to Daphnia grazing for several days. Algal population in ponds can also be reduced due to fungal parasitism and viral infection in short time by triggering changes in algal diversity and cell structure.

Zooplanktons may be controlled through application of some physical or chemical treatments such as filtration, centrifugation, low DO levels, chemicals/hormone mimics, increased pH, and free ammonia concentration. Most zooplanktons are not able to endure longer periods of low dissolved oxygen level and pH adjustment up to 11. These are perhaps the most practical methods for control of zooplanktons. Wastewater usually contains high levels of ammonia (30 mg/L) resulting in elevated pH level with toxic effects on zooplankton. There are various chemical, physical, and biological methods to inhibit growth of zooplanktons as detailed below:

#### 4.6.3.1 Physical Methods

(a) **Temperature**
High temperatures (up to 35 °C) can drastically reduce the survival of zooplankton and thus can be used to depress their populations in HRAPs. Very high temperature environment (60–100 °C) could be applied for short time periods, but would require large amounts of energy and can also have detrimental effect on algal community. However, moderate heat treatment (35–45 °C) is feasible, particularly when waste heat (e.g., coolant fluid from engines, power generators, or turbines) is available. However, heat treatment (35–38 °C for few hours) has been used to treat ship ballast water that proved to be effective in removing all zooplanktons but also killed a large proportion of microalgae.

(b) **Cavitation**
In this method, liquid is subjected to rapid pressure changes, resulting in empty cavities generation due to low pressures and the released energy can disrupt zooplanktons. Cavitation can be generated by sonication or hydrodynamic turbulences within liquid media. Sonication has been utilized and proved to be effective in treating zooplankton such as *Ceriodaphnia dubia*, *Brachionus plicatilis*, and *B. calyciflorus*, with very small effect on algae. But economics of using sonication is very high. Hydrodynamic cavitation can be produced mechanically by generating turbulence and pressure drops in a liquid flow using pumps or impellers to force the liquid through small openings. This is less expensive, easier to set up, and easy to maintain.

(c) **Shear stress**
Bead mills are applied using a rotating cylinder which is partially open at either end (e.g., perforated plate or metal mesh), holding large beads (dia. 10 mm) longitudinally to water flow. Algal suspension gently flows through the slow rotating device, with zooplankton killed by the friction generated between the beads. Various bead sizes and rotating speeds could be used to kill selective zooplankton species. This system is potentially applicable to HRAPs because it is simple to construct and user friendly. Furthermore, if operated at low rotating speed using large beads, cost of operation is projected to be less along with less damage to algae.

(d) **Filtration**
This is a cost-effective and simple technique which uses different sizes of filter mesh for the separation and capturing of zooplankton according to their size. Filtration using 50, 100, and 150 μm mesh size can be used to control large zooplankton grazers with removal rate between 79 and 89%. Rotating drum filters with 60 μm mesh size have been used to filter smaller zooplankton species. Unfortunately, mesh sizes $<200$ μm are not suitable for HRAPs because algal–bacterial flocs (~50–200 μm) growing in ponds can also be removed by such filters. Clogging of filters is another major issue increasing the operational cost.

#### 4.6.3.2 Chemical Methods

An ideal zooplankton control method should have minimal effect on growth of algae, i.e., it must be selective in nature. Chemical treatments are particularly used to reduce the abundance of smaller zooplankton, viz. rotifers, which are usually more challenging to be removed by means of mechanical methods.

(a) **pH alterations**
pH adjustment can be used to control/kill zooplanktons. Cladocerans can tolerate a wide pH range, but their mortality rate drastically increases at pH values greater than 10.5. In a highly eutrophic pond, the survival rate of *Ceriodaphnia reticulata* considerably decreases at pH 11.2, while in laboratory experiments the $LC_{50}$ of *D. magna* at pH 4.4 and pH 10.7 is reported to be 48 h. Some rotifers, including the *Brachionus* species, can tolerate pH between 3 and 10. *B. calyciflorus* exhibited less rate of reproduction with highest resting egg production at pH 10.5, whereas total mortality occurred at pH 11.5 within 24 h.

(b) **Ammoniacal effect**
Free ammonia ($NH_3$–N) toxicity can be used to control zooplankton in HRAPs at pH 9. At higher pH levels than 7.5, ammonium ions from wastewater get dissociated into unionized ammonia and the proportion further increases with increase in pH such as 50% at pH 9.5 and 80% at pH 10. The concentration of $NH^{4+}$ in wastewater generally ranges between 2 and 14 mg/L. Thus, at higher pH levels $NH_3$–N can potentially depress the zooplankton population. In laboratory-scale studies, 3 mg/L of $NH_3$–N showed no effect on the reproduction of *B. rubens*, whereas in range of 3–5 mg/L the population reversibly depreciated, and concentrations >5 mg/L destroyed all rotifers just within 2 days. Thus, concluding that increase in free ammonia in water can be proved to be effective in elimination of zooplanktons.

#### 4.6.3.3 Biological Methods

The introduction of zooplanktivorous (organisms surviving on zooplanktons) organisms to HRAPs could provide a natural zooplankton biocontrol method where a predator should be able to survive permanently in HRAPs but should not be transferred to downstream ponds. Natural parasites of cladocerans such as bacteria and fungi are not suitable because of the unavoidable contamination of maturation ponds. Fish, amphibians, and crustaceans have found to be effective; however, physicochemical parameters of HRAPs may limit the number of species. Their interactions with zooplankton have only been studied in natural and aquaculture systems.

(a) Notonectidae (Order: Hemiptera) are aquatic insects which swim backwards and usually called as backswimmers. These are predators of cladocerans and are found distributed worldwide, mainly dwelling in still and smoothly flowing freshwater ponds, lakes, and marshes.

(b) *Chaoborus* sp. is a type of midge whose larvae (~6–23 mm) can live up to several months in anoxic sediments of small ponds. They are found mainly in tropical and temperate climates excluding Antarctica and in the eutrophic conditions of standing water. *Cladocerans*, viz. *Cercopagis pengoi*, *Leptodora kindtii*, *Bythotrephes longimanus*, and *Polyphemus pediculus*, feed on smaller zooplankton species but not on algae.

Both larval and adult fish can prey on zooplanktons, affecting community composition and abundance in natural lakes. Fish are suggested as zooplankton predators in algae production ponds and can tolerate a wide range of pH and DO in aquaculture ponds. Species such as silver carp and Nile tilapia are proven to survive physicochemical conditions similar to those in HRAPs. Planktivorous common carp (*Cyprinus carpio*) is able to live in shallow, eutrophic, and turbid environments with pH ranging between 6.5 and 9.0 with very low (0.3–0.5 mg/L) or very high DO.

### 4.6.4 Commercially Available Compounds to Control Parasites

#### 4.6.4.1 Chemical Agents

Chemical agents typically have higher activity on cladocerans than rotifers. Some of the examples of these chemicals are as follows: (1) Cypermethrin is a neurotoxin insecticide, which at a concentration 6.1 μg/L completely kills the populations of Daphnia and Calanoid copepods in just 2 days. The half-life of cypermethrin is 25 days and can cost up to ~$15/kg (973.5 INR). (2) Temephos is an organophosphorus insecticide and at a concentration of 58.6 μg/L is potent towards all cladocerans and drastically reduces rotifer population. But recovery of the rotifer community after some time suggests their higher tolerance towards this agent. Half-life of Temephos is 28.7 days at pH 7 with rapid adsorption in suspended particles and sediments, with no detrimental effect to algae.

Carbaryl belongs to carbamate family and is an agricultural pesticide. At a concentration of 1 mg/L in small shallow and eutrophic ponds can potently kill all cladocerans, rotifers, and Chaoborus larvae. Studies suggest that absence of Chaoborus larvae (that can prey on zooplankton) can promote the reestablishment of cladocerans soon after treatment; in turn cladocerans will compete with rotifers and can probably reduce their population. This treatment has no detrimental effects on algae, but rapid recovery of cladocerans could probably be related to the rapid dissipation of chemical. Carbaryl half-life strongly depends on pH, i.e., 3.2 h at pH 9, 12.1 days at pH 7, and 1600 days at pH 5. Quinine sulfate is an antiprotozoal chemical which is usually used to control predatory ciliates in outdoor cultures of *D. salina*, with only minor damage to algae. Doses required to inactivate ciliates are expected to be similar to those required to inactivate rotifers. 24 h $LC_{100}$ (absolute lethal concentration) for the ciliates is reported to be 12–14 mg/L, while for algae, 72 h $EC_{50}$ (half-maximal effective concentration) is reported to be 14.5 mg/L.

Permethrin is another insecticide, which eradicated the population of *Daphnia rosea* at 10 μg/L concentration without reestablishment in the following month. It is reported that algal photosynthesis is not significantly affected by the use of this insecticide. Permethrin gets accumulated in the sediments at a concentration of 6 μg/L (wet weight), and after 24 h, permethrin concentration in water showed a sixfold decrease.

Menadione, the synthetic derivative of vitamins $K_1$ and $K_2$, also has biocidal activity. It is marketed as SeaKleen™ and selectively kills adult cladocerans and rotifers without affecting green microalgae. The use of chemicals to inactivate zooplankton in HRAPs seems to be a simple and effective approach, whereas several chemicals still need to be tested at hectare-scale HRAPs. Permethrin, Cypermethrin, and Carbaryl are effective even in low quantities with less inactivation time. More information is needed on persistence of these chemicals in the effluent water and sediments. Selection of particular chemical agent should be based on cost, availability, and effluent discharge limits for the specific chemical.

#### 4.6.4.2 Enzymes

Chitin is a structural polysaccharide, which is principal component of cladoceran exoskeletons. Chitin is hydrolyzed by chitinase, a group of enzymes produced by bacteria, fungi, plants, and insects during molting. Chitinases have been proposed as a bio-pesticide to degrade fungal cell walls. Cheaper option is usage of substances interfering with chitinase production and promoting the mortality of crustacean zooplankton, i.e., diflubenzuron, chitosan, and allosamidin. Chitosan and its derivatives repress chitinase activity through competitive inhibition preventing larval molting. Laboratory toxicity tests have shown that an artificial chitosan derivative when added to the diet of moth *Spodoptera littoralis* (Lepidoptera) larvae at a concentration of 0.625 g/kg resulted in 100% mortality. Allosamidin are pseudo-trisaccharides that exert inhibitory activity on chitinase at very low concentration. In laboratory experiments, 30 and 50 μg of allosamidin per ml of enzyme solution (200 units of chitinase/mL), respectively, inhibited 50 and 70% of chitinase activity.

#### 4.6.4.3 Infochemicals

Infochemicals are substances excreted by organisms that may modify the physiology, structure, and behavior of individuals of another species. Kairomone infochemicals have selective advantages for algae as they can induce signal pathway in algae against zooplanktons. For example, kairomones can induce defense mechanisms against zooplankton grazing by promoting colony formation, bio-flocculation, or generation of spines. Formation of colonies and spines helps in reduction of grazing effects of *B. patulus*, *B. calyciflorus*, *M. macrocopa*, and *C. dubia* on algae, which is directly proportional to the chemical concentration.

Some types of kairomones have been isolated and identified. For example, aliphatic sulfates and sulfamates as well as an artificial substitute (octyl sodium sulfate) have been experimented for induction of colony formation in *Desmodesmus* sp. and *Scenedesmus* sp.

## 5 Energy Costs

Energy necessities for operation of open pond reactors comprise pumping of wastewater, mixing, and bubbling carbon dioxide. Treatment of 12 MGD (Million Gallons per Day) wastewater requires 37 ha of algal pond/raceway pond (Benemann et al. 1978; Chisti 2007). Thus, to maintain flow, 75 HP is needed for centrifugal pumping of ~0 8333 GPM (Gallon Per Minute) with a 70% efficient pump requiring 1350 kW h/day or 9.7 GJ/ton biodiesel to ensure efficient working of wastewater treatment plants or other large-scale designs. But addition of $CO_2$ could add further load to energy requirements. Reports based on pilot raceway pond in Roswell, New Mexico, attained high biomass productivities of 50 $g/m^2$ day, having an average biomass of 10 $g/m^2$ day with above documented data (Hu et al. 2008). Numerous reports on life cycle analyses presumed algal productivities of 15–30 $g/m^2$ day after addition of $CO_2$ (Lardon et al. 2009; Batan et al. 2010; Campbell et al. 2011). Similarly, studies on WWTP based in Lawrence, USA, reported algal biomass productivities varying from 5 to 16 $g/m^2$ day with average lipid content up to 10% of dry mass without $CO_2$ enrichment, which is quite lower than laboratory-scale production. Carbon dioxide-enriched cultures are proposed to attain algal biomass with lipid content of 50–60% by weight in controlled conditions. With $CO_2$ addition, algal productivities of 25 $g/m^2$ day have been observed but can fluctuate in wastewater algal ponds. Thus, it could be concluded that addition of carbon dioxide in open ponds could help in generation of higher biomass and henceforth better phycoremediation (Sturm and Lamer 2011).

## 6 Immobilization of Algae

Immobilization is defined as the fixing/capturing of essential molecules or microbial cells on an inert and insoluble material providing resistance towards external conditions. In principle, presently there are different types of immobilization approaches available, i.e., affinity immobilization, covalent coupling, adsorption, capture behind semipermeable membrane, confinement in liquid–liquid emulsion, and entrapment, where adsorption and entrapment are the most common methods.

(a) Adsorption immobilization
Adsorption can be done either to external or internal surface of an inert carrier and is a reversible process with recoverable support after denaturation of catalyst. The most common adsorbents used are activated carbons, resins,

hydrous metal oxides, and glass. This has been successfully adopted in enzyme processes and also tried for algal species by some researchers. Elmahadi and Greenway (1991) used *Selenastrum capricornutum* immobilized on pore glass for preconcentration of heavy metals. Laurinavichene et al. (2006) immobilized *Chlorella reinhardtii* on fiber glass matrix for prolonged hydrogen production under sulfur-deprived condition. Shi et al. (2007) showed removal of ammonium, phosphate, and nitrate through *Scenedesmus rubescens* and *Chlorella vulgaris* cells entrapped in twin layer, and microalgal cells remained attached to nitrocellulose membrane, while receiving nutrient from macroporous glass fiber "source layer."

(b) Entrapment method

This is the most commonly used method and is dependent on retention of algae in gel lattice. Algal cells freely stay in their own compartments having pores for flow of materials to and from the cells. Diverse types of synthetic (polyurethane, acrylamide, polyvinyl, etc.) and natural (alginate, agar, collagen, agarose, cellulose, carrageenan, etc.) polymers are used by various researchers. Most frequently used natural gels for immobilization of algae are alginate and carrageenan.

## 6.1 Advantages of Immobilized Cell Systems

Immobilization of algal cells helps to overcome one of the most important issues in the exploitation of algae for wastewater treatment, i.e., separation of algal biomass from aqueous solution. Immobilization proposes numerous other benefits in batch or continuous mode as compared to free cells (Chevalier and De la-Noüe 1985; Rao and Hall 1984; De la Noüe and Proulx 1988). Entrapment of algae on a suitable inert support simplifies the treatment process and increases retention time (Travieso et al. 1992), further enhancing the efficiency of nitrate, ammonia, and phosphate removal. Chevalier and De la-Noüe (1985) revealed almost similar uptake rates for N and P with freely moving and k-carrageenan-immobilized *Scenedesmus*. Reports on hollow fiber-immobilized cyanobacterial systems suggest that these reactors are easy to construct, are less time-consuming, and are efficient in nutrient removal from wastewater (Sawayama et al. 1998). Hollow fiber immobilized cyanobacteria showed improved rates of hydrogen production along with better removal efficiency of inorganic nutrients (Markov et al. 1995). Immobilized cyanobacterial species such as *Phormidium* (Ochiai et al. 1983) and *Mastigocladus* (Ochiai et al. 1980) have also been tried for electricity generation successfully. Immobilized form of *Scenedesmus obliquus* and *Chlorella vulgaris* showed removal of phosphorus (86 and 81%, respectively) and ammonia (100 and 98.4%, respectively) (Kobbai et al. 2000). On the other hand, co-immobilization of *Azospirillum brasilense* and *C. vulgaris* in alginate beads removed ammonia and phosphate from simulated wastewater environment (De-Bashana et al. 2002). Co-immobilization of *Chlorella pyrenoidosa* and activated sludge removed nitrates and phosphates by Wang and Huang (2003). Studies also suggest potential heavy metal removal using immobilized cyanobacterial system from aqueous solution (Kiran et al. 2007; Kiran and Kaushik 2008).

## 7 Indian Scenario

India represents 16% of total world population but accounts for only 4 and 2.45% of water and land resources, respectively. Out of this, about 85% [688 BCM (Billion Cubic Meter)] of water is consumed for irrigation, which is supposed to be increased by 2050 to 1072 BCM. It has been further reported that by 2025, there may be an increase in demand for water usage in domestic and industrial sector to 29.2 BCM. It has been observed that with rising population and improved economic development, per capita annual availability of freshwater got reduced from 5177 to 1588 $m^3$ in India since 1951 till 2010. With this decreasing pattern, there is an imperative need for well-organized management of water resources, maybe through enhanced water use efficiency and recycling of wastewater.

As per reports (CPCB 2005a), India has 234 Sewage Treatment plants (STPs), typically settled under different river action plans and mainly located only in 5% cities situated along river banks. Most common technologies employed by existing STPs are oxidation ponds or Activated sludge process (59.5%), Waste Stabilization Ponds technology (28%), and Up-flow Anaerobic Sludge Blanket technology (26%) (Kaur et al. 2012). On the other hand, industries are also contributing a lot to wastewater pollution issues [about 13468 MLD (Million Liters per Day) wastewater]. For small-scale industries, which cannot afford installation cost of WWTPs, Common Effluent Treatment Plants were installed for a group of small-scale industries (CPCB 2005b). Here settable particles are removed through primary treatment such as sedimentation, grit removal, etc., followed by secondary and tertiary treatments which comprises of dual media filter, sludge drying beds, dissolved air floatation, activated carbon filter, flash mixer, tank stabilization, sand filtration, etc. These conventional methods are cost-intensive and time-consuming and require complex operations and maintenance efforts. As per collected data, the total cost estimated for the establishment and treatment of 100% wastewater generated is approximately Rs. 7560 crores for India (CPCB 2005a). Thus, to reduce per capita rate, scientists all over India have been working extensively on the development of wastewater treatment technologies. Wastewater treatment using algae seems to be promising due to various factors as explained earlier in this chapter. Laboratory-scale studies have confirmed that species such as *Chlorella*, *Scenedesmus*, *Oocystis*, *Chlamydomonas*, and *Botryococcus* can reduce the contaminants' level present in wastewater.

National Institutes such as DBT-ICT Centre for Energy Biosciences, Institute of Chemical Technology, Mumbai, focuses upon improvement of algal species through hybridization or adaptation. Genetic manipulation of algal strain, algal engineering for high metabolic production, modeling and designing of photobioreactors and raceway ponds, etc., are the major activities of DBT-ICT Centre. In 2011, Ministry of New and Renewable Energy (MNRE) developed an indigenous photobioreactor (PBR) for sequestration of waste $CO_2$. In the year 2010–2011, IISC Bangalore interpreted algae-based sewage treatment plant (STP) having capacity 67.65 MLD located at Mysore city for its treatment efficiency and found removal of total COD and BOD up to 60 and 82%, respectively. Phycoremediation with a retention time of 14.3 days also reported a sharp decline in nitrogen content. Studies at Indian Agriculture Research Institute (IARI) suggest that *Chlorella minutissima*, *Scenedesmus* sp., *Nostoc*, and their consortium could be effective for remediation purpose. Further studies reveal that *C. minutissima* having high biomass generation capability with high levels of nitrogen and phosphorus accumulation could also be a potential candidate for biofertilizer production (Bhatnagar et al. 2011). Various studies done for the removal of acid dyes at Central Leather Research Institute (CLRI), Chennai, reported that freshwater macroalga *Azolla filiculoides* is one of the efficient species in dye removal. Column reactor studies reported that an optimum bed height (25 cm), initial dye concentration (100 mg/L), and flow rate (5 mL/min) help in biosorption of Acid Green Dye up to 28.1 mg/g using *A. filiculoides*.

Phycospectrum Environmental Research Centre (PERC) works on the utilization of algal biomass for biogas, biodiesel, and bioethanol production along with remediation strategies. PERC has successfully cultivated *Chlorella* sp. in effluent of confectionery industry for correction of pH and sugar removal, and algal biomass produced high amount of lipids. For the treatment of industrial effluent using algae, first plant was installed at SNAP Natural and Alginate products at Ranipet, Tamil Nadu, in collaboration with PERC for the treatment of high acidic effluent having high TDS. The algal biomass hence generated is utilized for biofertilizer production. Furthermore, utilizing phycovolatilization for pH correction, PERC has installed its plants in companies such as Suntex Processing Mills Chennai, SVIS LABS Pvt LTD, Ultramarine and Pigments Ltd. Ranipet, Chemical Industry Ranipet, Perfetti van melle India, Sterlite Industries Mumbai, Arvind Mills Ahmedabad, Vani Pharma Hyderabad, TECHNO DRUGS, Ankhleshwar, and HMEL Mundra Gujarat for reduction of BOD, COD, and sludge from wastewater generated. Wheels India Ltd., Chennai, promoted by TVS Group has implanted pilot-scale system for remediation of effluent using microalgae. Pilot-scale plants implanted at Hindustan Coca Cola Ltd., Ahmedabad, have reported efficient phosphate and nitrate removal within 30 min.

Successful reduction in heavy metals was reported through phycovolatilization using *Chlorella vulgaris*, isolated from the effluent itself. Results also suggested that *Chlorella vulgaris* demonstrated an enhanced adsorption capacity in 8 h under open environmental conditions in comparison to laboratory studies (i.e., 30.6 vs. 10.5 mg/g dry weight). Other similar studies related to decrease in heavy metal levels and mass balance in high rate algal ponds revealed the role of phycovolatilization process through biotransformation in addition to complexation, adsorption, and entrapment mechanisms. Pycoremediation is a cost-effective, eco-friendly, nontoxic safe process with cheap construction and operational cost. Use of algal biomass for the treatment of effluent also helps in the removal of waste

$CO_2$, thus reducing greenhouse gases. Conventional chemical methods available for the treatment of effluents usually result in loads of toxic waste as sludge requiring large area for landfill sites to dispose of this sludge, whereas pycoremediation detoxifies and removes hazardous elements from the effluent. Therefore, sustainable treatment technology for STPs and ETPs with phycoremediation is a derivable solution with significant potential for nutrient recovery.

## 8 Conclusions

Environmental sustainability and economical development are linked to each other. Rise in population, urbanization, improper wastewater disposal and decomposition, and inefficient commercial wastewater treatment methodologies are leading to high levels of toxic and undesirable contaminants in water bodies, further resulting in decreased freshwater footprints across the nation. Shortcomings of existing methods led to the emergence of phycoremediation (algal wastewater treatment), an efficient and environmentally friendly approach. Lab-scale studies on algae for wastewater treatment have shown the reduction in levels of nitrogen, phosphorous, heavy metals, COD, BOD, coliforms, and various other disease-causing pathogens from simulated and real wastewater. Therefore, use of algae can be an effective solution for wastewater treatment. Various scale-up processes to treat wastewater have been tried and employed, and HRAPs (High Rate Algal Ponds) seem to be the most economical way for the treatment of wastewater which also produces valuable algal biomass (a feedstock for biofuel production) by utilizing nutrients from wastewater with minimum environmental impacts. Selection of robust algal species, $CO_2$ addition, natural bio-flocculation methods, operational parameters, local climatic conditions, control of grazers and parasites, and species control are currently the focus of researchers to further reduce the cost and environmental impact simultaneously improving the efficiency.

**Acknowledgment** Authors acknowledge financial assistance provided by Department of Science and Technology, Govt. of India under the INSPIRE Faculty Scheme (IFA12–EAS-01). The funding organization has not played any role in study design, decision to publish, or preparation of the manuscript.

**Conflict of Interest Statement** The authors declare that there are no conflicts of interest.

## References

Adey WH (1982) Algal turf scrubber. USA Patent No 4333263

Ahmad F, Khan AU, Yasar A (2012) Uptake of nutrients from municipal wastewater and biodiesel production by mixed algae culture. Pak J Nut 11:550–554. doi:10.3923/pjn.2012.648.652

Aziz MA, Ng WJ (1992) Feasibility of wastewater treatment using the activated-algae process. Bioresour Technol 40:205–208. doi:10.1016/0960-8524(92)90143-L

Badr SA, Ghazy ME, Moghazy RM (2010) Toxicity assessment of cyanobacteria in a wastewater plant. Egypt J Appl Sci Res 6:1511–1516

Batan L, Quinn J, Willson B, Bradley T (2010) Net energy and greenhouse gas emission evaluation of biodiesel derived from microalgae. Environ Sci Technol 44:7975–7980. doi:10.1021/es102052y

Benemann JR (2008) Open ponds and closed photobioreactors—comparative economics 5th annual world congress on industrial biotechnology and bioprocessing Chicago, 30 April

Benemann JR, Pursoff P, Oswald WJ (1978) Engineering design and cost analysis of a large-scale microalgae biomass system US Depart Energy

Bernal CB, Vazquez G, Quintal IB, Bussy AN (2008) Microalgal dynamics in batch reactors for municipal wastewater treatment containing dairy sewage water. Water Air Soil Pollut 190:259–270. doi:10.1007/s11270-007-9598-3

Bhatnagar A, Chinnasamy S, Singh M, Das KC (2011) Renewable biomass production by mixotrophic algae in the presence of various carbon sources and wastewaters. Appl Energy 88:3425–3431

Brown MR, Jeffrey SW, Volkman JK, Dunstan GA (1997) Nutritional properties of microalgae for mariculture. Aquacult 151:315–331. doi:10.1016/SOO44-8486(96)01501-3

Campbell PK, Beer T, Batten D (2011) Life cycle assessment of biodiesel production from microalgae in ponds. Bioresour Technol 102:50–56. doi:10.1016/j.biortech.2010.06.048

Cauchie HM, Hoffmann L, Jaspar-Versali MF, Salvia M, Thomé JP (1995) *Daphnia magna* straus living in an aerated sewage lagoon as a source of chitin: ecological aspects. J Zool 125:67–78

Chevalier P, De la-Noüe J (1985) Wastewater nutrient removal with microalgae immobilized in carrageenan. Enzyme Microb Technol 7:621–624. doi:10.1016/0141-0229(85)90032-8

Chinnasamy S, Bhatnagar A, Hunt RW, Das KC (2010) Biomass and bioenergy production potential of microalgae consortium in open and closed bioreactors using untreated carpet industry effluent as growth medium. Bioresour Technol 101:6751–6760. doi:10.1016/j.biortech.2010.03.094

Chisti Y (2007) Biodiesel from microalgae. Biotechnol Adv 25:294–306. doi:10.1016/j.biotechadv.2007.02.001

Cho S, Luong TT, Lee D, Oh YK, Lee T (2011) Reuse of effluent water from a municipal wastewater treatment plant in microalgae cultivation for biofuel production. Bioresour Technol 102:8639–8645. doi:10.1016/j.biortech.2011.03.037

CPCB (2005a) Parivesh sewage pollution—News Letter Central Pollution Control Board Ministry of Environment and Forests, Govt. of India

CPCB (2005b) Performance status of common effluent treatment plants in India Central Pollution Control Board, India

Craggs RJ (2005) Advanced integrated wastewater ponds In: Shilton A (ed) Pond treatment technology, IWA scientific and technical report series. IWA, London, pp 282–310

De la Noüe J, Proulx D (1988) Biological tertiary treatment of urban wastewaters with chitosan-immobilized *Phormidium*. Appl Microbiol Biotechnol 29:292–297. doi:10.1007/BF01982919

De-Bashana LE, Morenoa M, Hernandeza JP, Bashan Y (2002) Removal of ammonium and phosphorus ions from synthetic wastewater by the microalgae *Chlorella vulgaris* co-immobilized in alginate beads with the microalgae growth-promoting bacterium *Azospirillum brasilense*. Water Res 36:2941–2948. doi:10.1016/S0043-1354(01)00522-X

Dubey SK, Dubey J, Viswas AJ, Tiwali P (2011) Studies on cyanobacterial biodiversity in paper mill and pharmaceutical industrial effluents. Br Biotechnol J 1:61–67. doi:10.9734/BBJ/2011/395

Elmahadi HAM, Greenway GM (1991) Immobilized alga as a reagent for pre-concentration in trace element atomic absorption spectrometry. J Anal Atom Spectrom 6:643–646. doi:10.1039/JA9910600643

Fallowfield HD, Garrett MK (1985) The photosynthetic treatment of pig slurry in temperate climatic conditions: a pilot plant study. Agric Waste 12:111–136. doi:10.1016/0141-4607(85)90003-4

Fouilland E (2012) Biodiversity as a tool for waste phycoremediation and biomass production. Rev Environ Sci Biotechnol 11:1–4. doi:10.1007/s11157-012-9270-2

Hashimoto S, Furukawa K (1989) Nutrient removal from secondary effluent by filamentous algae. J Ferment Bioeng 67:62–69. doi:10.1016/0922-338X(89)90088-3

Hu Q, Sommerfeld M, Jarvis E, Ghirardi M, Posewitz M, Seibert M (2008) Microalgal triacylglycerols as feedstocks for biofuel production: perspectives and advances. Plant J 54:621–639. doi:10.1111/j.1365-313X.2008.03492.x

Huang W, Li B, Zhang C, Zhang Z, Lei Z, Lu B, Zhou B (2015) Effect of algae growth on aerobic granulation and nutrients removal from synthetic wastewater by using sequencing batch reactors. Bioresour Technol 179:187–192. doi:10.1016/j.biortech.2014.12.024

Hussein NR, Gharib SM (2012) Studies on spatio-temporal dynamics of phytoplankton in El-Umum drain in west of Alexandria. Egypt J Environ Biol 33:101–105

Jebali A, Acien FG, Gomez C, Fernandez-Sevilla JM, Mhiri N, Karray F, Dhouib A, Molina-Grima E, Sayadi S (2015) Selection of native Tunisian microalgae for simultaneous wastewater treatment and biofuel production. Bioresour Technol 198:424–430. doi:10.1016/j.biortech.2015.09.037

Kaur R, Wani SP, Singh AK, Lal K (2012) Wastewater production treatment and use in India. Country report, India

Kiran B, Kaushik A (2008) Cyanobacterial biosorption of Cr (VI): application of two parameter and Bohart Adams models for batch and column studies. Chem Eng J 144:391–399. doi:10.1016/j.cej.2008.02.003

Kiran B, Kaushik A, Kaushik CP (2007) Response surface methodological approach for optimizing removal of Cr (VI) from aqueous solution using immobilized cyanobacterium. Chem Eng J 126:147–153. doi:10.1016/j.cej.2006.09.002

Kiran B, Pathak K, Kumar R, Deshmukh D (2014a) Cultivation of *Chlorella* sp. IM-01 in municipal wastewater for simultaneous nutrient removal and energy feedstock production. Ecol Eng J 73:326–330. doi:10.1016/j.ecoleng.2014.09.094

Kiran B, Kumar R, Deshmukh D (2014b) Perspectives of microalgal biofuels as a renewable source of energy. Energy Conver Manage J 88:1228–1244. doi:10.1016/j.enconman.2014.06.022

Kobbai I, Dewedar AE, Hammouda O, Hameed MSA, May E (2000) Immobilized algae for wastewater treatment. Proc 1st Int Conf Biol Sci (ICBS) Fac Sci Tanta University 1:114–122

Kong QX, Li L, Martinez B, Chen P, Ruan R (2010) Culture of microalgae *Chlamydomonas reinhardtii* in wastewater for biomass feedstock production. Appl Biochem Biotechnol 160:9–18. doi:10.1007/s12010-009-8670-4

Lardon L, Helias A, Sialve B, Steyer JP, Bernard O (2009) Life-cycle assessment of biodiesel production from microalgae. Environ Sci Technol 43:6475–6481. doi:10.1021/es900705j

Laurinavichene TV, Fedorov AS, Ghirardi ML, Seibert M, Tsygankov AA (2006) Demonstration of sustained hydrogen photoproduction by immobilized sulfur-deprived *Chlamydomonas reinhardtii* cells. Int J Hydro Energy 31:659–667. doi:10.1016/j.ijhydene.2005.05.002

Lavoie A, De la Noüe J (1985) Hyperconcentrated cultures of *Scenedesmus obliquus*: a new approach for wastewater biological tertiary treatment. Water Res 19:1437–1442. doi:10.1016/0043-1354(85)90311-2

Li Y, Chen YF, Chen P, Min M, Zhou W, Matinez B, Zhou J, Roan R (2011) Characterization of a microalga *Chlorella* sp. well adapted to highly concentrated municipal wastewater for nutrient removal and biodiesel production. Bioresour Technol 102:5138–5144. doi:10.1016/j.biortech.2011.01.091

Liu J, Vyverman W (2015) Differences in nutrient uptake capacity of the benthic filamentous algae *Cladophora* sp. *Klebsormidium* sp. and *Pseudanabaena* sp. under varying N/P conditions. Bioresour Technol 179:234–242. doi:10.1016/j.biortech.2014.12.028

Liu W, Zhang Q, Liu G (2010) Lake eutrophication associated with geographic location lake morphology and climate in China. Hydrobiologia 644:289–299. doi:10.1007/s10750-010-0151-9

Markov SA, Bazin MJ, Hall DO (1995) Hydrogen photoproduction and carbon dioxide uptake by immobilized *Anabaena variabilis* in a hollow fibre photobioreactor enzyme. Microbiol Technol 17:306–310. doi:10.1016/0141-0229(94)00010-7

Marsot P, Cembella A, Houle L (1991) Growth kinetics and nitrogen-nutrition of the marine diatom *Phaeodactylum tricomutum* in continuous dialysis culture. J Appl Phycol 3:1–10. doi:10.1007/BF00003914

McGriff EC, McKinney RC (1972) The removal of nutrients and organics by activated algae. Water Res 6:1155–1158. doi:10.1016/0043-1354(72)90015-2

Morales J, De la Noüe J, Picard G (1985) Harvesting marine microalgae species by chitosan flocculation. Aquacult Eng 4:257–270. doi:10.1016/0144-8609(85)90018-4

Mulbry W, Kondrad S, Pizarro C (2006) Biofertilizers from algal treatment of dairy and swine manure effluents: characterization of algal Biomass as a slow release fertilizer. J Veget Sci 12:107–125. doi:10.1300/J484v12n04_08

Ochiai H, Shibata H, Sawa Y, Katoh T (1980) "Living electrode" as a long-lived photoconverter for bio-photolysis of water. Proc Natl Acad Sci U S A 77:2442–2444

Ochiai H, Shibata H, Sawa Y, Shoga M, Ohta S (1983) Properties of semiconductor electrodes coated with living films of cyanobacteria. Appl Biochem Biotechnol 8:289–303. doi:10.1007/BF02779496

Oswald WJ (1980) Algal production—problems achievements and potential In: Shelef G, Soeder CJ (eds) Algae biomass. Elsevier North/Holland/Biomedical Press, Amsterdam, pp 1–8

Pouliot Y, Buelna G, Racine C, De la Noüe J (1989) Culture of cyanobacteria for tertiary wastewater treatment and biomass production. Biol Waste 29:81–91. doi:10.1016/0269-7483(89)90089-X

Rao KK, Hall DO (1984) Photosynthetic production of fuels and chemicals in immobilized systems. Trends Biotechnol 2:124–129. doi:10.1016/0167-7799(84)90021-0

Renuka N, Sood A, Ratha SK, Prasanna R, Ahluwalia AS (2013) Nutrient sequestration biomass production by microalgae and phytoremediation of sewage water. Int J Phytorem 15:789–800. doi:10.1080/15226514.2012.736436

Renuka N, Sood A, Prasanna R, Ahluwalia AS (2015) Phycoremediation of wastewaters: a synergistic approach using microalgae for bioremediation and biomass generation. Int J Environ Sci Technol 12:1443–1460. doi:10.1007/s13762-014-0700-2

Ruiz-Marin A, Mendoza-Espinosa LG, Stephenson T (2010) Growth and nutrient removal in free and immobilized green algae in batch and semi-continuous cultures treating real wastewater. Bioresour Technol 101:58–64. doi:10.1016/j.biortech.2009.02.076

Sawayama S, Rao KK, Hall KK (1998) Nitrate and phosphate ion removal from water by *Phormidium laminosum* immobilized on hollow fibres in a photobioreactor. Appl Microbiol Biotechnol 49:463–468. doi:10.1007/s002530051199

Selvaratnam T, Pegallapati P, Montelya F, Rodriguez G, Nirmalakhandan N, Lammers PJ, Voorhies WV (2015) Feasibility of algal systems for sustainable wastewater treatment. Renew Energy 82:71–76. doi:10.1016/j.renene.2014.07.061

Sen B, Alp MT, Sonmez F, Kocer MAT, Canpolat O (2013) Relationship of algae to water pollution and waste water treatment. In: Elshorbagy W (ed) Water treatment. ISBN: 978-953-51-0928-0

Sharma GK, Khan SA (2013) Bioremediation of sewage wastewater using selective algae for manure production. Int J Environ Eng Manage 4:573–580

Shi J, Podola B, Melkonian M (2007) Removal of nitrogen and phosphorus from wastewater using microalgae immobilized on twin layers: an experimental study. J Appl Phycol 19:417–423. doi:10.1007/s10811-006-9148-1

Sturm BSM, Lamer SL (2011) An energy evaluation of coupling nutrient removal from wastewater with algal biomass production. Appl Energy 88:3499–3506. doi:10.1016/j.apenergy.2010.12.056

Su Y, Mennerich A, Urban B (2011) Municipal wastewater treatment and biomass accumulation with a wastewater-born and settleable algal-bacterial culture. Water Res 45:3351–3358. doi:10.1016/j.watres.2011.03.046

Sukacova K, Trtilek M, Rataj T (2015) Phosphorus removal using a microalgal biofilm in a new biofilm photobioreactor for tertiary wastewater treatment. Water Res 71:55–63. doi:10.1016/j.watres.2014.12.049

Tam NFY, Wong YS (1989) Wastewater nutrient removal by *Chlorella pyrenoidosa* and *Scenedesmus* sp. Environ Pollut 58:19–34. doi:10.1016/0269-7491(89)90234-0

Tam NFY, Wong YS (2000) Effect of immobilized microalgal bead concentrations on wastewater nutrient removal. Environ Pollut 7:145–151. doi:10.1016/S0269-7491(99)00118-9

Travieso L, Benitez F, Dupeiron R (1992) Sewage treatment using immobilized microalgae. Bioresour Technol 40:183–187. doi:10.1016/0960-8524(92)90207-E

Van Harmelen T, Oonk H (2006) Microalgae biofixation processes: applications and potential contributions to greenhouse gas mitigation options. Report for the international network on biofixation of $CO_2$ and greenhouse gas abatement with microalgae operated under the international energy agency greenhouse gas R&D programme, Netherlands, Order No 36562

Vasconcelos VM, Pereira E (2001) Cyanobacteria diversity and toxicity in a wastewater treatment plant (Portugal). Water Res 35:1354–1357. doi:10.1016/S0043-1354(00)00512-1

Vijayakumar S, Thajuddin N, Manoharan C (2007) Biodiversity of cyanobacteria in industrial effluents. Acta Bot Malacit 32:27–34

Voltolina D, Beatriz Cordero B, Nievesc M, Soto LP (1998) Growth of *Scenedesmus* sp. in artificial wastewater. Bioresour Technol 68:265–268. doi:10.1016/S0960-8524(98)00150-3

Wang Y, Huang G (2003) Nitrate and phosphate removal by co-immobilized *Chlorella pyrenoidosa* and activated sludge at different pH values. Water Qual Res J Can 38:541–551

Wang L, Mim M, Li Y, Chen P, Chen Y, Liu Y, Wang Y, Ruan R (2009) Cultivation of green algae *Chlorella* sp. in different wastewaters from municipal wastewater treatment plant. Appl Biochem Biotechnol 162:1174–1186. doi:10.1007/s12010-009-8866-7

Wei Q, Hu Z, Li G, Xiao B, Sun H, Tao M (2008) Removing nitrogen and phosphorus from simulated wastewater using algal biofilm technique. Front Environ Sci Eng China 2:446–451. doi:10.1007/s11783-008-0064-2

Xin L, Honh-ying H, Ke G, Ying-xue S (2010) Effects of different nitrogen and phosphorus concentrations on the growth nutrient uptake and lipid accumulation of a freshwater microalga *Scenedesmus* sp. Bioresour Technol 101:5494–5500. doi:10.1016/j.biortech.2010.02.016

Yang XE, Wu X, Hao HL, He ZL (2008) Mechanisms and assessment of water eutrophication. J Zhejiang Univ Sci B 9:197–209. doi:10.1631/jzus.B0710626

Zang TY, Wu HY, Hu HY (2014) Domestic wastewater treatment and biofuel production by using microalga *Scenedesmus* sp. ZTY1. Water Sci Technol 69:2492–2496. doi:10.2166/wst.2014.160

Zhu L, Hiltunen E, Shu Q, Zhou W, Li Z, Wang Z (2014) Biodiesel production from algae cultivated in winter with artificial wastewater through pH regulation by acetic acid. Appl Energy 128:103–110. doi:10.1016/j.apenergy.2014.04.039

# Microbial Conversion of Waste and Surplus Materials into High-Value Added Products: The Case of Biosurfactants

**Constantina Kourmentza, Filomena Freitas, Vitor Alves, and Maria A.M. Reis**

**Abstract** During the last decades, researchers have been focusing on the development of economically viable bioprocesses, in order to produce fine and platform chemicals and polymers that will eventually replace the ones deriving from the petrochemical industries. Waste and surplus materials are inexpensive raw materials that do not directly compete with food, while their utilization maximizes the efficiency of the overall production process. Microbial conversion of such materials, deriving from various industrial processes, into biosurfactants is considered beneficial from a societal, environmental, and economic point of view since it contributes toward environmental protection, sustainability, and reduction of costs. Biosurfactants produced by bacteria may eventually replace synthetic surfactants due to their biodegradability and environmental compatibility. Interestingly, the diversity of their chemical structures and their complexity provides them unique features. Therefore, they are characterized by remarkable physicochemical properties so their potential industrial applications are quite diverse as they can be used as emulsifiers, de-emulsifiers, wetting, spreading, and foaming agents, functional food ingredients, and detergents in various industrial sectors such as petroleum and petrochemicals, organic chemicals, food and beverages, cosmetics and pharmaceuticals, mining and metallurgy, agrochemicals and fertilizers, and environmental control and management.

**Keywords** Biosurfactants • Bioconversion • Waste valorization • Renewable substrates • Properties and applications

---

C. Kourmentza (✉) • F. Freitas • M.A.M. Reis
UCIBIO-REQUIMTE, Biochemical Engineering Group, Department of Chemistry, Faculdade de Ciências e Tecnologia, Universidade Nova de Lisboa (FCT/UNL), Quinta da Torre, Campus Universitário, 2829-516 Caparica, Portugal
e-mail: ckourmentza@gmail.com; c.kourmentza@fct.unl.pt

V. Alves
LEAF–Linking Environment Agriculture and Food, Instituto Superior de Agronomia, Universidade de Lisboa, Tapada da Ajuda, 1349-017 Lisboa, Portugal



V.C. Kalia, P. Kumar (eds.), *Microbial Applications Vol.1*,
DOI 10.1007/978-3-319-52666-9_2

# 1 Introduction

The term "surfactants" stands for "*sur*face-*act*ive *agents*" and includes various compounds, of both synthetic and biological origin, which are characterized by their tensioactive properties. Surfactants are amphiphilic molecules consisting of hydrophilic (their heads) and lipophilic (hydrophobic) groups (their tails).Thus, they can be found at interfaces between polar and nonpolar media and air–solid and liquid–solid surfaces. Plants, animals, and microorganisms produce surfactants of biological origin, viz. biosurfactants. Various types of bacteria, yeasts, and fungi produce a wide variety of surfactants with different chemical structures (Figs. 1 and 2). Biosurfactants are metabolites that are not directly associated with bacterial growth and therefore are characterized as secondary metabolites. The primary role and importance of biosurfactants that are secreted to the culture medium, or integrated into the cell wall of microorganisms, is their ability to lower the interfacial tension between water and hydrophobic compounds, allowing the micro-organisms to take up and grow on hydrophobic substrates. Due to that fact, hydrophobic substrates are most commonly used in order to induce biosurfactant production and also as the main carbon source. Although microbial biosurfactants exhibit several important advantages over their chemically-derived counterparts, their extensive use in industry is still limited (Fontes et al. 2012). The industrial development and commercialization of microbial biosurfactants has been hampered by their high production costs, which are due to the low product yields and/or long fermentation periods, the use of expensive substrates like oleic acid, and the formation of a mixture of components instead of pure compounds (Rocha et al. 2007; Syldatk and Hausmann, 2010). In addition, the cost regarding their

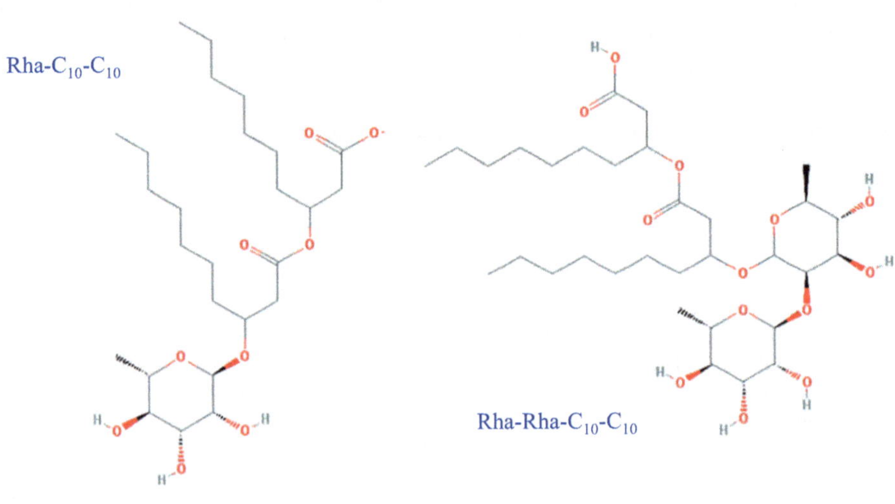


**Fig. 1** Characteristic structure of the mono-rhamnolipid Rha-$C_{10}$-$C_{10}$ and the di-rhamnolipid Rha-Rha-$C_{10}$-$C_{10}$ (obtained by PubChem Compound Database: http://www.ncbi.nlm.nih.gov/pccompound/)

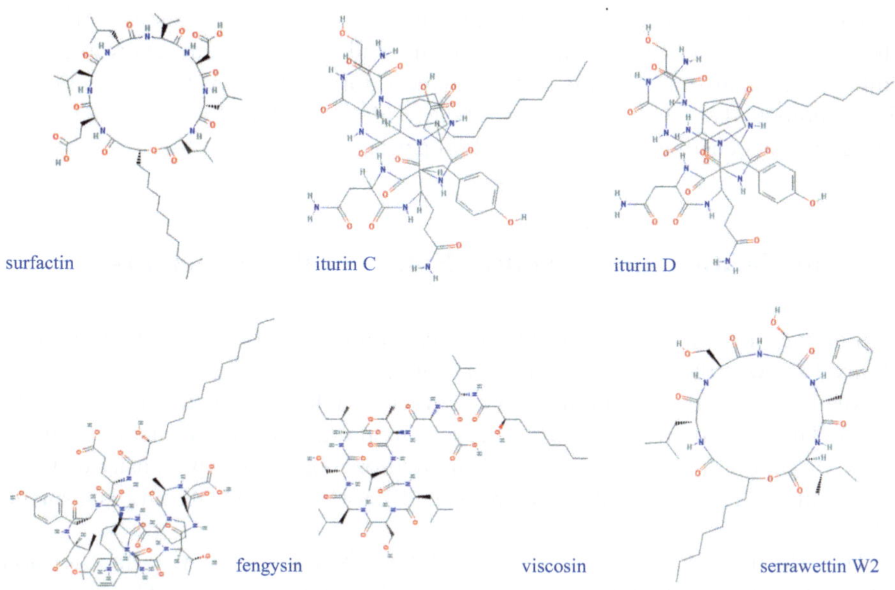


**Fig. 2** Chemical structure of the lipopeptide surfactants surfactin, iturin C, iturin D, fengysin, viscosin, and serrawettin W2 (obtained by PubChem Compound Database: http://www.ncbi.nlm.nih.gov/pccompound/)

downstream processing, in order to separate and recover these microbial surfactants, is considerably high (Banat et al. 2014).

Substrate cost accounts for up to 30–50% of the overall economy of the bioprocess (Bednarski et al. 2004; Nitschke and Pastore 2004; Rufino et al. 2014). In order to address this issue, a lot of researchers worldwide are focusing on the exploitation and biological conversion of low- or no-cost substrates towards the production of biosurfactants. Waste and by-product streams occurring during food processing and consumption, mainly from oil processing and agricultural industries, are ideal for the production of biosurfactants. Gradually, even more food industries in Europe, as well as worldwide, are interested in managing their waste streams, due to strict legislation concerning their environmental disposal, and converting them to high-value added chemicals. Evidence of their interest can be found in their participation in relevant research projects and the development of their R&D (Research and Development) section. As a result, in the near future the concept of Integrated BioRefinery (IBR) is expected to grow and become increasingly popular through the development of bioprocesses that produce minimal waste with limited impact on the environment. As it regards biosurfactants, according to recent reports, their global market is expected to grow at a Compound Annual Growth Rate (CAGR) of 6.14% in 2015–2018, reaching US$2210.5 million and a market volume of around 476,000 tons by 2018–2019, with Europe's market revenue share forecasted at 53.3%, followed by North America (Transparency Market Research 2012; TechNavio 2015).

The aim of this chapter is to provide an overview on the types and different chemical structures of bacterial biosurfactants and an insight into renewable agricultural and industrial substrates that may be valorized towards the production of those compounds as well as on the wide range of potential applications due to their unique physicochemical and biological properties.

## 2 Classification and Chemical Nature of Biosurfactants

One of the most important physicochemical properties of biosurfactants is their ability to lower the surface tension of water, as well as the interfacial tension between two immiscible liquids, due to their amphiphilic nature (Banat et al. 2010). The hydrophilic part may consist of mono-/di-/or polysaccharides, amino acids, or peptides. On the other hand, the hydrophobic part may consist of long-chain saturated, unsaturated, or hydroxylated fatty acids (Syldatk and Hausmann 2010).

The effectiveness of a biosurfactant is defined by the minimum surface tension achieved at a certain concentration, called the Critical Micelle Concentration (CMC), above which no further reduction in the surface tension occurs, by the ability to form water–hydrocarbon emulsions and by enhancing the apparent water solubility of hydrophobic compounds. Biosurfactants are molecules that are characterized by complex and various chemical structures. Classification of microbial surfactants can be based on their functional groups: glycolipids, lipopeptides, polymeric surfactants, as well as particulate surfactants. They can also be categorized as low- and high-molecular mass biosurfactants. In general, low-molecular mass biosurfactants are more effective in reducing the surface and interfacial surface tension, whereas high-molecular mass biosurfactants are more efficient in forming stable hydrocarbon–water emulsions. Characteristic examples of biosurfactants, their producers, applications, and related patents are presented in Table 1.

### *2.1 Glycolipids*

Glycolipids are biosurfactants of low-molecular mass, comprising fatty acids esterified with a carbohydrate moiety. The most extensively studied members of this group include rhamnolipids, sophorolipids, and trehalose lipids.

#### 2.1.1 Rhamnolipids

Rhamnolipids are predominantly produced by strains of the opportunistic pathogen *Pseudomonas aeruginosa*, and they are classified as mono- and di-rhamnolipids

**Table 1** Classification of microbial biosurfactants and their applications

| Type of biosurfactant | Microorganism | Application | Patent(s) |
|---|---|---|---|
| Rhamnolipids | *Pseudomonas aeruginosa*, *Pseudomonas* sp., *Burkholderia* sp. | Biological control, pharmaceuticals and cosmetics, petroleum recovery, pesticides, enhanced oil recovery, toiletries and household cleaners, food products, bioremediation, agriculture, water treatment | Piljac and Piljac (1995)<br>Kim et al. (1997)<br>Van Haesendonck and Vanzeveren (2005)<br>Bensaci et al. (2007)<br>De Santo (2009)<br>Parry et al. (2013)<br>Rhamnopharma Inc. USA (2013)<br>Kuppert et al. (2014) |
| Sophorolipids | *Candida bombicola*, *Candida apicola*, *Candida petrophilum* | Detergents, secondary oil recovery, removal of heavy metals, sophorolipid-based cosmetics for body and skin, antimicrobial agents, protein inducers, and inhibitors in fermentation media | Futura et al. (2002)<br>Pesce (2002)<br>Concaix (2003)<br>Gross and Vishai (2004)<br>Gross et al. (2007)<br>Giessler-Blank et al. (2012)<br>Parry et al. (2012)<br>Gross and Schofield (2014) |
| Trehalose lipids | *Mycobacterium* sp., *Nocardia* sp., *Arthrobacter paraffineus*, *Corynebacterium* sp. *Rhodococcus erythropolis*, *Rhodococcus wratislaviensis* | Detergents, bioremediation, pharmaceuticals formulation, biocontrol agents, therapeutic agents | Fukushi (1984)<br>Sawai Pharmaceutical Co. (1991)<br>Lindoerfer et al. (1992)<br>Unilever Plc. (1992)<br>Lopes and Rodrigues (2003)<br>Volke and Meiche (2005)<br>Ojha et al. (2013) |
| Lipopeptides | *Bacillus subtilis*, *Bacillus pumilus*, *Bacillus licheniformis*, *Serratia marcescens*, *Pseudomonas fluorescens* | Cleaning agents, food and nutrition, biocontrol agents, fungicides, pharmaceuticals, agriculture | Heins et al., (2001)<br>Nero (2007)<br>Krohn and Zinke (2011)<br>Bralkowski et al. (2013)<br>Keith et al. (2014) |
| Polymeric biosurfactants | *Acinetobacter calcoaceticus*, *Acinetobacter radioresistens*, *Pseudomonas* sp., *Candida lypolytica*, *Candida tropicalis* | Cosmetic and pharmaceutical compositions, detergents, personal care products, drug delivery vehicles, emulsifiers, dispersants, stabilizers, wound healing, drug delivery | Hayes and Holzner (1986)<br>Eigen and Simone (1990)<br>Rosenberg and Ron (1990)<br>Rosenberg and Ron (1998)<br>Fuhrman et al. (2001)<br>Gutnick and Bach (2002)<br>Castro et al. (2009)<br>Buthe et al. (2012) |
| Particulate biosurfactants | *Acinetobacter* sp. | Laundry detergent | Unilever Plc. (2015) |

according to the number of rhamnose molecules present (Maier and Soberón-Chávez 2000; Gong et al. 2015). Rhamnose units are linked glycosidically to long-chain hydroxyl fatty acids of varying lengths representing the hydrophobic tail of rhamnolipids. The most common fatty acid constituent is β-hydroxydecanoic acid, although hydroxyl fatty acids consisting of 8–16 carbon atoms may be present depending on the bacterial species employed and growth conditions. Examples of mono- and di-rhamnolipid consisting of two C10 fatty acids, viz. Rha-$C_{10}$-$C_{10}$ and Rha-Rha-$C_{10}$-$C_{10}$, are illustrated in Fig. 1.

Apart from *P. aeruginosa*, other bacteria that belong to *Pseudomonas* genus have been reported in the literature as rhamnolipid producers, i.e., *P. chlororaphis* (Gunther et al. 2005), *P. putida* (Wittgens et al. 2011), and *P. fluorescens* (Toribio et al. 2010; Vasileva-Tonkova et al. 2011). However, a recent study revealed that there are incorrect claims in the literature concerning the ability and productivity of certain bacteria to produce rhamnolipids mainly due to inaccurate qualitative and quantitative analysis and the lack of information concerning the identification of genes involved in rhamnolipid biosynthesis through molecular biology techniques (Marchant et al. 2015). On the other hand, bacteria that belong to *Burkholderia* genus, which is a genus closely related to *Pseudomonas*, as it used to be a branch of it since 1992, have been identified to produce rhamnolipids. Several studies have described the production of rhamnolipids from nonpathogenic, to humans and animals, bacteria of these species such as *B. thailandensis* (Dubeau et al. 2009), *B. kururiensis* (Tavares et al. 2013), *B. glumae* (Costa et al. 2011), and *B. plantarii* (Hoermann et al. 2010), however, in lower rates compared to *P. aeruginosa*. Despite the advantage of employing a Biosafety Level 1 microorganism for the production of rhamnolipids, further work is required in order to optimize the fermentation conditions so as to increase productivity and design a cost-effective industrial process.

Rhamnolipids produced by *Burkholderia* strains are comprised of fewer and less complex rhamnolipid congeners than *P. aeruginosa* and are mostly abundant in di-rhamnolipids. Another main difference lies in the length of the hydrophobic moiety, because unlike *P. aeruginosa*, *Burkholderia* strains produce rhamnolipids consisting mainly of C14 hydroxy fatty acids (Costa et al. 2011).

Rhamnolipids can be produced in high yields within relatively short fermentation periods (from 48 to 120 h). They have been demonstrated to reduce the surface tension to 25–30 mN/m and the interfacial tension, against hexadecane, to 1 mN/m, whereas their CMC ranges between 20 and 225 mg/L (Dubeau et al. 2009). They are able to form emulsions with various hydrocarbons, from hexane to octadecane, that have been shown to be stable for up to 48 h (Patel and Desai 2003). Rhamnolipids are highly applicable, i.e., in areas such as bioremediation, cosmetics and pharmaceuticals, enhanced oil recovery, etc. (Table 1) However, there is still limited biosurfactant production at a marketable scale by companies such as TeeGene Biotech (UK), AGAE Technologies LLC (USA), Jeneil Biosurfactant Co. LLC (USA), Paradigm Biomedical Inc. (USA), Henkel (Germany), BioFuture Ltd. (Dublin), Ecochem Ltd. (Canada), and Rhamnolipid Companies Inc. (USA) (Connolly et al. 2010; Randhawa and Rahman 2014). Their current market price, depending on their purity and composition, varies between \$234–409/10 mg (Sigma-Aldrich) and \$200–350/10 mg (AGAE technologies, USA).

### 2.1.2 Sophorolipids

Sophorolipids, or else sophorosides, are mainly produced by yeast species, and in particular by *Candida bombicola*, *C. petrophilum*, and *C. apicola* species (Kurtzman et al. 2010). They are considered the most promising biosurfactants since they are produced at high rates from nonpathogenic species. Their hydrophilic moiety consists of the sugar sophorose, a diglucose with an unusual β-D-Glc-(1→2)-D-Glc linkage and the 6- and 6′-hydroxyl groups generally acetylated. Sophorolipids are a mixture that consists of six to nine different hydrophobic sophorosides (Desai and Banat 1997) with differences in the fatty acid moiety regarding the chain length, saturation, and position of hydroxylation as well as the lactonization and acetylation pattern; thereof they are categorized into acidic and lactonic sophorolipids. Their hydrophobic part consists of a terminal or subterminal hydroxylated fatty acid, linked β-glycosidically to sophorose. In case the carboxylic end of the fatty acid is free acidic sophorolipid is formed. Otherwise, when the carboxylic end is internally esterified a lactonic sophorolipid is formed. The hydroxyl fatty acid consists of 16 or 18 carbon atoms containing one or more unsaturated bonds (Asmer et al. 1988; Davila et al. 1993; van Bogaert and Soetaert 2011). Due to their structural differences, they possess different physicochemical properties. For example, acidic sophorolipids are characterized by enhanced solubility and foam production; however, lactonic sophorolipids are more efficient as surface-active and antimicrobial agents. Sophorolipids can reduce the surface tension of water to 30–40 mN/m, and their CMCs range between 40 and 100 mg/L (Hirata et al. 2009a, b). Both acidic and lactonic sophorolipids can lower the interfacial tension against *n*-hexadecane from 40 to 5 mN/m and show stability during pH and temperature changes (Desai and Banat 1997).

Sophorolipids are mostly used in personal care products, cosmetics and pharmaceuticals, and cleaning products in general (Table 1). Among sophorolipid-biosurfactant producing companies are Saraya Co. Ltd (Japan), Ecover (Belgium), Groupe Soliance (France), MG Intobio Co. Ltd. (South Korea), ACS Ltd. (Japan), Kaneca Co. (Japan), and Henkel (Germany).

### 2.1.3 Trehalolipids

Trehalolipids, or else trehalose lipids, are mainly produced by Actinomycetales, an order of Actinobacteria that includes *Anthrobacter*, *Rhodococcus*, *Corynebacterium*, *Nocardia*, *Gordonia*, *Micrococcus*, *Brevibacterium*, and *Mycobacterium* species (Asselineau and Asselineau 1978). Trehalose is a nonreducing diglucose linked by an α,α-1,1-glycosidic bond and represents the hydrophilic moiety of this glycolipid. It is also the basic component present in cell wall glycolipids in

Mycobacteria and Corynebacteria. This disaccharide is acylated with long-chain α-branched 3-hydroxyl fatty acids, called mycolic acids (Hausmann and Syldatk 2015). They are diverse in their chemical structures and generally divided into two categories: monomycolates, where acyl groups are linked to the C6 position, and dimycolates, where acyl groups are linked to C6 and C6′ positions of trehalose. Depending on the microorganism, trehalolipids vary due to the complexity of the mycolic acids, which differ in length, structure, degree of unsaturation, and the number of carbon atoms, resulting in almost 500 distinct molecular species.

An example of a well-known trehalose lipid is 6,6′-dimycolytrehalose, also called cord factor. It is produced by mycobacteria and it has been suggested that the mycolic moiety is related with potential virulence mechanisms (Guidry et al. 2007; Ishikawa et al. 2009). Most of trehalose lipids possess strong surface activity since they are able to reduce water surface tension to 19–43 mN/m, the interfacial tension against hydrocarbons like hexadecane, decane, and kerosene to 5 mN/m, and in some cases even less than 1 mN/m. Their CMCs are quite low and can reach 0.7 mg/L while in general they range between 0.7 and 37 mg/L (Shao 2011). In addition when a purified trehalose tetraester, produced by *Rhodococcus wratislaviensis*, was studied as regards its emulsifying properties against several hydrophobic substrates, it was shown that it could efficiently emulsify aliphatic and aromatic hydrocarbons as well as carbon mixtures and several oils and was characterized by higher activity than the ones of chemical surfactants such as Triton X-100, Tween 20, and Tween 80 (Tuleva et al. 2008). Furthermore, they have been reported for their biomedical and therapeutic properties such as their ability to inhibit HSV and influenza virus (Uchida et al. 1989a, b; Inaba et al. 2013).

Despite their advanced characteristics, trehalose lipids have not been successfully commercialized yet. Compared to the other glycolipid biosurfactants, the yield of trehalose lipids is lower, usually below 3 g/L (Christova and Stoineva 2014). Research towards the optimization of culture conditions, bioprocess, as well as the development of engineered strains, so as to enhance yield and productivity, will favor the production of these biosurfactants with such unique features.

## 2.2 *Lipopeptides*

Microbial lipopeptide biosurfactants are in general cyclic compounds. Their hydrophilic part consists of peptides, comprising of seven to ten amino acids, linked to their hydrophobic fatty acid moiety. They are produced by many microorganisms, of both Gram-positive, i.e., *Bacillus* and *Streptomyces*, and Gram-negative species, i.e., *Pseudomonas* and *Serratia* (Raaijmakers et al. 2006). Lipopeptides are versatile and well-known antibiotics because they possess remarkable antimicrobial activity. The lipopeptide antibiotics Polymyxins, produced by *Bacillus polymyxa*, have been reported for their strong surface activities (Marchant and Banat 2012a). Polymyxins are cyclic decapeptides consisting of eight amino acids and linked to a branched fatty acid hydrophobic moiety.

Cyclic lipopeptides produced by *Bacillus subtilis* are classified into three groups: surfactin, iturin, and fengycin (Fig. 2). Surfactin is the most studied lipopeptide biosurfactant and consists of a seven-amino acid cyclic sequence linked to C13-C16 3-hydroxyl fatty acids (Kakinuma et al. 1969a, b). Iturin is also a heptapeptide, different to surfactin, interlinked with a β-amino acid fatty acid consisting of C14-C17, while fengycin consists of ten amino acids connected to C14-C18 fatty acids (Lang 2002). According to the variety of amino acids in the peptide part, iturins have been distinguished as follows: iturin A, iturin C, iturin D, iturin E, bacillomycin D, bacillomycin F, bacillomycin L, bacillomycin Lc, and mycosubtilin (Pecci et al. 2010). Surfactin is considered to be one of the most powerful biosurfactants as it can reduce the surface tension of water from 72 to 27.9 mN/m at low concentrations of 0.005% (Desai and Banat 1997). In addition, lichenysin, produced by *Bacillus licheniformis*, bears similar structure and physicochemical properties to those of surfactin. The certain biosurfactant has been reported to exhibit excellent stability over a wide range of temperature, pH, and salinity (McInerney et al. 1990).

Lipopeptides from *Pseudomonas* species have been also identified and are distinguished as follows: viscosin, amphisin, tolaasin, and syringomycin (Raaijmakers et al. 2006), from which viscosin and amphisisn groups are the best known. Viscosin produced by *P. fluorescens*, *P. libanensis*, and *P. viscosa* is composed of a peptide of nine amino acids, usually linked to C10 fatty acid, while in the amphisin group the peptide consists of eleven amino acids (Saini et al. 2008). Another group of cyclodepsipeptides is serrawettin, produced by *Serratia marcescens*. Serrawettin W1 or Serratamolide is rotationally symmetric and consists of two serine and two 3-hydroxydecanoic acids. Serrawettin W2 is a pentapeptide linked to 3-hydroxydecanoic acid and is responsible for the strain's swarming mobility (Pradel et al. 2007).

## *2.3 Polymeric Biosurfactants*

Polymeric biosurfactants are high-molecular mass biopolymers produced by various bacterial species. They can be composed of lipoproteins, proteins, polysaccharides, and lipopolysaccharides or complex mixtures of those compounds. The best-characterized polymeric biosurfactants are emulsan and alasan produced by *Acinetobacter* species (Rosenberg et al. 1979; Navon-Venezia et al. 1995). Both biosurfactants are characterized of molecular weights of around 1000 kDa. Two types of emulsan have been extensively studied. RAG-1 emulsan produced by *Acinetobacter* ATCC 31012 is composed of a mixture of a protein, anionic exopolysaccharides, as well as lipopolysaccharides with their polysaccharide part consisting of D-galactosamine, D-galactosaminuronic acid, and di-amino-6-deoxy-D-glucose at a ratio of 1:1:1 (Dams-Kozlowska et al. 2008; Hausmann and Syldatk 2015). The lipid part of emulsan corresponds to 23% of its mass and comprises of C10-C18 unsaturated fatty acids (Zhang et al. 1999). Due to their long-chain fatty acids, linked to the polysaccharide backbone, and the protein associated, emulsan is characterized by strong surface and emulsification activities even at low concentrations, 0.001–0.01% (Zosim et al. 1982). BD4 emulsan produced by *Acinetobacter calcoaceticus* BD4 is a rather complex mixture of protein and polysaccharide. Its polysaccharide moiety is composed of a repeating heptapolysaccharide unit that consists of L-rhamnose, D-glucose, D-glucuronic acid, and D-mannose at a ratio of 4:1:1:1. Interestingly, the protein and the polysaccharide do not bear any surface or emulsification activity when purified. However, after mixing both purified parts surface and emulsification activities are reconstituted (Kaplan and Rosenberg 1982; Gudiña et al. 2015; Hausmann and Syldatk 2015).

Alasan is a complex of anionic heteropolysaccharides and proteins produced by *Acinetobacter radioresistens* KA53 (Navon-Venezia et al. 1995, 1998). The polysaccharide part is rather unusual because of the covalently bound alanine. According to previous studies, it became evident that the protein content of alasan is responsible for its strong emulsification activity (Toren et al. 2001). Biodispersan is another polymeric biosurfactant produced by *A. calcoaceticus* A2 that has been known by its ability to bind to inorganic salts, like limestone, and is capable of dispersing limestone in water (Rosenberg et al. 1988; Towner 1996). Polymeric surfactants, even at low concentrations, can alter the rheological properties of aqueous solutions; thereof they are mostly applied as thickeners, emulsion stabilizers, and dispersion and suspension agents. Due to their high viscosity and shear resistance, they also have a wide range of industrial uses in pharmaceuticals and cosmetics formulations and food industries (Table 1).

## 2.4 Particulate Biosurfactants

Membrane or extracellular vesicles are associated with hydrocarbon uptake and act as particulate biosurfactants, able to form microemulsions (Rosenberg 1986). They are composed of proteins, phospholipids, and lipopolysaccharides, and their content in phospholipid and polysaccharides is 5 and 350 times higher than the respective ones found in the outer membrane of the same microorganism. It has been reported that when grown on hexadecane *Acinetobacter* sp. HO1-N is accumulating extracellular membrane vesicles characterized by a diameter of 20–50 nm, while their buoyant density reaches up to 1.16 g/cm$^3$ (Kappeli and Finnerty 1979). In general, bacteria with high cell surface hydrophobicity, such as microorganisms able to degrade hydrocarbons as well as some pathogenic ones, have a strong resemblance to hydrocarbon–water and air–water interfaces due to several cell surface components. Such component structures include M protein and lipoteichoic acid in *Streptococcus*, protein A in *Staphylococcus aureus*, layer A in *Aeromonas salmonicida*, gramicidins in *Brevibacillus brevis*, and fimbriae in *A. calcoaceticus* RAG-1 (Wilkinson and Galbraith 1975; Fattom and Shilo 1985; Rosenberg 1986; Desai 1987; Lang et al. 1989; Desai and Banat 1997).

# 3 Renewable Substrates for Biosurfactant Production

Several industrial and agro-food wastes and by-products have been proposed for cultivation of biosurfactant-producing microorganisms, including hydrophobic and hydrophilic raw materials, like oil refinery wastes, used frying oil, vegetable and fruit processing wastes, glycerol, etc. (Tables 2 and 3). The use of water-soluble substrates is an advantage since it eliminates the problems associated with the use of water-immiscible substrates, like oils and fats. The selection of suitable substrates for production of each type of biosurfactant must take into consideration the composition of the waste materials in terms of nutrients and other components that might interfere with cell growth and/or biosurfactant synthesis. Suitable wastes must have the adequate macro- and micronutrients' composition to allow cell growth and product synthesis. The effect of the constituents of such substrates on the final products' purity and properties must also be considered (Banat et al. 2014). Therefore, a compromise must be made between the cost reduction by the use of low-cost waste substrates and the costs associated with the more complex extraction and purification downstream steps. For applications that require highly pure biosurfactant products, like pharmaceuticals or food products, waste substrates may not be suitable choices because more intensive downstream processing will be necessary.

## *3.1 Hydrophobic Raw Materials*

### 3.1.1 Vegetable Oil Processing Wastes

Vegetable oil processing industries generate large amounts of by-products, including soapstocks, oil seed cakes, and fatty acids residues, which are major sources of water and land pollution (Dumont and Narine 2007). The lipid content of these wastes difficults their biodegradation and makes their treatment costly (Cammarota and Freire 2006). Notwithstanding the chemical complexity of these wastes that renders those challenging raw materials, their high content of fats and oils as well as other nutrients makes them suitable as feedstock in microbial cultivation (Kourmentza et al. 2015). Soapstock is an alkaline waste produced by the seed oil refinery industry during the neutralization process that represents 2–3% of the total oil produced (Benincasa et al. 2002; Benincasa and Accorsini 2008). It is mainly composed of linoleic (50%) and oleic (25%) acids, with minor contents of other fatty acids, such as palmitic (7%) and stearic (4%) acids (Benincasa et al. 2002; Benincasa and Accorsini 2008). Another major waste generated by the oil refining industry is acid water. The treatment costs of this waste are very high due to the large volumes that are generated (10 L of wastewater per kg of oil produced) and its high chemical oxygen demand concentration (2700 mg/L) (Benincasa and Accorsini 2008).

**Table 2** Examples of the use of different hydrophobic wastes for the production of microbial biosurfactants

| Substrate | Type of biosurfactant | Microorganism | Cultivation mode | Production (g/L) | Productivity (g/L day) | Yield (g/g) | References |
|---|---|---|---|---|---|---|---|
| Soapstock | Glycolipid | *Candida antarctica* ATCC 20509 | Batch shake flask | 13.4 | 2.23 | n.a. | Bednarski et al. (2004) |
| | | *Candida apicola* ATCC 96134 | Batch shake flask | 10.3 | 1.72 | n.a. | Bednarski et al. (2004) |
| | Rhamnolipid | *Pseudomonas aeruginosa* LBI | Fed-batch bioreactor | 15.9 | 7.07 | 0.70 | Benincasa et al. (2002) |
| | | | Batch shake flask | 7.3 | 2.43 | 0.22–0.23 | Benincasa and Accorsini (2008) |
| | | | Fed-batch shake flask | 11.72 | 1.95 | n.a. | Nitschke et al. (2005) |
| | | *Pseudomonas aeruginosa* MR01 | Batch shake flask | 13.0 | 1.30 | 0.54 | Partovi et al. (2013) |
| Oil seed cake | Lipopolysaccharide | *Enterobacter* sp. MS16 | Batch shake flask | 1.0–1.5 | 0.14–0.21 | n.a. | Jadhav et al. (2011) |
| | Glycolipopeptide | *Corynebacterium kutscheri* | Batch bioreactor | 6.4 | 1.16 | n.a. | Thavasi et al. (2007) |
| | Glycolipid | *Bacillus megaterium* | Batch bioreactor | 7.8 | 1.42 | n.a. | Thavasi et al. (2008) |
| | | *Ochrobactrum anthropi* 2/3 AB542934 | Batch shake flask | 4.52 | 1.13 | n.a. | Noparat et al. (2014) |
| Fatty acids residues | Sophorolipid | *Candida bombicola* ATCC 22214 | Fed-batch bioreactor | 40–120 | 6.0–12.0 | 0.15–0.40 | Felse et al. (2007) |
| | Glycolipid | *Candida antarctica* ATCC 20509 | Batch shake flask | 10.3 | 1.72 | n.a. | Bednarski et al. (2004) |
| | | *Candida apicola* ATCC 96134 | Batch shake flask | 8.2 | 1.37 | n.a. | Bednarski et al. (2004) |

| | | | | | | | |
|---|---|---|---|---|---|---|---|
| Oil refinery residue | Glycolipid | *Candida sphaerica* UCP0995 | Batch shake flask | 1.6–9.0 | 0.26–1.5 | n.a. | Sobrinho et al. (2008); Luna et al. (2013) |
| | Polysaccharide–protein–lipid complex | *Candida lipolytica* UCP 0988 | Batch shake flask | ~4.5 | 0.75 | n.a. | Rufino et al. (2007) |
| | Lipopeptide | *Candida lipolytica* UCP 0988 | Batch shake flask | 8.0 | 2.67 | n.a. | Rufino et al. (2014) |
| Deodorizer distillate | Rhamnolipid | *Pseudomonas aeruginosa* MR01 | Batch shake flask | 21.3 | 2.13 | 0.51 | Partovi et al. (2013) |
| Two-phase olive mill waste | Surfactin | *Bacillus subtilis* DSM3256 | Batch shake flask | 0.25 | 0.17 | n.a. | Maass et al. (2016) |
| | | *Bacillus subtilis* N1 | Batch shake flask | 0.003 | 0.0003 | n.a. | Ramírez et al. (2015) |
| | Rhamnolipids | *Pseudomonas aeruginosa* PAO1 | Batch shake flask | 0.19 | 0.02 | n.a. | Ramírez et al. (2015) |
| Waste frying oil | Rhamnolipids | *Pseudomonas aeruginosa* LBI | Batch shake flask | 7.6 | 1.27 | n.a. | Nitschke et al. (2005) |
| | | *Pseudomonas aeruginosa* DG30 | Batch bioreactor | 15.56 | 5.19 | n.a. | Zheng et al. (2011) |
| | | *Pseudomonas* SWP-4 | Batch shake flask | 13.93 | 5.57 | n.a. | Lan et al. (2015) |
| | | *Pseudomonas aeruginosa* MTCC 2297 | Batch shake flask | 2.26 | 0.57 | n.a. | George and Jayachandran (2013) |
| | | *Pseudomonas aeruginosa* D | Batch shake flask | 3.55 | 0.89 | n.a. | George Jayachandran (2013) |
| | Glycolipid | *Pseudomonas cepacia* CCT6659 | Batch shake flask | 5.2 | 0.87 | n.a. | Silva et al. (2014a, b) |
| | Surfactin | *Bacillus subtilis* MTCC 2423 | Batch shake flask | 0.45–0.65 | 0.09–0.13 | n.a. | Vedaraman and Venkatesh (2011) |
| | Sophorolipid | *Candida bombicola* ATCC 22214 | Batch shake flask | 5.13–8.23 | 0.62–0.99 | n.a. | Wadekar et al. (2012) |

(continued)

**Table 2** (continued)

| Substrate | Type of biosurfactant | Microorganism | Cultivation mode | Production (g/L) | Productivity (g/L day) | Yield (g/g) | References |
|---|---|---|---|---|---|---|---|
| Animal fats and grease | Rhamnolipid | *Pseudomonas aeruginosa* LBI | Batch shake flask | 6.8 | 1.13 | n.a. | Nitschke et al. (2005) |
| | | *Pseudomonas aeruginosa* ATCC 10145 | Batch bioreactor | 3.84 | 1.92 | n.a. | Borges et al. (2012) |
| | Sophorolipid | *Candida bombicola* ATCC 22214 | Batch bioreactor | 120 | 42.35 | n.a. | Deshpande and Daniels (1995) |
| Waste motor lubricant oil | Glycolipopeptide | *Corynebacterium kutscheri* | Batch bioreactor | 3.85 | 0.70 | n.a. | Thavasi et al. (2007) |
| | Glycolipid | *Bacillus megaterium* | Batch bioreactor | ~4.0 | 0.73 | n.a. | Thavasi et al. (2008) |
| Crude oil | Glycolipid | *Bacillus megaterium* | Batch bioreactor | ~3.0 | 0.55 | n.a. | Thavasi et al. (2008) |

*n.a.* data not available

**Table 3** Examples of the use of different hydrophilic wastes for the production of microbial biosurfactants

| Substrate | Type of biosurfactant | Microorganism | Cultivation mode | Production (g/L) | Productivity (g/L day) | Yield (g/g) | References |
|---|---|---|---|---|---|---|---|
| Glycerol by-product | Rhamnolipid | *Pseudomonas aeruginosa* LBI | Batch shake flask | 8.05 | 1.34 | n.a. | Nitschke et al. (2005) |
| | | *Pseudomonas aeruginosa* MSIC02 | Batch shake flask | 1.2 | 0.30 | 0.57 | Sousa et al. (2011) |
| | Sophorolipid | *Candida bombicola* ATCC 22214 | Fed-batch bioreactor | 60.0 | 12.0 | n.a. | Ashby et al. (2005) |
| | Glycolipid | *Pseudozyma antarctica* JCM 10317 | Batch shake flask | 12.6 | 0.60 | n.a. | Morita et al. (2007) |
| Cashew apple juice | Rhamnolipid | *Pseudomonas aeruginosa* ATCC 10145 | Batch shake flask | 3.86 | 1.93 | n.a. | Rocha et al. (2007) |
| | Surfactin | *Bacillus subtilis* LAMI008 | Batch shake flask | 0.0035–0.32 | 0.0012–0.11 | n.a. | Rocha et al., (2009); Oliveira et al. (2013) |
| Banana peel | Lipopeptpide | *Halobacteriaceae archaeon* AS65 | Batch shake flask | 5.3 | 2.65 | n.a. | Chooklin et al. (2014) |
| Orange peel | Rhamnolipid | *Pseudomonas aeruginosa* MTCC 2297 | Batch shake flask | 9.2 | 1.15 | n.a. | George and Jayachandran (2009) |
| | Lipopeptide | *Bacillus subtilis* SPB1 | Batch shake flask | 2.5–4.4 | 1.2–2.2 | n.a. | Ghribi et al. (2011) |
| Lime peelings | Rhamnolipid | *Pseudomonas aeruginosa* MTCC 2297 | Batch shake flask | 4.4 | 0.55 | n.a. | George and Jayachandran (2009) |
| Pineapple peel | Glycolipopeptide | *Pantoea* sp. | Batch shake flask | 1.5 | 0.50 | n.a. | Almeida et al. (2012) |
| Carrot peel waste | Rhamnolipid | *Pseudomonas aeruginosa* MTCC 2297 | Batch shake flask | 5.7 | 0.71 | n.a. | George and Jayachandran (2009) |
| Potato process waste streams | Polysaccharide | *Klebsiella* sp. strain RJ-03 | Batch shake flask | 9.6 | 3.20 | n.a. | Jain et al. (2013) |
| | Surfactin | *Bacillus subtilis* ATCC 21332 | Chemostat | 0.6 | 0.6 | n.a. | Noah et al. (2005) |
| | | | Batch shake flask | 0.87–4.10 | 0.29–1.37 | 0.01–0.16 | Thompson et al. (2000) |

(continued)

**Table 3** (continued)

| Substrate | Type of biosurfactant | Microorganism | Cultivation mode | Production (g/L) | Productivity (g/L day) | Yield (g/g) | References |
|---|---|---|---|---|---|---|---|
| Cassava wastewater | Surfactin | *Bacillus subtilis* LB5a | Batch bioreactor | 2.4 | 1.6 | n.a. | Barros et al. (2008) |
| | | | Batch shake flask | 3.0 | 1.5 | n.a. | Nitschke and Pastore (2004) |
| | | *Bacillus subtilis* ATCC 21332 | Batch shake flask | 2.2 | 1.47 | n.a. | Nitschke and Pastore (2004) |
| Sugarcane bagasse | Polysaccharide | *Klebsiella* sp. strain RJ-03 | Batch shake flask | 6.2 | 2.07 | n.a. | Jain et al., (2013) |
| Lignocellulose hydrolysates | Sophorolipids | *Candida bombicola* ATCC 22214 | Batch shake flask | 3.6–84.6 | 0.45–10.58 | 0.55 | Samad et al. (2015) |
| | | *Wickerhamiella domercqiae var. sophorolipid* CGMCC 1576 | Batch shake flask | 39.08–42.06 | 13.03–14.02 | n.a. | Ma et al. (2014) |
| Molasses | Rhamnolipid | *Pseudomonas luteola* B17 | Batch shake flask | 0.40–0.51 | 0.40–0.51 | n.a. | Onbasli and Aslim (2009) |
| | | *Pseudomonas putida* B12 | Batch shake flask | 0.44–0.50 | 0.44–0.50 | n.a. | Onbasli and Aslim (2009) |
| | Sophorolipids | *Candida bombicola* NRRL Y-17069 | Batch shake flask | 12.67 | 2.11 | n.a. | Daverey and Pakshirajan (2009a) |
| | | *Candida bombicola* ATCC22214 | Batch shake flask | 21.0 | 4.20 | n.a. | Solaiman et al. (2004) |
| | | | Fed-batch bioreactor | 53.0 | 5.89 | n.a. | Solaiman et al. (2007) |
| | Lipopolysaccharide | *Enterobacter* sp. MS16 | Batch shake flask | 1.0 | 0.14 | n.a. | Jadhav et al. (2011) |
| Whey | Surfactin | *Bacillus subtilis* ATCC 6633 | Batch shake flask | 0.18–0.29 | 0.06–0.10 | n.a. | Cagri-Mehmetoglu et al. (2012) |
| | Sophorolipids | *Candida bombicola* NRRL Y-17069 | Batch bioreactor | 33.0 | 4.12 | n.a. | Daverey and Pakshirajan (2010) |

*n.a.* data not available

Soapstock has been studied as substrate for the cultivation of the yeasts *Candida antarctica* and *Candida apicola* (Bednarski et al. 2004). Both strains were able to grow on the oil refinery waste and produce glycolipids (between 8.2 and 13.4 g/L) within 144 h of cultivation (Table 2). Soapstock was also tested as substrate for the production of rhamnolipds by *P. aeruginosa* strains that reached productions between 10.3 and 15.9 g/L (Benincasa et al. 2002; Nitschke et al. 2005; Partovi et al. 2013) (Table 2). In another study, acidic water and soapstock from sunflower oil refinery were used for the production of rhamnolipids (Benincasa and Accorsini 2008). Although a lower production was obtained (7.3 g/L), the use of both the wastes, one alkaline and the other acidic, allowed to more easily adjust the pH of the cultivation medium to physiological values. Oil seed cake is the solid waste obtained after mechanical extraction of oil from the seeds (Sharma et al. 2013). Depending on the type of oil seed and extraction procedure, such wastes may still contain considerable amounts of oil (up to 26%), together with variable amounts of proteins (6–50%), fiber (3–40%), and ashes (4–12%) (Ramachandran et al. 2007; King et al. 2010), which render their disposal and treatment problematic. In view of this, the conversion of oil seed cake into high-value added microbial products is of great interest. A few examples of the use of such wastes as substrates for the production of biosurfactants include the production of a lipopolysaccharide by the bacterium *Enterobacter* sp. MS16 using sunflower oil cake and groundnut oil cake (Jadhav et al. 2011), the production of a glycolipopeptide by *Corynebacterium kutscheri* grown on peanut oil cake (Thavasi et al. 2007), and the production of glycolipids by *Bacillus megaterium* and *Ochrobactrum anthropi* from peanut oil cake (Thavasi et al. 2008) and palm oil decanter cake (Noparat et al. 2014), respectively (Table 2).

The olive oil extraction process generates huge amounts of a semisolid sludge waste that accounts for almost 80% of the processed olive, the two-phase olive mill waste, composed of the vegetation water of the olive, husks, pits, and oil (Maass et al. 2016). The high organic matter concentration, including a lipid content of 0.03–4.25%, renders this waste great potential for use as feedstock for the production of many high-value added products. Recently, it was demonstrated that two-phase olive mill waste can be used as carbon source for the production of surfactin by *B. subtilis* DSM 3256 and *B. subtilis* N1, respectively (Maass et al. 2016; Ramírez et al. 2015). This waste was also tested as substrate for the production of rhamnolipids by *P. aeruginosa* PAO1 (Ramírez et al. 2015) (Table 2). Further examples of the use of by-products and wastes generated by the oil processing industry as substrates for the cultivation of biosurfactant-producing strains include the production of glycolipids by several species of the Genus *Candida*, namely, *C. bombicola* ATCC 22214 (Felse et al. 2007), *C. antarctica* ATCC 20509, and *C. apicola* ATCC 96134 (Bednarski et al. 2004) grown on fatty acid residues and *C. sphaerica* UCP0995 grown on groundnut oil refinery residue (Sobrinho et al. 2008; Luna et al. 2013). *C. lipolytica* UCP 0988 was also reported to produce a polysaccharide–protein–lipid complex and a lipopeptide from groundnut oil refinery residue (Rufino et al. 2007) and soybean oil refinery residue (Rufino et al. 2014), respectively (Table 2).

### 3.1.2 Waste Frying Oils

During the frying process for food products, oils undergo certain degradation reactions, including hydrolysis, peroxidation, and polymerization (Knothe and Steidley 2009; Wadekar et al. 2012). The major components of used frying oils are triglycerides and di-glycerides, while monoglycerides and free fatty acids represent minor fractions (6–8%) (Choe and Min 2007; Martino et al. 2014). The large amounts of waste frying oils make their disposal a growing problem (Vedaraman and Venkatesh 2011). Although waste frying oils are commonly used as raw material for biodiesel production (Knothe and Steidley 2009; Cruz et al. 2016), there is a surplus of this waste that can be efficiently converted into biosurfactants. Waste frying oils have been tested as substrate for the production of rhamnolipids by several *P. aeruginosa* strains (Nitschke et al. 2005; Zheng et al. 2011; George and Jayachandran 2013; Lan et al. 2015), glycolipids (Wadekar et al. 2012; Silva et al. 2014a, b), and the lipopetide surfactin (Vedaraman and Venkatesh 2011) (Table 2). High volumetric productivities have been reported for the production of rhamnolipids by *P. aeruginosa* SWP-4 (5.19 g/L day) (Zheng et al. 2011) and *P. aeruginosa* SWP-4 (5.57 g/L day) (Lan et al. 2015), with used vegetable oils as the sole carbon source (Table 2).

### 3.1.3 Animal Fats and Grease

Meat processing industries generate large amounts of waste animal fat, tallow, and lard that are sources of fatty acids, soaps, and lubricants (Deshpande and Daniels 1995). Edible fats are mainly used for cooking foods. However, recently, this market was shifted toward the use of vegetable oils (Santos et al. 2013). In view of this, large amounts of cheap animal fat are being generated, making their utilization and disposal problematic (Banat et al. 2014). In view of the limited number of available applications, they are potential candidates for microbial processes. High production of sophorolipids, 120 g/L in a 3 days cultivation, has been achieved by using animal fat as substrate for the cultivation of *C. bombicola* ATCC 22214 (Deshpande and Daniels 1995). The use of chicken fat and the greasy effluent from the treatment station of a slaughterhouse were also tested for the production of rhamnolipids by *P. aeruginosa* LB1 (Nitschke et al. 2005) and *P. aeruginosa* ATCC 10145 (Borges et al. 2012), respectively (Table 2). Bovine grease was also tested for the production of a glycolipid by *Candida lipolytica* UCP 0988 (Santos et al. 2013). These studies demonstrated the potential of such wastes for use as substrates for the production of different biosurfactants.

### 3.1.4 Crude Oil and Refined Petroleum Products

The production of biosurfactants utilizing crude oil and refined petroleum products could help to tackle the environmental problem caused by their spillage. This would allow simultaneously cleaning the oil spill and converting the spilled oil into valuable bioproducts. The production of biosurfactants from these wastes could help to ease the financial burden required for mechanical cleaning of oil spillage. Diesel oil recovered from the groundwater at a petrol station has been tested for the cultivation of different biosurfactant-producing strains, namely, *Staphylococcus hominis*, *Kocuria palustris*, and *P. aeruginosa* LBI, but no biosurfactant production was detected (Mariano et al. 2008). On the other hand, waste motor lubricant oil was successfully used for the production of a glycolipopeptide by *Corynebacterium kutscheri* (Thavasi et al. 2007) and a glycolipid by *B. megaterium* (Thavasi et al. 2008) (Table 2). *B. megaterium* was also able to use crude oil for the production of glycolipid (Thavasi et al. 2008). Although low volumetric productivities were attained (0.55–0.73 g/L day), the strains' ability to use these types of substrates for cell growth and biosurfactant production became evident. The use of diesel and kerosene as substrates for the production of biosurfactant by a halo-thermophilic bacterium, *B. licheniformis* B-4, isolated from a petrochemical waste dump site has also been demonstrated (Igbonekwu et al. 2014). These biosurfactants, as well as the producing strains, can be used in the bioremediation of sites contaminated with hydrocarbons.

## *3.2 Hydrophilic Raw Materials*

### 3.2.1 Glycerol from Biodiesel Production

Glycerol is a by-product of many industrial processes, mainly from biodiesel production. Besides glycerol, this by-product contains residual amounts of methanol, NaOH, fat/oil, esters, sulfur compounds, proteins, and minerals (Hu et al. 2012). The presence of these contaminants prevents the use of this by-product in many of the traditional glycerol applications (Freitas et al. 2010). Moreover, glycerol by-product is generated in quantities that exceed its current consumption, thus making urgent the development of new routes for its valorization (Çelik et al. 2008). The use of glycerol as feedstock for cultivation of microorganisms contributes to make the bioprocesses more cost-effective (Hilliou et al. 2009). The use of glycerol, which is a water-soluble substrate, also avoids the problems related to the presence of water-immiscible substrates, like oils and fats, in culture broth (Nitschke et al. 2005).

The suitability of glycerol for the production of different biosurfactants was demonstrated in several studies. High rhamnolipid production, up to 8.05 g/L in 6 days of cultivation of *P. aeruginosa* LB1 in glycerol, has been reported (Nitschke et al. 2005). More recently, *P. aeruginosa* MSIC02, a strain isolated from oil-contaminated soils, was reported to produce rhamnolipids from crude glycerol, but only 1.2 g/L were obtained in 4 days batch shake flask cultivations (Sousa et al. 2011). The use of glycerol as substrate was also reported for the production of other glycolipids by *C. bombicola* ATCC 22214 (Ashby et al. 2005) and *Pseudozyma antarctica* JCM 10317 (Morita et al. 2007). *C. bombicola* ATCC 22214 reached a high production of sophorolipids (60 g/L) with high volumetric productivity (12.0 g/L day) (Ashby et al. 2005). Lower production was reported for *P. antarctica* JCM 10317, but the produced glycolipids, mannosyl erythritol lipids, are highly promising biosurfactants, because they exhibit surface-active properties and biological activity. The use of glycerol is advantageous over lipidic substrates because it eliminates the problem of free fatty acids and mono- or diacylglycerols being carried over to the final product (Morita et al. 2007).

### 3.2.2 Fruit and Vegetable Processing Wastes

Many wastes generated by fruit and vegetable processing industries can also be exploited for the production of biosurfactants. Such sugar-rich raw materials can provide many of the adequate macro- and micronutrients for microbial cultivation. Examples of the utilization of fruit and vegetable wastes include cashew apple juice, peels from banana, orange, or potato, cassava wastewater, and corn powder (Table 3). Cashew apple (*Anacardium occidentale* L.) is a tropical pseudofruit in which the real fruit is a nut, the cashew nut (Rocha et al. 2006; Oliveira et al. 2013). The edible portion represents 90% and is rich in vitamin C, flavour, and aroma (Rocha et al. 2007, 2009). Besides cashew nut production, cashew apples can be consumed in products such as juice and ice cream. Since only 12% of the pseudofruit is used industrially, large amounts of wastes are generated and usually left to spoil in soil, causing environmental pollution (Rocha et al. 2007). The carbohydrate-rich composition of cashew apple juice turns it into a valuable and low-cost raw material for several potential applications (Rocha et al. 2006).

Cashew apple juice was used as feedstock for the cultivation of the yeast *Yarrowia lipolytica* IMUFRJ 50682. The microorganism could grow and produce biosurfactant, achieving a production of 6.9 g/L within 4 days of cultivation (Fontes et al. 2012). Exploitation of cashew apple juice as substrate for *A. calcoaceticus* growth and production of emulsan, an extracellular polyanionic amphipathic heteropolysaccharide, has been investigated (Rocha et al. 2006). Biosurfactant production was evidenced by the reduction of the surface tension by 11% and the emulsification activity of the supernatant against kerosene. Cashew apple juice was also tested as substrate for the biosynthesis of rhamnolipids and surfactin (Table 3). Production of 3.86 g/L of rhamnolipids was achieved within 2 days of cultivation of *P. aeruginosa* ATCC 10145 on medium supplemented with cashew apple juice (Rocha et al. 2007). Surfactin production by *B. subtilis* LAMI008 was considerably lower (0.0035–0.32 g/L in 3 days cultivations) (Rocha et al. 2009; Oliveira et al. 2013). The increasing demand and consumption of processed fruits and vegetables generates large quantities of wastes, including peels, stalks, and pomaces. These wastes often represent an economic burden to the processing industries that have to deal with their disposal. Hence, there is a growing interest in the utilization of wastes like fruit or vegetable peels for microbial cultivation and production of high-value added products.

Banana peel, which accounts for almost 30% of the fruit, is the main by-product of the banana processing industry (Saisa-Ard et al. 2013). This by-product is rich in fiber (50%), proteins (7%), amino acids, polyunsaturated fatty acids, and potassium (Emaga et al. 2007; Chooklin et al. 2014). Due to its high water content and nitrogen and phosphorus contents, banana peel constitutes an environmental problem. Banana peel has recently been tested as substrate for the production of a lipopetpide by *Halobacteriaceae archaeon* AS65 (Chooklin et al. 2014), and a high production (5.30 g/L) was obtained within 2 days of cultivation. The produced lipopeptide had high surface tension reduction and small CMC value, as well as thermal and pH stability, and a high salt tolerance.

Orange peelings, carrot peel waste, and lime peelings have been also valorized for the production of rhamnolipids by *P. aeruginosa* MTCC 2297 (George and Jayachandran 2013). Volumetric productivity values between 0.55 and 1.15 g/L day were reported, with orange peelings being the best substrate among the tested ones (Table 3). Orange peel was also used for the production of a lipopeptide by *B. subtilis* SPB1, and volumetric productivities between 1.2 and 2.2 g/L day were reported (Ghribi et al. 2011). The suitability of pineapple peel as substrate for the production of a glycolipopeptide by *Pantoea* sp. (Table 3) has been evident (Almeida et al. 2012). Dried potato peels were tested for the production of a thermostable lipopeptide by two *B. subtilis* strains in solid-state and submerged fermentations. Between 20 and 80 mg of lipopeptide was obtained per g of dry substrate (Das and Mukherjee 2007). In addition, the ability of *Klebsiella* sp. strain RJ-03 to use potato peel powder as substrate for the production of a polysaccharide with a high volumetric productivity, 3.20 g/L day, has been studied (Jain et al. 2013).

Other potato processing wastes, besides potato peels, like potato pulp and effluents, have also been screened for their suitability to support the growth of biosurfactant-producing microorganisms. Production of surfactin by *B. subtilis* using high-solids (HS) and also low-solids (LS) potato process effluents has been evaluated (Thompson et al. 2000). In another study, surfactin production by *B. subtilis* has also been reported by using an airlift reactor (Noah et al. 2002). The produced surfactin was recovered from the foam that overflowed the effluent air filter. Although both studies were successful when purified starch was used, the presence of recalcitrant indigenous bacteria hampered the production of surfactin since they outcompeted *B. subtilis*. Spores of the indigenous bacteria remained even after heat treatment of the wastes by autoclaving. Latter studies reduced the cultivation time from about 48 to 12–24 h and also prevented contamination, reaching a production of 0.6 g/L surfactin (Noah et al. 2005).

### 3.2.3 Molasses

Molasses is a low-cost sucrose-rich by-product of sugar production. Its content in sucrose, as well as other compounds like minerals, organic compounds, and vitamins, makes molasses a major raw material for production of several microbial products, including baker's yeast, citric acid, and amino acids, among others (Makkar and Cameotra 1997). Molasses resulting from the processing of different products, such as sugarcane, beet, date, and soy, have also been tested as substrates for the production of different biosurfactants (Table 3). For example, beet molasses have been used for the production of biosurfactants by *Pseudomonas luteola* B17 and *P. putida* B12 (Onbasli and Aslim 2009) and *B. subtilis* B20 (Al-Bahry et al. 2013). Production of sophorolipids in high productivities (2.11–5.89 g/L day) by *C. bombicola* strains using molasses as substrates was also demonstrated (Solaiman et al. 2004, 2007; Daverey and Pakshirajan 2009a).

### 3.2.4 Other Wastes

Other unconventional wastes, such as cassava wastewater, cheese whey, lignocellulosic residues, and distillery wastes, among others, have been proposed as low-cost raw materials for the production of different types of biosurfactants (Table 3). A two-stage fed-batch process has been proposed based on the use of deproteinized whey for the cultivation of the yeast *Cryptococcus curvatus* ATCC 20509 and production of single-cell oil that was then used together with rapeseed oil for the production of sophorolipid by *C. bombicola* (Daniel et al. 1999). Agricultural lignocellulosic residues, including straws, hulls, and stalks, are abundant, inexpensive, and carbon-rich resources with great potential for biotechnological upgrade. Nevertheless, their use is still limited to animal feed or soil incorporation, being the most part burned or sent to landfills with associated pollution issues (Mussatto 2014; Rafiqul and Sakinah 2013). Lignocellulose hydrolysates have been used for the production of sophorolipids by *C. bombicola* ATCC 22214 and *Wickerhamiella domercqiae* var. *sophorolipid* CGMCC 1576, reaching productivity values as high as 10.48–14.02 g/L day (Ma et al. 2014; Samad et al. 2015). Cassava wastewater has also been used for the cultivation of *B. subtilis* strains and production of 2.2–3.0 g/L of surfactin was achieved (Nitschke and Pastore 2004; Barros et al. 2008) (see Table 3).

The wide diversity of wastes that have been screened reveals the versatility of microbial biosurfactant producers. Depending on the type of biosurfactant, the producing strain, and the envisaged application, several wastes may be suitable for use as raw material in the bioprocesses. However, up to date, the use of wastes as substrates has not reached industrial scale, mainly because most of the proposed strategies of cultivation, as well as upstream and downstream procedures, are still under optimization. Nevertheless, the suitability of many of the tested wastes was demonstrated and opened up the opportunity for their valorization toward the production of many different types of biosurfactants.

# 4 Properties and Applications of Biosurfactants

## *4.1 Properties*

Biosurfactants, as amphiphilic molecules having both hydrophobic and hydrophilic moieties, are able to orient themselves and accumulate at interfaces, decreasing the interfacial tension of liquids and surface tension. Like other surfactants, besides their chemical composition, other physical-chemical properties are assessed in order to foresee their functional properties. The hydrophilic–lipophilic balance (HLB) may be calculated from the chemical composition, according to Eq. (1) (Griffin 1954):

$$HLB = 20\left(\frac{MW_{HP}}{MW_{SA}}\right) \tag{1}$$

where $MW_{HP}$ and $MW_{SA}$ refer to the molecular mass of the hydrophilic part and of the whole surface agent, respectively. The HLB value varies between 0 and 20. It enables the prediction of the behavior of surface-active compounds as follows: 0–3 antifoaming agents, 4–6 water-in-oil emulsifiers (W/O), 7–9 wetting agents, 8–18 oil-in-water emulsifiers (O/W), 13–15 typical detergents, and 10–18 solubilizers/ hydrotropes. Even though the HLB value is a good starting point to estimate the surfactant properties, experimental characterization must be further carried out. The emulsification index ($E_{24}\%$) is a simple method generally used to evaluate the emulsification capacity of a surfactant even before its chemical characterization (Freitas et al. 2011). It is based on mixing vigorously the surfactant with a hydrophobic compound, and the ratio between the emulsion phase heights by the overall height is measured after 24 h. Decrease of interfacial and surface tension, and CMC, may be measured by tensiometric methods, i.e, Du-Noüy-Ring or Wilhelmy plate method. Table 4 presents the indicative values of these parameters for some of the most studied biosurfactants.

## *4.2 Applications*

The diversity of microbial biosurfactants enables a variety of functional properties, such as emulsification, foaming and surface activity, phase separation, and biological activity. In addition, they are biodegradable, biocompatible, highly specific, and stable at extreme conditions of pH, temperature, and salinity. All these

**Table 4** Indicative values of surface and interfacial tension, CMC, and emulsification index for selected biosurfactants

| Microrganism | Biosurfactant | ST (mN/m) | IFT (mN/m) | CMC (mg/L) *or* CMD | $E_{24}$% | Reference(s) |
|---|---|---|---|---|---|---|
| *Rhamnolipids* | | | | | | |
| *Pseudomonas alcaligenes* | Rha-$C_{10}$-$C_{10}$: 42.2%<br>Rha-$C_{12}$-$C_{10}$: 13.5%<br>Rha-Rha-$C_8$-$C_{10}$: 2.3%<br>Rha-Rha-$C_{10}$-$C_{10}$: 6.2%<br>Rha-Rha-$C_{10}$-$C_{12}$: 14.6%<br>Rha-Rha-$C_{12}$-$C_{10}$: 13.8% | 31 | | 30 mg/L | *Cell-free supernatant containing 2.3 g/L*<br>*S/H ratio 1:1*<br>Hexane: 85%<br>KAV: 88 %<br>Crude oil: 70% | Oliveira et al. (2009) |
| *Pseudomonas aeruginosa* D | Rha-$C_{12}$-$C_{10}$<br>Rha-$C_{10}$-$C_8$/Rha-$C_8$-$C_{10}$<br>Rha-Rha-$C_{10}$-$C_{10}$ | 24.02 | | | *Cell-free supernatant containing 3.55 g/L*<br>*S/H ratio 6:4*<br>Kerosene: 71.7 | George and Jayachandran (2013) |
| *Burkholderia thailandensis* | Rha-$C_{14}$-$C_{14}$: 6.5%<br>Rha-Rha-$C_{12}$-$C_{12}$: 3.4%<br>Rha-Rha-$C_{12}$-$C_{14}$: 12.3%<br>Rha-Rha-$C_{14}$-$C_{14}$: 69.9%<br>Rha-Rha$C_{14}$-$C_{16}$: 5.6% | 43 | | 250 | | Dubeau et al. (2009) |
| *Burkholderia glumae* | Rha-Rha-$C_{12}$-$C_{12}$: 18.8%<br>Rha-Rha-$C_{12}$-$C_{14}$: 18.2%<br>Rha-Rha$C_{14}$-$C_{14}$: 62.8% | 28.5 | Hexadecane: 1.8 | 25–27 mg/L | *solution at CMC*<br>*S/H ratio 1:1*<br>Canola oil: 100%<br>Motor oil: 100%<br>Hexadecane: 44.5%<br>Toluene: 61.5%<br>Pentane: 50%<br>Cannot emulsify cyclohexane | Costa et al. (2011) |

| | | | | | | |
|---|---|---|---|---|---|---|
| *Sophorolipids* | | | | | | |
| *Candida bombicola* | | 34.15 | | 59.43 mg/L | | Daverey and Pakshirajan (2009b) |
| *Candida bombicola* ATCC22214 | | 33.5 | | 9.5 mg/L | | Joshi-Navare et al. (2013) |
| *Candida bombicola* NRRL Y-17069 | (17-hydroxyoctadecanoic)-1′4-″-lactone-6′6″-diacetate sophorolipid | 34.18 | *n*-hexane: 0.99<br>Sunflower oil: 3.44<br>Olive oil: 4.46 | 27.17 mg/L | | Daverey and Pakshirajan (2010) |
| *Trehalose lipids* | | | | | | |
| *Rhodococcus* sp. SD-74 | 3,4-di-*O*-alkanoyl-2-*O*-succinoyl-alpha-D-glucopyranosyl-2′-*O*-succinoyl-alpha-D-glucopyranoside, major fatty acid of was $C_{16}$ | 19 | | $5.6 \times 10^{-6}$ M | | Tokumoto et al. (2009) |
| *Rhodococcus erythropolis* 51T7 | Trehalose tetraester consisting of six components, most abundant (44%) TH-suc-$C_9$-$C_9$-$C_{10}$ or TH-suc-$C_{11}$-$C_{10}$-$C_7$ | 27.9 | Hexadecane: 5 | 37 mg/L | | Marqués et al. (2009) |
| *Tsukamurella spumae* DSM 44113 | 1-α-glucopyranosyl-1-α-glucopyranoside carrying two acyl chains varying of C4 to C6 and C16 to C18 at the 2′ and 3′ carbon atom of one sugar unit. | 32.77 | | 50 mg/L | *Cell-free supernatant containing 50 mg/L*<br>*S/H Ratio 1:1*<br>Kerosene: ~46% | Kügler et al. (2014) |
| *Rhodococcus wratislaviensis* BN38 | 2,3,4,2′-trehalose tetraester with molecular mass of 876 g/mol | 24.4 | Hexadecane: 1.3 | 5 mg/L | *Purified, 0.5% w/v*<br>*S/H Ratio 1:1*<br>Toluene: 54%<br>Benzene: 68%<br>Xylene 62%<br>Hexadecane 65%<br>N-alkanes ($C_{12-22}$): 47%<br>Kerosene: 50% | Tuleva et al. (2008) |

(continued)

**Table 4** (continued)

| Microrganism | Biosurfactant | ST (mN/m) | IFT (mN/m) | CMC (mg/L) *or* CMD | $E_{24}$% | Reference(s) |
|---|---|---|---|---|---|---|
| | | | | | Crude oil: 23%<br>Mineral oils: 52%<br>Olive oil: 69%<br>Sunflower oil: 63%<br>Almond oil: 70% | |
| *Lipopeptides* | | | | | | |
| *Bacillus subtilis* | Surfactin | 36 | | 15.6 | *Cell-free supernatant containing 1.12 g/L*<br>*S/H Ratio 3:2*<br>Hexadecane: 68%<br>Querosene: 53%<br>Diezel: 67%<br>Motor oil: 6% | Abdel-Mawgoud et al. (2008) |
| *Bacillus amyloliquefaciens* AR2 | Surfactin, iturin, fengysin mixture | 30–37 | | 80–110 | *Cell-free supernatant containing*<br>*173–641 mg/L*<br>*S/H ratio 1:1*<br>Kerosene: 32–66% | Singh et al. (2014) |
| *B. licheniformis* | Lichenysin | 27 | Decane: 0.006 | 10 | | Nerurkar (2010) |
| *P. fluorescens* | Viscosin | 26.5 | | 150 | | Neu and Poralla (1990) |
| *Pseudomonas libanensis* M9-3 | Viscosin | 27.5 | | 54 | *Purified, 7.5 mg/L*<br>*S/H ratio 1:1*<br>10% (v/v) Pennzoil 10W-40: 100% | Saini et al. (2008) |

| | | | | | | |
|---|---|---|---|---|---|---|
| *Serratia marcescens* NS25 | Serrawettin W2 | 33.9 | | | | Matsuyama et al. (1986) |
| *Polymeric biosurfactants* | | | | | | |
| *Acinetobacter calcoaceticus* BU03 | | 38.4 | | 152.4 | | Zhao and Wong (2009) |
| *A. venetianus* RAG-1 | Emulsan | ~50 | | 20 | *Cell-free supernatant containing 65 mg/L S/H ratio 1:10*<br>Mineral oil: 60% | Su et al. (2009) |

ST, surface tension; IFT, interfacial tension, CMC, critical micelle concentration; CMD, critical micelle dilution; $E_{24}$%, emulsification index at 24 h; S/H ratio, solution[a]/hydrocarbon ratio

[a]Water solution or cell-free supernatant containing biosurfactants

properties make them suitable for diverse biological, industrial, and ionic strength applications. These include food, agriculture, medicine and pharmaceuticals, cosmetics, enhanced oil recovery, bioremediation, soil washing, paints, and detergents, among others (Banat et al. 2010; Shekhar et al. 2015).

### 4.2.1 Medicine and Pharmaceuticals

The use of biosurfactants in the medical field with potential commercial applications is firstly related to their low or inexistent toxicity, being generally considered safer than the synthetic ones (Marchant and Banat 2012a). Furthermore, the large spectrum of biosurfactant molecules with diverse chemical and physical properties leads to a broad range of potential applications, especially in biomedicine: (1) antibacterial, antifungal, and antiviral activities, which makes them important for combating many diseases; (2) ability to partition at the interfaces, affecting adhesion of microorganisms; (3) biological activities such as anti-inflammatory, antitumor, immune modulatory action, human cells' stimulation and differentiation, cell-to-cell signaling, interaction with stratum corneum lipids, and hemolytic activity; (4) genetic manipulation; and (5) drug delivery (Fracchia et al. 2012; Shekhar et al. 2015).

**Antimicrobial and Antiviral Activity**

Lipopeptides are referred to as the type of surfactacts having the most potent bacteriocidal activity. This is attributed to the ability of these biomolecules to self-associate and interact with lipid membranes. For example, surfactin can cause membrane disruptions through hydrophobic interactions with both Gram-positive and Gram-negative bacteria (Lu et al. 2007). A lipopeptide fraction, identified to contain surfactin- and lichenysin-like molecules produced by the marine *Bacillus circulans*, presented antimicrobial action against various Gram-positive and Gram-negative pathogenic and semi-pathogenic bacteria including *Escherichia coli*, *Alcaligenes faecalis*, *Enterobacter cloacae*, *Micrococcus flavus*, and *Proteus vulgaris* (Das et al. 2008). An example of an antimicrobial lipopeptide under commercial development is daptomycin (Cubicin®). It is produced by *Streptomyces roseosporus* and has shown to be highly active against clinically relevant resistant bacteria, such as vancomycin-resistant *Enterococci*, penicillin-resistant *Streptococcus pneumoniae*, as well as methicillin-resistant *Staphylococcus aureus* (Tally et al. 1999).

Regarding antifungal activity, many studies have also been reported. For instance, pathogenic yeasts associated with human mycoses such as *Candida lusitaniae*, *Trichosporon asahii*, *Cryptococcus neoformans*, and *Candida albicans* have shown to be sensitive to the cellobiose lipid flocculosin isolated from *Pseudozyma flocculosa* (Mimee et al. 2009). In addition, the synthetically modified lipopeptides, caspofungin, micafungin, and anidulafungin, have also been used to treat fungal related diseases. While the first and the second were applied in the treatment of invasive fungal infections by *Candida and Aspergillus* species (Emiroglu 2011; Ngai et al. 2011), the third has shown clinical efficacy in the treatment of candidemia and other forms of candidiasis (George and Reboli 2012).

Some lipopeptides have presented antiviral activity, mostly against enveloped viruses, indicating that inhibitory action may be related to biosurfactants' interactions with the virus envelope (Vollenbroich et al. 1997). As examples, surfactin and fengycin produced by *B. subtilis* have shown in vitro inactivation of pseudorabies virus, porcine parvovirus, bursal disease virus, and newcastle disease virus (Huang et al. 2006). In addition to lipopeptides, sophorolipids were reported to possess activity against human immunodeficiency virus (Shah et al. 2005).

**Antiadhesive Capacity**

The adhesion of microorganisms to surfaces with the formation of microbial biofilms on the equipment and devices used in the medical field (e.g., urinary catheters, heart valves, hip prostheses, and intrauterine devices) is a quite important safety issue, especially because bacteria of those biofilms tend to become highly resistant to antibiotics (Fracchia et al. 2012). Due to their surface-active properties against several pathogens, biosurfactants present a potential effectiveness as coating and antiadhesive agents for medical devices, which eventually may lead to a reduction of hospital infections without the need of using synthetic drugs.

Lunasan, a biosurfactant produced by the yeast *Candida sphaerica* UCP0995, completely inhibits adhesion of several bacteria such as *Streptococcus agalactiae*, *Streptococcus sanguis* 12, *Streptococcus mutans*, *Streptococcus mutans* NS, *Staphylococcus epidermidis*, *Staphylococcus aureus*, and *Candida albicans* on plastic tissue culture plates (Luna et al. 2011). In addition, biosurfactants produced by *Lactobacillus* species have been reported to inhibit bacterial pathogens' adherence on medical devices, particularly for avoiding urogenital infection in mammals (Bruce et al. 2000). Moreover, in a recent study it was shown that the lipopeptide biosurfactant V9T14 when combined with antibiotics results in a synergistic increase in the efficacy of *E. coli* CFT073 biofilm inhibition, while total eradication of the uropathogenic strain biofilm was observed in several other combinations (Rivardo et al. 2011).

**Antitumor and Immune Modulatory Action**

It has been observed that some types of surfactants are able to participate in several molecular recognitions in cells, such as signal transduction, cell differentiation, and cell immune response, which make them potential biomolecules to be used in related medical applications, such as in cancer and autoimmune diseases.

As examples, surfactin has shown to induce apoptosis in human breast cancer MCF-7 cells (Cao et al. 2010), viscosin has demonstrated to inhibit migration regarding the metastatic prostate cancer cell line PC-3M (Saini et al. 2008), and the sophorolipid produced by the yeast *Wickerhamiella domercqiae* was able to induce apoptosis in H7402 human liver cancer cells (Chen et al. 2006). Furthermore, from the work of Park and Kim (2009), it is envisaged that surfactin is a potent immunosuppressive, suggesting important therapeutic responses for transplantation and autoimmune diseases like allergy, arthritis, and diabetes.

**Gene Transfer**
Gene transfection, i.e., introduction of exogenous genes into target cells, is a fundamental technology for clinical gene therapy. This strategy requires the implementation of efficient and safe methods without side effects. Due to the physicochemical properties of cationic liposomes, such as lipid packing density, shape, and zeta-potential, lipofection using those structures is considered to be a promising methodology (Inoh et al. 2010; Gharaei-Fathabad, 2011).

### 4.2.2 Food Industry

Biosurfactants properties, such as low CMC, high surface activity, low toxicity, stability under large variations of temperature, pH and salt concentrations, as well as biological activity, make them interesting molecules to be used in the food industry. They can eventually substitute conventional synthetic surfactants since they are characterized by functional advantages. Furthermore, as they are bioproducts, and depending on the microorganism and the production process involved in their production, it may be easier to confer them the status of natural additives, which is in line with the actual market needs. In the area of food produc*ts*' formulation and processing, biosurfactants have been applied mostly as emulsifiers, biofilm, and antimicrobial agents.

**Food Emulsifiers**
Emulsions are formed when two immiscible liquids are mixed and one of them is dispersed in small drops into the other. They may be classified as O/W or W/O emulsions, when oil droplets are dispersed in an aqueous continuous phase or water droplets are dispersed in an oil phase, respectively. These systems are thermodynamically unstable. In order to prevent droplet coalescence followed by phase separation, emulsifiers (surfactants) are added, which will create a stabilizing film at droplets surfaces. The most used emulsifiers in food products are phospholipids, proteins, and synthetic surfactants. From those, lecithin is one of the most important. It can be obtained from either egg yolk or soybean. Taking into account the different consumers' preferences, lecithin from egg yolk is not tolerated by vegans, as well as by Jewish and Islamic communities that demand Kosher-certified raw materials. Lecithin soybean may not be accepted for food requiring non-genetically modified ingredients. As such, emulsifiers from microbial sources are alternatives to those traditionally used, with some of them being more tailored to modern food processing requirements. There are many studies regarding the ability of biosurfactants to stabilize O/W emulsions based on edible oils. As examples, high emulsification indexes have been reported for a biosurfactant produced by a *B. subtilis* strain with sunflower, mustard, gingelly, and coconut oils (Chander et al. 2012). In another study, emulsion formation was presented with vegetable oils (e.g., peanut, olive, sunflower, rice bran, and sesame oils) and surfactants from

marine bacteria, which remained stable for more than 3 months (Radhakrishnan et al. 2011). Mayonnaise, butter, salad dressing, chocolates, whippable toppings, and hotdogs are examples of food processes based on emulsions (Nitschke and Costa 2007; Kralova and Sjöblom, 2009). The application of a bioemulsifier produced by *Candida utilis* in salad dressings has also been reported (Shepherd et al. 1995). Beyond emulsions, biosurfactants are also used in food products for foam stabilization, control of texture and shelf-life of products containing starch, modification of rheological properties of wheat dough, improvement of fat-based product texture, solubilization of flavor oils and as fat stabilizers in bakery and ice cream, enhancement of dough texture, as well as conservation of bakery products (Kralova and Sjöblom 2009; Marchant and Banat 2012b).

**Biofilm Control and Antimicrobial Activity**

Upon food processing, it is essential to produce safe food products without compromising their nutritional content and sensory attributes, in order to possess an overall quality meeting the consumers' requirements. Microbial growth has to be controlled as much as possible in order to guarantee food safety. There are microorganisms that have the ability to attach to surfaces, such as polystyrene, glass, rubber, and stainless steel, mostly found in the food processing environment, by forming biofilms (Kalia et al. 2015). Many of those microorganisms, like *E. coli* O157:H7, *Listeria monocytogenes*, *Salmonella typhimurium*, *Staphylococcus aureus*, and *Cronobacter sakazakii*, are pathogenic bacterial species (Bae et al. 2012). Physical methods (e.g., extensive washing and high-pressure sprays), and also chemical compounds (e.g., hypochlorite), are usually applied to minimize and eliminate biofilm formation, although there is evidence that bacteria can still remain at equipment surfaces and survive for long periods of time even using standard cleaning procedures.

Hydrophobic surfaces may be preconditioned with biosurfactants in order to turn them more hydrophilic. The molecules align with the hydrophobic part to the surface and the hydrophilic group to water phase. Subsequently, a reduction of microbial cell attachment is expected, which eventually decreases or inhibits biofilm development. From the studies reported so far, biosurfactants present a great potential to be used in biofilm control. There are several works regarding preconditioning polystyrene surfaces with rhamnolipids and surfactin. Reduced adhesion of different strains of *L. monocytogenes*, lower than when the synthetic sodium-dodecyl sulfate surfactant was used, has been observed (Araujo et al. 2011). In addition, a lower adhesion of *S. aureus*, *L. monocytogenes*, and *M. luteus* has been reported (Zeraik and Nitschke 2010). Using only surfactin, similar behavior with *L. monocytogenes* ATCC 19112 cells (Gomes and Nitschke 2012) and also with *Listeria monocytogenes* and *Salmonella enteritidis* has been observed (Nitschke et al. 2009).

Beyond its ability to decrease biofilm formation, due to their surface active properties, biosurfactants have also showed antimicrobial activity against bacteria, fungi, yeasts, algae, as well as viruses (Nitschke and Costa 2007). The antimicrobial potential of rhamnolipids against *L. monocytogenes* testing 32 cultures has been evaluated (Magalhães and Nitschke 2013). From that, 90.6% were susceptible to the biosurfactants. Moreover, rhamnolipids showed a synergistic effect when combined with nisin. It has also been reported that a *S. enteritidis* strain was very sensitive to surfactin in milk (Huang et al. 2011). In addition, surfactin from *B. subtilis* was reported to possess a strong anti-Legionella activity (Loiseau et al. 2015).

### 4.2.3 Environmental Applications

Biological processes offer several advantages over conventional technologies for bioremediation, because they are often more environmentally friendly, economic, and versatile, and they can reduce the concentration and toxicity of a large number of contaminants. Since the rate of the biodegradation of a contaminant strongly depends on its bioavailability, the bottleneck of these processes relies on the low solubility in water of most contaminant molecules (e.g., hydrocarbons and other hydrophobic compounds), limiting their availability to microorganisms (Bustamante et al. 2012).

The bioavailability of contaminants can be improved by the addition of biosurfactants, which reduce surface and interfacial tension increasing their solubility. They have been used in bioremediation of petroleum hydrocarbons (Thavasi et al. 2011, 2015; Zeng et al. 2011; Kumari et al. 2012), polyaromatic hydrocarbons (Husain 2008; Zhang et al. 2010), pesticides (Zhang et al. 2011), and other hydrophobic compounds (Kardena et al. 2015). Additionally, biosurfactants and biosurfactant-producing microorganisms have been applied to contaminated sites, i.e., soils, regarding the removal of heavy metals, either by biosorption or metal ions complexation with surfactant's active chemical groups (e.g., hydroxyl, carbonyl, or amine) (Juwarkar et al. 2007; Huang and Liu 2013).

### 4.2.4 Microbial Enhanced Oil Recovery

Up to two-thirds of the oil present in reservoirs is not removed after applying primary and secondary recovery technologies. Subsequently, enhanced oil recovery (EOR) technologies are employed, which involve both chemical and microbial-based methods (MEOR). The latter include the use of biosurfactants, taking advantage of their feature to reduce the oil/water interfacial tension and to produce an oil-in-water emulsion, leading to an improvement in the mobility of the oil through rock fractures (Khan et al. 2015).

The application of biosurfactants in MEOR may be carried out using different methods: (1) ex situ production, followed by injection into the oil reservoir, and (2) in situ biosurfactant production, either by injection of bacteria or by stimulation of indigenous microorganisms (Banat et al. 2010).

Emulsan and lipopeptides, such as surfactin and lichenysin, have shown to be effective when used in the first method (Sen 2008). The bottleneck of this method relies on the biosurfactant production and purification costs. The costs involved in the second methodology are apparently lower; however, it requires the identification and isolation of microbial strains adapted to the extreme environmental conditions of the reservoirs (high pressure, ionic strength, temperature, and pH values). There are reports on the isolation of *B. subtilis* strains from oil reservoirs that are able to produce biosurfactants (Simpson et al. 2011; Gudiña et al. 2012) and also regarding the injection of biosurfactant-producing indigenous microorganisms (She et al. 2011).

### 4.2.5 Agriculture

Agricultural productivity is very important to meet the increasing food demand of human population, and it is a major concern worldwide. This productivity is quite dependent on diverse strategies, like the use of chemical fertilizers to improve soil fertility and pesticides for eradication of plant pathogens, which are applied in the form of agrochemical formulations that include traditionally synthetic surfactants. Because of being environmentally friendly, biosurfactants have been tested in such formulations. Furthermore, besides their low toxicity and biodegradability, biosurfactants have demonstrated functional properties that make them important biological agents to be used also in plant disease control and immunity enhancement, pest/insect control, and improvement of plant–microbes interaction (Sachdev and Cameotra 2013).

**Agrochemical Formulations**

From the wide range of biosurfactants' properties of biosurfactants, those of particular interest for agrochemical formulations are thickening, dispersion, emulsification, and penetrating properties. Biosurfactants (e.g., rhamnolipids, sophorolipids, and other glycolipids) are used in agrochemical formulations mainly with the purpose of (1) increasing the permeation of active agents into the plants by controlling their solubility in water (Wattanaphon et al. 2008; Awada et al. 2011), (2) acting as adjuvants to enhance the effectiveness of active ingredients (Giessler-Blank et al. 2012), (3) serving as cleaning agents of herbicide and pesticide residues from fruits and vegetables (Cheowtirakul and Linh 2010), (4) increasing leaf's surface wettability, enabling to increase the residence time of active compounds (e.g., pesticides) on the leaf surface aiming prolonged protection, and (5) solubilizing nutrients in fertilizer formulations (Thavasi et al. 2015).

**Plant Disease Control and Immunity Enhancers**

Some biosurfactants have shown antimicrobial activity against plant pathogens that makes them excellent candidates to substitute chemical and toxic fungicides and bactericides, aiming at achieving sustainable agriculture. According to a report presented recently (Schofield et al. 2013), the potential of microbial surfactants as plant disease controllers was shown. The authors reported that sophorolipids and sophorolipid derivatives have shown significant antimicrobial activity for a large number of fungal (e.g., *Alternaria tomatophila*, *Botrytis cinerea*, and *A. alternata*) and bacterial (e.g., *Ralstonia solanacearum*, *Erwinia amylovora*, *Acidovorax carotovorum*) pathogens. In addition, *P. aeruginosa* has been reported to be inhibited by biosurfactants produced by *Staphylococcus* sp., and *Colletotrichum gloeosporioides*, which causes anthracnose on papaya leaves, has also been reported to be inhibited by a biosurfactant produced by *Bacillus subtilis* isolated from soil (Kim et al. 2010; Eddouaouda et al. 2012). Another important aspect being studied is the identification and isolation of biosurfactant-producing microorganisms present in the plant rhizosphere environment. Several reports have been presented regarding antimicrobial activity of biomolecules produced from those strains. As an example, plant growth-promoting *P. putida* strain 267, which has been isolated from black pepper's rhizosphere, produces cyclic lipopeptide biosurfactants that are responsible for the lysis of zoospores of the pathogen *Phytophthora capsici* and also restrict growth of the pathogenic fungi *Botrytis cinerea* and *Rhizoctonia solani* (Kruijt et al. 2009). Additionally, purified cyclic lipopeptide massetolide A produced by *P. fluorescens* SS101 provided significant control of *Phytophthora infestans* in tomato plants (Tran et al. 2007).

Beyond their antimicrobial activity, biosurfactants present the ability of inducing plant disease resistance, which is called induced systemic resistance (ISR) (Thavasi et al. 2015). This ability may be explored to trigger plant's permanent defense mechanisms against specific diseases, thus reducing frequent application of pesticides to combat those diseases. An ISR-mediated protective effect of surfactins and fengycins, produced by a *B. subtilis* strain, on tomato plants has been reported (Ongena et al. 2007). Moreover, an induced protection of grapevine leaves against *B. cinerea* by rhamnolipids produced by *P. aeruginosa* and *B. plantarii* has been observed (Varnier et al. 2009).

### 4.2.6 Nanotechnology

The role of biosurfactants in the field of nanotechnology relies on their functional properties that enable them to participate in the formation of structures at the nanoscale with fine-tuned properties. The attention has been driven into research regarding the synthesis of metallic nanoparticles by using biosurfactants as green alternatives for promoting nanoparticle synthesis and also stabilization. One of the modes is avoiding aggregation by adsorbing onto nanoparticles surface. Different nanoparticle (sphere-like) biosurfactant systems have been reported, such as gold and silver nanoparticles with surfactin (Reddy et al. 2009a, b), cadmium sulfide nanoparticles using surfactin (Singh et al. 2011), and rhamnolipids for capping ZnS nanoparticles (Narayanan et al. 2010). Structures other than spheres, like nanofibers and nanotubes, have also been synthesized with biosurafactants (rhamnolipids) acting as soft templates (Worakitsiri et al. 2011).

Nanoemulsion formulation for the encapsulation and subsequent delivery of functional compounds is also an emerging field of nanotechnology applied in the food industry. Encapsulation of hydrophobic molecules (e.g., flavors, antioxidants, and antimicrobial agents) in nanoemulsions may improve their solubility, bioavailability, and functionality, as those nanostructures may act as carriers and delivery systems (Silva et al. 2012). Nanoemulsions have themselves been tested as antimicrobial systems for decontamination of food or food contact materials. As example, surfactin–sunflower-based nanoemulsions showed antibacterial activity against *Salmonella typhi*, *L. monocytogenes*, and *S. aureus*. In addition, those nanoemulsions decreased significantly the native bacterial and fungal populations upon in situ evaluation of their antimicrobial activity on raw chicken, apple juice, milk, etc. (Manoharan et al. 2012). In a recent study, lipid–polymer hybrid nanoparticles using poly(lactic-*co*-glycolic) acid as the polymer nanoparticle core and phosphatidylcholine as lipid coat were formulated in order to encapsulate antibiotics to control biofilm formation by *P. aeruginosa* (Cheow and Hadinoto 2012). They have shown an antibiotic trigger release from hybrid nanoparticles in response to interaction with rhamnolipids present in biofilm colonies, which disrupted the nanoparticles' phosphatidylcholine coating.

## 5 Perspectives

The singular properties of microbial surfactants make them unquestionable important molecules to be used in a wide range of applications, some of them already available commercially due to the variety of their chemical structures. Nevertheless, there is still potential for new biosurfactant-producing strain isolation and production process optimization, not only aiming at costs reduction but also to obtain novel molecules with tailored properties for specific applications. Biodegradability, biological activity, and lack of toxicity and stability under a variety of physicochemical environments are the major advantages claimed for the use of biosurfactants. However, the addition of these molecules to the environment has to be carried out conscientiously, because it may affect negatively some ecosystems due to their antimicrobial activity against identified groups of microorganisms.

**Acknowledgments** Dr Constantina Kourmentza would like to thank the European Commission for providing support through the FP7-PEOPLE-2013-IEF-Marie-Curie Action: Intra-European Fellowships for Career Development (grant no. 625774). Dr Filomena Freitas acknowledges FCT-MEC for fellowship SFRH/BPD/72280/2010. The authors acknowledge Fundação para a Ciência e Tecnologia, Ministério da Educação e Ciência (FCT-MEC), Portugal, through projects PEst-C/EQB/LA0006/2013 and PEst-OE/AGR/UI0245/2014.

## References

Abdel-Mawgoud AM, Aboulwafa MM, Hassouna NA (2008) Characterization of surfactin produced by *Bacillus subtilis* isolate BS5. Appl Biochem Biotechnol 150:289–303. doi:10.1007/s12010-008-8153-z

AGAE technologies, USA. http://www.agaetech.com/new-product-catalog-2/, accessed on October 2015

Al-Bahry SN, Al-Wahaibi YM, Elshafie AE, Al-Bemani AS, Joshi SJ, Al-Makhamari HS, Sulaimani HS (2013) Biosurfactant production by *Bacillus subtilis* B20 using date molasses and its possible application in enhanced oil recovery. Int Biodeter Biodegr 81:141–146. doi:10.1016/j.ibiod.2012.01.006

Almeida FCG, Lins CIM, Vieira AM, Vilar CJ, Mota Lins MC, Campos-Takaki GM, Tambourgi EB (2012) Biosurfactant production by *Pantoea* sp in submerged fermentation using pineapple peel as an alternative medium. In: Mendez-Vilas A (ed) Microbes in applied research-current advances and challenges. World Scientific, Singapore, pp 348–352

Araujo LV, Abreu F, Lins U, Santa Anna LMM, Nitschke M, Freire DMG (2011) Rhamnolipid and surfactin inhibit *Listeria monocytogenes* adhesion. Food Res Int 44:481–488. doi:10.1016/j.foodres.2010.09.002

Ashby RD, Nuñez A, Solaiman DKY, Foglia TA (2005) Sophorolipid biosynthesis from a biodiesel co-product stream. J Am Oil Chem Soc 82:625–630. doi:10.1007/s11746-005-1120-3

Asmer HJ, Lang S, Wagner F, Wray V (1988) Microbial-production, structure elucidation and bioconversion of sophorose lipids. J Am Oil Chem Soc 65:1460–1466. doi:10.1007/BF02898308

Asselineau C, Asselineau J (1978) Trehalose containing glycolipids. Prog Chem Fats Other Lipids 16:59–99. doi:10.1016/0079-6832(78)90037-X

Awada SM, Awada M, Spendlove RS (2011) Method of controlling pests with biosurfactant penetrants as carriers for active agents. Patent: US2011/0319341 A1

Bae Y, Baek S, Lee S (2012) Resistance of pathogenic bacteria on the surface of stainless steel depending on attachment form and efficacy of chemical sanitizers. Int J Food Microbiol 153:465–473. doi:10.1016/j.ijfoodmicro.2011.12.017

Banat IM, Franzetti A, Gandolfi I, Bestetti G, Martinotti MG, Fracchia L, Smyth TJ, Marchant R (2010) Microbial biosurfactants production, applications and future potential. Appl Microbiol Biotechnol 87:427–444. doi:10.1007/s00253-010-2589-0

Banat IM, Satpute SK, Cameotra SS, Patil R, Nyayanit NV (2014) Cost effective technologies and renewable substrates for biosurfactants' production. Front Microbiol 5:697. doi:10.3389/fmicb.2014.00697

Barros FFC, Ponezi AN, Pastore GM (2008) Production of biosurfactant by *Bacillus subtilis* LB5a on a pilot scale using cassava wastewater as substrate. J Ind Microbiol Biotechnol 35:1071–1078. doi:10.1007/s10295-008-0385-y

Bednarski W, Adamczak M, Tomasik J, Plaszczyk M (2004) Application of oil refinery waste in the biosynthesis of glycolipids by yeast. Bioresour Technol 95:15–18. doi:10.1016/j.biortech.2004.01.009

Benincasa M, Accorsini FR (2008) *Pseudomonas aeruginosa* LBI production as an integrated process using the wastes from sunflower-oil refining as a substrate. Bioresour Technol 99:3843–3849. doi:10.1016/j.biortech.2007.06.048

Benincasa M, Contiero J, Manresa MA, Moraes IO (2002) Rhamnolipid production by *Pseudomonas aeruginosa* LBI growing on soapstock as the sole carbon source. J Food Eng 54:283–288. doi:10.1016/S0260-8774(01)00214-X

Bensaci MF, Ganghi NR, Palmer SVL, Takemoto JY (2007) Antimycotic rhamnolipid compositions and related methods of use. Patent: WO2007095259 A2

Borges WS, Cardoso VL, Resende MM (2012) Use of a greasy effluent floater treatment station from the slaughterhouse for biosurfactant production. Biotechnol Appl Biochem 59:238–244. doi:10.1002/bab.1018

Bralkowski MP, Brooks SA, Hinton SM, Wright DM, Yang SH (2013) Improved poultry farm practices. Patent: WO2013110133 A1

Bruce AW, Busscher HJ, Reid G, Van der Mei HC (2000) *Lactobacillus* therapies. U.S. Patent US6051552 A

Bustamante M, Durán N, Diez MC (2012) Biosurfactants are useful tools for the bioremediation of contaminated soil: a review. J Soil Sci Plant Nutr 12:667–687. doi:10.4067/S0718-95162012005000024

Buthe A, Hafner A, Kaufmann F, Fiedler B, Meurer G, Bradley G (2012) Compositions, use and method for the use of surface active proteins in topical drug delivery across keratin. Patent: EP2400988 A1

Cagri-Mehmetoglu A, Kusakli S, van de Venter M (2012) Production of polysaccharide and surfactin by *Bacillus subtilis* ATCC 6633 using rehydrated whey powder as the fermentation medium. J Dairy Sci 95:3643–3649. doi:10.3168/jds.2012-5385

Cammarota MC, Freire DMG (2006) A review on hydrolytic enzymes in the treatment of wastewater with high oil and grease content. Bioresour Technol 97:2195–2210. doi:10.1016/j.biortech.2006.02.030

Cao XH, Wang AH, Wang CL, Mao DZ, Lu MF, Cui YQ, Jiao RZ (2010) Surfactin induces apoptosis in human breast cancer MCF-7 cells through a ROS/JNK-mediated mitochondrial/caspase pathway. Chem Biol Interact 183:357–362. doi:10.1016/j.cbi.2009.11.027

Castro GR, Kaplan DL, Panilaitis B (2009) Emulsan-alginate microspheres and methods of use thereof. Patent: WO2006028996 A3

Çelik E, Ozbay N, Oktar N, Çalık P (2008) Use of biodiesel byproduct crude glycerol as the carbon source for fermentation processes by recombinant *Pichia pastoris*. Ind Eng Chem Res 47:2985–2990. doi:10.1021/ie071613o

Chander CRS, Lohitnath T, Kumar DJM, Kalaichelvan PT (2012) Production and characterization of biosurfactant from *Bacillus subtilis* MTCC441 and its evaluation to use as bioemulsifier for food bio–preservative. Adv Appl Sci Res 3:1827–1831

Chen J, Song X, Zhang H, Qu Y, Miao J (2006) Sophorolipid produced from the new yeast strain *Wickerhamiella domercqiae* induces apoptosis in H7402 human liver cancer cells. Appl Microbiol Biotechnol 72:52–59. doi:10.1007/s00253-005-0243-z

Cheow WS, Hadinoto K (2012) Lipid-polymer hybrid nanoparticles with rhamnolipid-triggered release capabilities as anti-biofilm drug delivery vehicles. Particuology 10:327–333. doi:10.1016/j.partic.2011.08.007

Cheowtirakul C, Linh ND (2010) The study of biosurfactant as a cleaning agent for insecticide residue in leafy vegetables. AU J T 14(2):75–87

Choe E, Min DB (2007) Chemistry of deep-fat frying oils. J Food Sci 72:77–86. doi:10.1111/j.1750-3841.2007.00352.x

Chooklin CS, Maneerat S, Saimmai A (2014) Utilization of banana peel as a novel substrate for biosurfactant production by *Halobacteriaceae archaeon* AS65. Appl Biochem Biotechnol 173:624–645. doi:10.1007/s12010-014-0870-x

Christova N, Stoineva I (2014) Trehalose biosurfactants. In: Mulligan CN, Sharma SK, Mudhoo A (eds) Biosurfactants—recent trends and applications. CRC, Taylor & Francis, Boca Raton, FL, pp 197–216

Concaix FB (2003) Use of sophorolipids comprising diacetyl lactones as agent for stimulating skin fibroblast, metabolism. Patent: US6596265 B1

Connolly HE, Rahman PKSM, Banat IM, Lord RA (2010) Resource recovery and reduction of oily hazardous wastes via biosurfactant washing and bioremediation. In: Grazyna P (ed) Trends in bioremediation and phytoremediation. Research Signpost, Trivandrum, India, pp 157–172

Costa SGVAO, Deziel E, Lepine F (2011) Characterization of rhamnolipid production by *Burkholderia glumae*. Lett Appl Microbiol 53:620–627. doi:10.1111/j.1472-765X.2011.03154.x

Cruz MV, Freitas F, Paiva A, Mano F, Dionísio M, Ramos AM, Reis MAM (2016) Valorisation of fatty acids-containing wastes and byproducts into short- and medium-chain length polyhydroxyalkanoates. New Biotechnol 33(1):206–215. doi:10.1016/j.nbt.2015.05.005

Dams-Kozlowska H, Mercaldi MP, Panilaitis BJ, Kapla DJ (2008) Modifications and applications of the *Acinetobacter venetianus* RAG-1 exopolysaccharide, the emulsan complex and its components. Appl Microbiol Biotechnol 81:201–210. doi:10.1007/s00253-008-1664-2

Daniel H-J, Otto RT, Binder M, Reuss M, Syldatk C (1999) Production of sophorolipids from whey: development of a two-stage process with *Cryptococcus curvatus* ATCC 20509 and *Candida bombicola* ATCC 22214 using deproteinized whey concentrates as substrates. Appl Microbiol Biotechnol 51:40–45. doi:10.1023/A:1005332605003

Das K, Mukherjee AK (2007) Comparison of lipopeptide biosurfactants production by *Bacillus subtilis* strains in submerged and solid state fermentation systems using a cheap carbon source: some industrial applications of biosurfactants. Proc Biochem 42:1191–1199. doi:10.1016/j.procbio.2007.05.011

Das P, Mukherjee S, Sen R (2008) Antimicrobial potential of a lipopeptide biosurfactant derived from a marine *Bacillus circulans*. J Appl Microbiol. 104:1675–1684. doi:10.1111/j.1365-2672.2007.03701.x

Daverey A, Pakshirajan K (2009a) Production of sophorolipids by the yeast *Candida bombicola* using simple and low cost fermentative media. Food Res Int 42:499–504. doi:10.1016/j.foodres.2009.01.014

Daverey A, Pakshirajan K (2009b) Production, characterization and properties of sophorolipids from the yeast *Candida bombicola* using a low-cost fermentative medium. Appl Biochem Biotechnol 158:663–674. doi:10.1007/s12010-008-8449-z

Daverey A, Pakshirajan K (2010) Sophorolipids from *Candida bombicola* using mixed hydrophilic substrates: production, purification and characterization. Colloids Surf B Biointerfaces 79:246–253. doi:10.1016/j.colsurfb.2010.04.002

Davila AM, Marchal R, Monin N, Vandecasteele JP (1993) Identification and determination of individual sophorolipids in fermentation products by gradient elution high-performance liquid-chromatography with evaporative light-scattering detection. J Chromatogr 648:139–149. doi:10.1016/0021-9673(93)83295-4

Desai JD (1987) Microbial surfactants: evaluation, types and future applications. J Sci Ind Res 46:440–449

Desai JD, Banat IM (1997) Microbial production of surfactants and their commercial potential. Microb Mol Biol Rev 61(1):47–64

De Santo K (2009) Rhamnolipid-based formulations. Patent: WO2008013899 A2

Deshpande M, Daniels L (1995) Evaluation of sophorolipid biosurfactant production by *Candida bombicola* using animal fat. Bioresour Technol 54:143–150. doi:10.1016/0960-8524(95)00116-6

Dubeau D, Deziel E, Woods DE, Lepine F (2009) *Burkholderia thailandensis* harbors two identical *rhl* gene clusters responsible for the biosynthesis of rhamnolipids. BMC Microbiol 9:263. doi:10.1186/1471-2180-9-263

Dumont M-J, Narine SS (2007) Soapstock and deodorizer distillates from North American vegetable oils: review on their characterization, extraction and utilization. Food Res Int 40:957–974. doi:10.1016/j.foodres.2007.06.006

Eddouaouda K, Mnif S, Badis A, Younes SB, Cherif S, Ferhat S, Mhiri N, Chamkha M, Sayadi S (2012) Characterization of a novel biosurfactant produced by *Staphylococcus* sp. strain 1E with potential application on hydrocarbon bioremediation. J Basic Microbiol 52:408–418. doi:10.1002/jobm.201100268

Eigen A, Simone AJ (1990) Control of dental plaque and caries. Patent: CA1269333 A1

Emaga TH, Andrianaivo RH, Wathelet B, Tchango JT, Paquot M (2007) Effects of the stage of maturation and varieties on the chemical composition of banana and plantain peels. Food Chem 103:590–600. doi:10.1016/j.foodchem.2006.09.006

Emiroglu M (2011) Micafungin use in children. Expert Rev Anti Infect Ther 9:821–834. doi:10.1586/eri.11.91

Fattom A, Shilo M (1985) Production of emulcyan by *Phormidium* J-1: its activity and function. FEMS Microbiol Lett 31:3–9. doi:10.1111/j.1574-6968.1985.tb01125.x

Felse PA, Shah V, Chan J, Rao KJ, Gross RA (2007) Sophorolipid biosynthesis by *Candida bombicola* from industrial fatty acid residues. Enzyme Microb Technol 70:316–323. doi:10.1016/j.enzmictec.2006.04.013

Fontes GC, Ramos NM, Amaral PFF, Nele M, Coelho MAZ (2012) Renewable resources for biosurfactant production by *Yarrowia lipolytica*. Braz J Chem Eng 29:483–493. doi:10.1590/S0104-66322012000300005

Fracchia L, Cavallo M, Martinotti MG, Banat IM (2012) Biosurfactants and bioemulsifiers: biomedical and related applications-present status and future potentials. In: Ghista DN (ed) Biomedical science, engineering and technology. InTech, Rijeka, pp 325–370

Freitas F, Alves VD, Pais J, Carvalheira M, Costa N, Oliveira R, Reis MAM (2010) Production of a new exopolysaccharide (EPS) by *Pseudomonas oleovorans* NRRL B-14682 grown on glycerol. Proc Biochem 45:297–305. doi:10.1016/j.procbio.2009.09.020

Freitas F, Alves VD, Torres AV, Cruz M, Sousa I, Melo MJ, Ramos AM, MAM (2011) Fucose-containing exopolysaccharide produced by the newly isolated *Enterobacter* strain A47 DSM 23139. Carbohydr Polymer 83:1582–1590. doi:10.1016/j.carbpol.2010.07.034

Fuhrman J, Gross RA, Kaplan DL (2001) Emulsan and emulsan analogs immunization formulations and use. Patent: WO2000051635 A3

Fukushi K (1984) Therapeutic agents. Patent: US4454119 A

Futura T, Igarashi K, Hirata Y (2002) Low-foaming detergent compositions. World patent 03/002700

George S, Jayachandran K (2009) Analysis of rhamnolipid biosurfactants produced through submerged fermentation using orange fruit peelings as sole carbon source. Appl Biochem Biotechnol 158:694–705. doi:10.1007/s12010-008-8337-6

George S, Jayachandran K (2013) Production and characterization of rhamnolipid biosurfactant from waste frying coconut oil using a novel *Pseudomonas aeruginosa* D. J Appl Microbiol 114:373–383. doi:10.1111/jam.12069

George J, Reboli AC (2012) Anidulafungin: when and how? The clinician's view. Mycoses 55:36–44. doi:10.1111/j.1439-0507.2011.02052.x

Gharaei-Fathabad E (2011) Biosurfactants in pharmaceutical industry (a mini review). Am J Drug Discov Dev 1:58–69. doi:10.3923/ajdd.2011.58.69

Ghribi D, Mnif I, Boukedi H, Kammoun R, Ellouze-Chaabouni S (2011) Statistical optimization of low-cost medium for economical production of *Bacillus subtilis* biosurfactant, a biocontrol agent for the olive moth *Prays oleae*. Afr J Microbiol Res 5:4927–4936. doi:10.5897/AJMR11.1125

Giessler-Blank S, Schilling M, Thum O, Sieverding E (2012) Use of sophorolipids and derivatives thereof in combination with pesticides as adjuvant/additive for plant protection and the industrial non-crop field. Patent: US20120220464 A1

Gomes MZV, Nitschke M (2012) Evaluation of rhamnolipid and surfactin to reduce the adhesion and remove biofilms of individual and mixed cultures of food pathogenic bacteria. Food Control 25:441–447. doi:10.1016/j.foodcont.2011.11.025

Gong Z, Peng Y, Wang Q (2015) Rhamnolipid production, characterization and fermentation scale-up by *Pseudomonas aeruginosa* with plant oils. Biotechnol Lett 37:2033–2038. doi:10.1007/s10529-015-1885-2

Griffin W (1954) Calculation of HLB values of non-ionic surfactants. J Soc Cosmet Chem 5:235–249

Gross RA, Schofield MH (2014) Sophorolipid analog compositions. Patent: US20140024816 A1

Gross RA, Vishai S (2004) Antimicrobial properties of various forms of sophorolipids. Patent: WO2004044216 A1

Gross RA, Shah V, Nerud F, Madamwars D (2007) Sophorolipids as protein inducers and inhibitors in fermentation medium. Patent: WO2007/073371

Gudiña EJ, Pereira JFB, Rodrigues LR, Coutinho JAP, Teixeira JA (2012) Isolation and study of microorganisms from oil samples for application in microbial enhanced oil recovery. Int Biodeter Biodegr 68:56–64. doi:10.1016/j.ibiod.2012.01.001

Gudiña EJ, Pereira JFB, Costa R, Evtuguin DV, Coutinho JAP, Teixeira JA, Rodrigues LR (2015) Novel bioemulsifier produced by a *Paenibacillus* strain isolated from crude oil. Microb Cell Fact 14:14. doi:10.1186/s12934-015-0197-5

Guidry TV, Hunter RL, Actor JK (2007) Mycobacterial glycolipid trehalose 6,6′-dimycolate-inducedhypersensitive granulomas: contribution of $CD4^+$ lymphocytes. Microbiology 153:3360–3369. doi: 10.1099/mic.0.2007/010850-0

Gunther NW, Nunez A, Fett W, Solaiman DK (2005) Production of rhamnolipids by Pseudomonas chlororaphis, a nonpathogenic bacterium. Appl Environ Microbiol 71:2288–2293. doi:10.1128/AEM.71.5.2288-2293.2005

Gutnick DL, Bach HR (2002) Compositions containing bioemulsifiers and a method for their preparation. Patent: WO2002048327

Hausmann R, Syldatk C (2015) Types and classification of microbial surfactants. In: Kosaric N, Vardar-Sukan F (eds) Biosurfactants, production and utilization-processes, technologies, and economics. CRC, Taylor & Francis, Boca Raton, FL, pp 3–18

Hayes ME, Holzner G (1986) Cosmetic and pharmaceutical compositions containing bioemulsifiers. Patent: EP0178443 A2

Heins SD, Manker DC, Jimerez DR, McCoy RJ, Marrone PG, Orjala JE (2001) Strain of *Bacillus* for controlling plant diseases and corn rootworm. Patent: US6291426 B1

Hilliou L, Freitas F, Oliveira R, Reis MAM, Lespineux D, Grandfils C, Alves VD (2009) Solution properties of exopolysaccharide from a *Pseudomonas* strain obtained using glycerol as single carbon source. Carbohydr Polymer 78:526–532. doi:10.1016/j.carbpol.2009.05.011

Hirata Y, Ryu M, Igarashi K, Nagatsuka A, Furuta T, Kanaya S, Sugiura M (2009a) Natural synergism of acid and lactone type mixed sophorolipids in interfacial activities and cytotoxicities. J Oleo Sci 58:565–572. doi:10.5650/jos.58.565

Hirata Y, Ryu M, Oda Y, Igarashi K, Nagatsuka A, Furuta T, Sugiura M (2009b) Novel characteristics of sophorolipids, yeast glycolipid biosurfactants, as biodegradable low-foaming surfactants. J Biosci Bioeng 108:142–146. doi:10.1016/j.jbiosc.2009.03.012

Hoermann B, Mueller MM, Syldatk C, Hausmann R (2010) Rhamnolipid production by *Burkholderia plantarii* DSM 9509$^{T}$. Eur J Lipid Sci Technol 112:674–680. doi:10.1002/ejlt.201000030

Hu S, Luo X, Wan C, Li Y (2012) Characterization of crude glycerol from biodiesel plants. J Agric Food Chem 60:5915–5921. doi:10.1021/jf3008629

Huang W, Liu Z-M (2013) Biosorption of Cd(II)/Pb(II) from aqueous solution by biosurfactant-producing bacteria: isotherm kinetic characteristic and mechanism studies. Colloids Surf B 105:113–119. doi:10.1016/j.colsurfb.2012.12.040

Huang X, Lu Z, Zhao H, Bie X, Lü FX, Yang S (2006) Antiviral activity of antimicrobial lipopeptide from *Bacillus subtilis* fmbj against pseudorabies virus, porcine parvovirus, newcastle disease virus and infectious bursal disease virus in vitro. Int J Pept Res Ther 12:373–377. doi:10.1007/s10989-006-9041-4

Huang X, Suo J, Cui Y (2011) Optimization of antimicrobial activity of surfactin and polylysine against *Salmonella enteritidis* in milk evaluated by a response surface methodology. Foodborne Pathog Dis 8:439–443. doi:10.1089/fpd.2010.0738

Husain S (2008) Effect of surfactants on pyrene degradation by *Pseudomonas fluorescens* 29 L. World J Microbiol Biotechnol 24:2411–2419. doi:10.1007/s11274-008-9756-9

Igbonekwu CJ, Osonwa UE, Umeyor CE, Ikegbunam MN, Uronnachi EM, Esimone CO (2014) Biosurfactant produced by novel halo-thermophilic bacterial strain from unconventional substrates. Int J Pharmacogn 9:73–81. doi:10.13040/IJPSR.0975-8232.IJP

Inaba T, Tokumoto Y, Miyazaki Y, Inoue N, Maseda H, Nakajima-Kambe T, Uchiyama H, Nomura N (2013) Analysis of genes for succinoyl trehalose lipid production and increasing production in *Rhodococcus* sp strain SD-74. Appl Environ Microbiol 79:7082–7090. doi:10.1128/AEM.01664-13

Inoh Y, Furuno T, Hirashima N, Kitamoto D, Nakanishi M (2010) The ratio of unsaturated fatty acids in biosurfactants affects the efficiency of gene transfection. Int J Pharm 398:225–230. doi:10.1016/j.ijpharm.2010.07.042

Ishikawa E, Ishikawa T, Morita YS, Toyonaga K, Yamada H, Takeuchi O, Kinoshita T, Akira S, Yoshikai Y, Yamasaki S (2009) Direct recognition of the mycobacterial glycolipid, trehalose dimycolate, by C-type lectin Mincle. J Exp Med 206:2879–2888. doi:10.1084/jem.20091750

Jadhav M, Kagalkar A, Jadhav S, Govindwar S (2011) Isolation, characterization, and antifungal application of a biosurfactant produced by *Enterobacter* sp. MS16. Eur J Lipid Sci Technol 113:1347–1356. doi:10.1002/ejlt.201100023

Jain RM, Mody K, Joshi N, Mishra A, Jha B (2013) Effect of unconventional carbon sources on biosurfactant production and its application in bioremediation. Int J Biol Macromol 62:52–58. doi:10.1016/j.ijbiomac.2013.08.030

Joshi-Navare K, Khanvilkar P, Prabhune A (2013) Jatropha oil derived sophorolipids: production and characterization as laundry detergente additive. Biochem Res Int 2013:169797. doi:10.1155/2013/169797

Juwarkar AA, Nair A, Dubey KV, Singh SK, Devotta S (2007) Biosurfactant technology for remediation of cadmium and lead contaminated soils. Chemosphere 68:1996–2002. doi:10.1016/j.chemosphere.2007.02.027

Kakinuma A, Hori M, Isono M, Tamura G, Arima K (1969a) Determination of amino acid sequence in surfactin, a crystalline peptidolipid surfactant produced by *Bacillus subtilis*. Agric Biol Chem 33:971–972. doi:10.1271/bbb1961.33.971

Kakinuma A, Sugino H, Isono M, Tamura G, Arima K (1969b) Determination of fatty acid in surfactin and elucidation of the total structure of surfactin. Agric Biol Chem 33:973–976. doi:10.1080/00021369.1969.10859409

Kalia VC, Kumar P, Pandian SK, Sharma P (2015) Biofouling control by quorum quenching. In: Kim SK (ed) Handbook of marine biotechnology. Springer, Berlin, pp 431–440. doi:10.1007/978-3-642-53971-8_15, isbn:978-3-642-53970-1

Kaplan N, Rosenberg E (1982) Exopolysaccharide distribution and bioemulsifier production by *Acinetobacter calcoaceticus* BD4 and BD413. Appl Environ Microbiol 44:1335–1341

Kappeli O, Finnerty WR (1979) Partition of alkane by an extracellular vesicle derived from hexadecane-grown *Acinetobacter*. J Bacteriol 140:707–712

Kardena E, Helmy Q, Funamizu N (2015) Biosurfactants and soil bioremediation. In: Kosaric N, Vardar-Sukan F (eds) Biosurfactants, production and utilization-processes, technologies, and economics. CRC, Taylor & Francis, Boca Raton, FL, pp 327–360

Keith D, Lai JJ, Govardhan C, Khalaf N (2014) Method for preparing purified lipopeptides. Patent: US8697638 B2

Khan MSA, Singh B, Cameotra SS (2015) Biological applications of biosurfactants and strategies to potentiate commercial production. In: Kosaric N, Vardar-Sukan F (eds) Biosurfactants, production and utilization-processes, technologies, and economics. CRC, Taylor & Francis, Boca Raton, FL, pp 270–294

Kim DH, Miler RM, Rasmussen SL, Stanghellini MR, Zhang Y (1997) Microbially produced rhamnolipids (biosurfactants) for the control of plant pathogenic zoosporic fungi. Patent: US1997/000701

Kim PI, Ryu J, Kim YH, Chi YT (2010) Production of biosurfactant lipopeptides iturin A, fengycin and surfactin A from *Bacillus subtilis* CMB32 for control of *Colletotrichum gloeosporioides*. J Microbiol Biotechnol 20:138–145. doi:10.4014/jmb.0905.05007

King JW, Srinivas K, Zhang D (2010) Advances in critical fluid processing. In: Proctor A (ed) Alternatives to conventional food processing. Royal Society of Chemistry, London

Knothe G, Steidley KR (2009) A comparison of used cooking oils: a very heterogeneous feedstock for biodiesel. Bioresour Technol 100:5796–5801. doi:10.1016/j.biortech.2008.11.064

Kourmentza C, Ntaikou I, Lyberatos G, Kornaros M (2015) Polyhydroxyalkanoates from *Pseudomonas* sp. using synthetic and olive mill wastewater under limiting conditions. Int J Biol Macromol 74:202–210. doi:10.1016/j.ijbiomac.2014.12.032

Kralova I, Sjöblom J (2009) Surfactants used in food industry: a review. J Dispers Sci Technol 30:1363–1383. doi:10.1080/01932690902735561

Krohn M, Zinke H (2011) Cyclic lipopeptides for use as taste modulators. Patent: 20110256291

Kruijt M, Tran H, Raaijmakers JM (2009) Functional, genetic and chemical characterization of biosurfactants produced by plant growth-promoting *Pseudomonas putida* 267. J Appl Microbiol 107:546–556. doi:10.1111/j.1365-2672.2009.04244.x

Kügler JH, Muhle-Goll C, Kühl B, Kraft A, Heinzler R, Kirschhofer F, Henkel M, Wray C, Luy B, Brebber-Weiss G, Lang S, Syldatk C, Hausmann R (2014) Trehalose lipid biosurfactant produced by the actinomycetes *Tsukamurella spumae* and *T. pseudospumae*. Appl Microbiol Biotechnol 98:8905–8915. doi:10.1007/s00253-014-5972-4

Kumari B, Singh SN, Singh DP (2012) Characterization of two biosurfactant producing strains in crude oil degradation. Process Biochem 47:2463–2471. doi:10.1016/j.procbio.2012.10.010

Kuppert D, Kottke U, Lattich J, Volk M, Wenk HH, Cabirol F, Schilling M, Schaffer S, Allef P (2014) Detergent formulation for textiles, comprising rhamnolipids with a predominant content of di-rhamnolipids. Patent: 20140296125 A1

Kurtzman CP, Price NP, Ray KJ, Kuo TM (2010) Production of sophorolipid biosurfactants by multiple species of the *Starmerella* (*Candida*) *bombicola* yeast clade. FEMS Microbiol Lett 311:140–146. doi:10.1111/j.1574-6968.2010.02082.x

Lan G, Fan Q, Liu Y, Chen C, Li G, Liu Y, Yin X (2015) Rhamnolipid production from waste cooking oil using *Pseudomonas* SWP-4. Biochem Eng J 101:44–54. doi:10.1016/j.bej.2015.05.001

Lang S (2002) Biological amphiphiles (microbial biosurfactants). Curr Opin Colloid 7:12–20. doi:10.1016/S1359-0294(02)00007-9

Lang S, Katsiwela E, Wagner F (1989) Antimicrobial effects of biosurfactants. Lipid/Fett 91:363–366. doi:10.1002/lipi.19890910908

Lindoerfer W, Sewe KU, Oberbremer A, Mueller-Hurtig R, Wagner F (1992) Microbial decontamination of soils contaminated with hydrocarbons, in particular mineral oils by microbial oxidation. Patent: US5128262 A

Loiseau C, Schlusselhuber M, Bigot R, Bertaux J, Berjeaud J, Verdon J (2015) Surfactin from *Bacillus subtilis* displays an unexpected anti-Legionella activity. Appl Microbiol Biotechnol 99:5083–5093. doi:10.1007/s00253-014-6317-z

Lopes SC, Rodrigues JM (2003) Immunogenic compositions containing antigens, gene vectors and adjuvants-loaded biodegradable microspheres. Patent: WO2001007869 A2

Lu JR, Zhao XB, Yaseen M (2007) Biomimetic amphiphiles: biosurfactants. Curr Opin Colloid Interface Sci 12:60–67. doi:10.1016/j.cocis.2007.05.004

Luna JM, Rufino RD, Sarubbo LA, Rodrigues LR, Teixeira JA, de Campos-Takaki GM (2011) Evaluation antimicrobial and antiadhesive properties of the biosurfactant lunasan produced by *Candida sphaerica* UCP 0995. Curr Microbiol 62:1527–1534. doi:10.1007/s00284-011-9889-1

Luna JM, Rufino RD, Sarubbo LA, Campos-Takaki GM (2013) Characterization, surface properties and biological activity of a biosurfactant produced from industrial waste by *Candida sphaerica* UCP0995 for application in the petroleum industry. Coll Surf B Biointerfaces 102:202–209. doi:10.1016/j.colsurfb.2012.08.008

Ma X-J, Li H, Wang D-X, Song X (2014) Sophorolipid production from delignined corncob residue by *Wickerhamiella domercqiae* var. *sophorolipid* CGMCC 1576 and *Cryptococcus curvatus* ATCC 96219. Appl Microbiol Biotechnol 98:475–483. doi:10.1007/s00253-013-4856-3

Maass D, Ramírez IM, Román MG, Alameda EJ, de Souza AAU, Valle JAB, Vaz DA (2016) Two-phase olive mill waste (alpeorujo) as carbon source for biosurfactant production. J Chem Technol Biotechnol 91:1990–1997. doi:10.1002/jctb.4790

Magalhães L, Nitschke M (2013) Antimicrobial activity of rhamnolipids against *Listeria monocytogenes* and their synergistic interaction with nisin. Food Control 29:138–142. doi:10.1016/j.foodcont.2012.06.009

Maier RM, Soberón-Chávez G (2000) *Pseudomonas aeruginosa* rhamnolipids: biosynthesis and potential applications. Appl Microbiol Biotechnol 54:625–633. doi:10.1007/s002530000443

Makkar RS, Cameotra SS (1997) Utilization of molasses for biosurfactant production by two *Bacillus* strains at thermophilic conditions. JAOCS 74:887–889. doi:10.1007/s11746-997-0233-7

Manoharan MJ, Bradeeba K, Parthasarathi R (2012) Development of surfactin based nanoemulsion formulation from selected cooking oils: evaluation for antimicrobial activity against selected food associated microorganisms. J Taiwan Inst Chem E 43:172–180. doi:10.1016/j.jtice.2011.08.008

Marchant R, Banat IM (2012a) Biosurfactants: a sustainable replacement for chemical surfactants? Biotechnol Lett 34:1597–1605. doi:10.1007/s10529-012-0956-x

Marchant R, Banat IM (2012b) Microbial biosurfactants: challenges and opportunities for future exploitation. Trends Biotechnol 30:558–565. doi:10.1016/j.tibtech.2012.07.003

Marchant R, Funston S, Uzoigwe C, Rahman PKSM, Banat IM (2015) Production of biosurfactants from nonpathogenic bacteria. In: Kosaric N, Vardar-Sukan F (eds) Biosurfactants, production and utilization-processes, technologies, and economics. CRC, Taylor & Francis, Boca Raton, FL, pp 73–82

Mariano AP, Bonotto DM, Angelis DF, Pirôllo MPS, Contiero J (2008) Use of weathered diesel oil as a low-cost raw material for biosurfactant production. Braz J Chem Eng 25:269–274. doi:10.1590/S0104-66322008000200006

Marqués AM, Pinazo A, Farfan M, Aranda FJ, Teruel JA, Ortiz A, Manresa A, Espuny MJ (2009) The physicochemical properties and chemical composition of trehalose lipids produced by *Rhodococcus erythropolis* 51T7. Chem Phys Lipids 158:110–117. doi:10.1016/j.chemphyslip.2009.01.001

Martino L, Cruz MV, Scoma A, Freitas F, Bertin L, Scandola M, Reis MAM (2014) Recovery of amorphous polyhydroxybutyrate granules from *Cupriavidus necator* cells grown on used cooking oil. Int J Biol Macromol 71:117–123. doi:10.1016/j.ijbiomac.2014.04.016

Matsuyama T, Murakami T, Fujita M, Fujita S, Yano I (1986) Extracellular vesicle formation and biosurfactant production by *Serratia marcescens*. J Gen Microbiol 132:865–875. doi:10.1099/00221287-132-4-865

McInerney MJ, Javaheri M, Nagle DP (1990) Properties of the biosurfactant produced by *Bacillus licheniformis* strain JF-2. J Ind Microbiol 5:95–102. doi:10.1007/BF01573858

Mimee B, Pelletier R, Bélanger RR (2009) In vitro antibacterial activity and antifungal mode of action of flocculosin, a membrane-active cellobiose lipid. J Appl Microbiol 107:989–996. doi:10.1111/j.1365-2672.2009.04280.x

Morita T, Konishi M, Fukuoka T, Imura T, Kitamoto D (2007) Microbial conversion of glycerol into glycolipid biosurfactants, mannosylerythritol lipids, by a Basidiomycete yeast, *Pseudozyma antarctica* JCM 10317T. J Biosci Bioeng 104:78–81. doi:10.1263/jbb.104.78

Mussatto SI (2014) Brewer's spent grain: a valuable feedstock for industrial applications. J Sci Food Agric 94:1264–1275. doi:10.1002/jsfa.6486

Narayanan J, Ramji R, Sahu H, Gautam P (2010) Synthesis, stabilisation and characterisation of rhamnolipid-capped ZnS nanoparticles in aqueous medium. IET Nanobiotechnol 4:29–34. doi:10.1049/iet-nbt.2009.0010

Navon-Venezia S, Zosim Z, Gottlieb A, Legmann R, Carmeli S, Ron EZ, Rosenberg E (1995) Alasan, a new bioemulsifier from *Acinetobacter radioresistens*. Appl Environ Microbiol 61:3240–3244

Navon-Venezia S, Banin E, Ron EZ, Rosenberg E (1998) The bioemulsifier alasan: role of protein in maintaining structure and activity. Appl Microbiol Biotechnol 49:382–384. doi:10.1007/s002530051186

Nero MD (2007) Systems and methods for spot cleaning materials. Patent: US7291585 B2

Nerurkar AS (2010) Structural and molecular characteristics of lichenysin and its relationship with surface activity. In: Sen R (ed) Biosurfactants, Advances in experimental medicine and biology, vol 672. Springer, New York, pp 304–315

Neu TR, Poralla K (1990) Emulsifying agent from bacteria isolated during screening for cells with hydrophobic surfaces. Appl Microbiol Biotechnol 32:521–525. doi:10.1007/BF00173721

Ngai AL, Bourque MR, Lupinacci RJ, Strohmaier KM, Kartsonis NA (2011) Overview of safety experience with caspofungin in clinical trials conducted over the first 15 years: a brief report. Int J Antimicrob Agents 38:540–544. doi:10.1016/j.ijantimicag.2011.07.008

Nitschke M, Costa SG (2007) Biosurfactants in food industry. Trends Food Sci Technol 18:252–259. doi:10.1016/j.tifs.2007.01.002

Nitschke M, Pastore GM (2004) Biosurfactant production by *Bacillus subtilis* using cassava-processing effluent. Appl Biochem Biotechnol 112:163–172. doi:10.1016/j.biortech.2006.12.010

Nitschke M, SGVAO C, Haddad R, LAG G, Eberlin MN, Contiero J (2005) Oil wastes as unconventional substrates for rhamnolipid biosurfactant production by *Pseudomonas aeruginosa* LBI. Biotechnol Prog 21:1562–1566. doi:10.1021/bp050198x

Nitschke M, Araújo LV, Costa SGVAO, Pires RC, Zeraik AE, Fernandes ACLB, Freire DMG, Contiero J (2009) Surfactin reduces the adhesion of food-borne pathogenic bacteria to solid surfaces. Lett Appl Microbiol 49:241–247. doi:10.1111/j.1472-765X.2009.02646.x

Noah KS, Fox SL, Bruhn DF, Thompson DN, Bala GA (2002) Development of continuous surfactin production from potato process effluent by *Bacillus subtilis* in an airlift reactor. Appl Biochem Biotechnol 98-100:803–813. doi:10.1385/ABAB:98-100:1-9:803

Noah KS, Bruhn DF, Bala GA (2005) Surfactin production from potato process effluent by *Bacillus subtilis* in a chemostat. Appl Biochem Biotechnol 121-124:465–473. doi:10.1385/ABAB:122:1-3:0465

Noparat P, Maneerat S, Saimmai A (2014) Utilization of palm oil decanter cake as a novel substrate for biosurfactant production from a new and promising strain of *Ochrobactrum anthropi* 2/3. World J Microbiol Biotechnol 30:865–877. doi:10.1007/s11274-013-1493-z

Ojha AK, Yang Y, Zhang P (2013) Use of *Mycobacterium smegmatis* trehalose dimycolate hydrolase. Patent: US20130323224

Oliveira FJS, Vazquez L, de Campos NP, de Franca FP (2009) Production of rhamnolipids by a *Pseudomonas alcaligenes* strain. Process Biochem 44(4):383–389. doi:10.1016/j.procbio.2008.11.014

Oliveira DWF, França ÍWL, Félix AKN, Martins JJL, Giro MEA, Melob VMM, Gonçalves LRB (2013) Kinetic study of biosurfactant production by *Bacillus subtilis* LAMI005 grown in clarified cashew apple juice. Colloids Surf B Biointerfaces 101:34–43. doi:10.1016/j.colsurfb.2012.06.011

Onbasli D, Aslim B (2009) Biosurfactant production in sugar beet molasses by some *Pseudomonas* spp. J Environ Biol 30:161–163

Ongena M, Jourdan E, Adam A, Paquot M, Brans A, Joris B, Arpigny JL, Thonart P (2007) Surfactin and fengycin lipopeptides of *Bacillus subtilis* as elicitors of induced systemic resistance in plants. Environ Microbiol 9(4):1084–1090. doi:10.1111/j.1462-2920.2006.01202.x

Park SY, Kim Y (2009) Surfactin inhibits immunostimulatory function of macrophages through blocking NK-κB, MAPK and Akt pathway. Int Immunopharmacol 9:886–893. doi:10.1016/j.intimp.2009.03.013

Parry AJ, Parry NJ, Peilow AC, Stevenson PS (2012) Detergent compositions comprising biosurfactant and lipase. Patent: WO2012010407 A1

Parry AJ, Parry NL, Peilow AC, Stevenson PS (2013) Combinations of rhamnolipids and enzymes for improved cleaning. Patent: WO2012010406 A1

Partovi M, Lotfabad TB, Roostaazad R, Bahmaei M, Tayyebi S (2013) Management of soybean oil refinery wastes through recycling them for producing biosurfactant using *Pseudomonas aeruginosa* MR01. World J Microbiol Biotechnol 29:1039–1047. doi:10.1007/s11274-013-1267-7

Patel RM, Desai AJ (2003) Biosurfactant production by *Pseudomonas aeruginosa*GS3 from molasses. Lett Appl Microbiol 25:91–94. doi:10.1046/j.1472-765X.1997.00172.x

Pecci Y, Rivardo F, Martinotti MG, Allegrone G (2010) LC/ESI-MS/ MS characterization of lipopeptide biosurfactants produced by *Bacillus licheniformis* V9T14 strain. J Mass Spectrom 45:772–778. doi:10.1002/jms.1767

Pesce L (2002) A biotechnological method for the regeneration of hydrocarbons from dregs and muds, on the base of biosurfactants. World patent 02/062495

Piljac G, Piljac V (1995) Pharmaceutical preparation based on rhamnolipid. Patent: US5455232 A

Pradel E, Zhang E, Pujol N, Matsuyama T, Bargmann CI, Ewbank JJ (2007) Detection and avoidance of a natural product from the pathogenic bacterium *Serratia marcescens* by *Caenorhabditis elegans*. Proc Natl Acad Sci USA 104:2295–2300. doi:10.1073/pnas.0610281104

PubChem Compound Database. http://www.ncbi.nlm.nih.gov/pccompound/, accessed on October 2015

Raaijmakers JM, de Bruijn I, de Kock MJ (2006) Cyclic lipopeptide production by plant-associated *Pseudomonas* spp.: diversity, activity, biosynthesis, and regulation. Mol Plant Microbe Interact 19:699–710. doi:10.1094/MPMI-19-0699

Radhakrishnan N, Kavitha V, Madhavacharyulu E, Gnanamani A, Mandal AB (2011) Isolation, production and characterization of bioemulsifiers of marine bacteria of coastal Tamil Nadu. Indian J Geo Mar Sci 40:76–82

Rafiqul ISM, Sakinah AMM (2013) Processes for the production of xylitol—a review. Food Rev Int 29:127–156. doi:10.1080/87559129.2012.714434

Ramachandran S, Singh SK, Larroche C, Soccol CR, Pandey A (2007) Oil cakes and their biotechnological applications—a review. Bioresour Technol 98:2000–2009. doi:10.1016/j.biortech.2006.08.002

Ramírez IM, Tsaousi K, Rudden M, Marchant R, Jurado AE, García RM, Banat IM (2015) Rhamnolipid and surfactin production from olive oil mill waste as sole carbon source. Bioresour Technol 198:231–236. doi:10.1016/j.biortech.2015.09.012

Randhawa KKS, Rahman PKSM (2014) Rhamnolipid biosurfactants—past, present, and future scenario of global market. Front Microbiol 5:454. doi:10.3389/fmicb.2014.00454

Reddy AS, Chen C, Chen C, Jean J, Fan C, Chen H, Wang J, Nimje VR (2009a) Synthesis of gold nanoparticles via an environmentally benign route using a biosurfactant. J Nanosci Nanotechnol 9:6693–6699. doi:10.1166/jnn.2009.1347

Reddy AS, Chen C, Baker SC, Chen C, Jean J, Fan C, Chen H, Wang J (2009b) Synthesis of silver nanoparticles using surfactin: a biosurfactant as stabilizing agent. Mater Lett 63:1227–1230. doi:10.1016/j.matlet.2009.02.028

Rhamnopharma Inc. USA (2013) Method for treating rhinitis and sinusitis by rhamnolipids. Patent: US20140080771 A1

Rivardo F, Martinotti MG, Turner RJ, Ceri H (2011) Synergistic effect of lipopeptide biosurfactant with antibiotics against *Escherichia coli* CFT073 biofilm. Int J Antimicrob Agents 37:324–331. doi:10.1016/j.ijantimicag.2010.12.011

Rocha MVP, Oliveira AHS, Souza MCM, Gonçalves LRB (2006) Natural cashew apple juice as fermentation medium for biosurfactant production by *Acinetobacter calcoaceticus*. World J Microbiol Biotechnol 22:1295–1299. doi:10.1007/s11274-006-9175-8

Rocha MV, Souza MC, Benedicto SC, Bezerra MS, Macedo GR, Pinto GA, Gonçalves LR (2007) Production of biosurfactant by *Pseudomonas aeruginosa* grown on cashew apple juice. Appl Biochem Biotechnol 137–140:185–194. doi:10.1007/s12010-007-9050-6

Rocha MVP, Barreto RVG, Melo VMM, Gonçalves LRB (2009) Evaluation of cashew apple juice for surfactin production by *Bacillus subtilis* LAMI008. Appl Biochem Biotechnol 155:366–378. doi:10.1007/s12010-008-8459-x

Rosenberg E (1986) Microbial surfactants. CRC Crit Rev Biotechnol 3:109–132. doi:10.3109/07388558509150781

Rosenberg E, Ron EZ (1990) Bacterial process for the production of dispersants. Patent: US4921793 A

Rosenberg E, Ron EZ (1998) Bioemulsifiers. Patent: US5840547 A

Rosenberg E, Zuckerberg A, Rubinovitz C, Gutnick DL (1979) Emulsifier of *Arthrobacter* RAG-1: isolation and emulsifying properties. Appl Environ Microbiol 37:402–408

Rosenberg E, Rubinovitz C, Legmann R, Ron EZ (1988) Purification and chemical properties of *Acinetobacter calcoaceticus* A2 biodispersan. Appl Environ Microbiol 54:323–326

Rufino RD, Sarubbo LA, Campos-Takaki GM (2007) Enhancement of stability of biosurfactant produced by *Candida lipolytica* using industrial residue as substrate. World J Microbiol Biotechnol 23:729–734. doi:10.1007/s11274-006-9278-2

Rufino RD, Luna JM, Takaki GMC, Sarubbo LA (2014) Characterization and properties of the biosurfactant produced by *Candida lipolytica* UCP 0988. Electron J Biotechnol 17:34–38. doi:10.1016/j.ejbt.2013.12.006

Sachdev DP, Cameotra SS (2013) Biosurfactants in agriculture. Appl Microbiol Biotechnol 97:1005–1016. doi:10.1007/s00253-012-4641-8

Saini HS, Barragan-Huerta BE, Lebron-Paler A, Pemberton JE, Vazquez RR, Burns AM, Marron MT, Seliga CJ, Gunatilaka AA, Maier RM (2008) Efficient purification of the biosurfactant viscosin from *Pseudomonas libanensis* strain M9–3 and its physicochemical and biological properties. J Nat Prod 71:1011–1015. doi:10.1021/np800069u

Saisa-Ard K, Maneerat S, Saimmai A (2013) Isolation and characterization of biosurfactants-producing bacteria isolated from palm oil industry and evaluation for biosurfactants production using low-cost substrates. J Biotechnol Comput Biol Bionanotechnol 94:275–284. doi:10.5114/bta.2013.46421

Samad A, Zhang J, Chen D, Liang Y (2015) Sophorolipid production from biomass hydrolysates. Appl Biochem Biotechnol 175:2246–2257. doi:10.1007/s12010-014-1425-x

Santos DKF, Rufino RD, Luna JM, Santos VA, Salgueiro AA, Sarubbo LA (2013) Synthesis and evaluation of biosurfactant produced by *Candida lipolytica* using animal fat and corn steep liquor. J Petrol Sci Eng 105:43–50. doi:10.1016/j.petrol.2013.03.028

Sawai Pharmaceutical Co., LTD (1991) Trehalose trimycolate and medicinal composition. Patent: EP0261248 B1

Schofield MH, Thavasi R, Gross RA (2013) Modified sophorolipids for the inhibition of plant pathogens. Patent: US20130085067 A1

Sen R (2008) Biotechnology in petroleum recovery: the microbial EOR. Prog Energy Combust 34:714–724. doi:10.1016/j.pecs.2008.05.001

Shah V, Doncel GF, Seyoum T, Eaton KM, Zalenskaya I, Hagver R, Azim A, Gross R (2005) Sophorolipids, microbial glycolipids with anti-human immunodeficiency virus and sperm-immobilizing activities. Antimicrob Agents Chemother 49:4093–4100. doi:10.1128/AAC.49.10.4093-4100.2005

Shao Z (2011) Trehalolipids. In: Soberón-Chávez G (ed) Biosurfactants—from genes to applications, Microbiology monographs, vol 20. Springer, Berlin, pp 121–144

Sharma S, Verma M, Sharma A (2013) Utilization of non edible oil seed cakes as substrate for growth of *Paecilomyces lilacinus* and as biopesticide against termites. Waste Biomass Valoriz 4:325–330. doi:10.1007/s12649-012-9134-6

She YH, Zhang F, Xia JJ, Kong S, Wang Z, Shu F, Hu J (2011) Investigation of biosurfactant-producing indigenous microorganisms that enhance residue oil recovery in an oil reservoir after polymer flooding. Appl Biochem Biotechnol 163:223–234. doi:10.1007/s12010-010-9032-y

Shekhar S, Sundaramanickam A, Balasubramanian T (2015) Biosurfactant producing microbes and their potential applications: a review. Crit Rev Environ Sci Technol 45:1522–1554. doi:10.1080/10643389.2014.955631

Shepherd R, Rockey J, Shutherland IW, Roller S (1995) Novel bioemulsifiers from microorganisms for use in foods. J Biotechnol 40:207–217. doi:10.1016/0168-1656(95)00053-S

Sigma Aldrich (Company website). http://www.sigmaaldrich.com/life-science/biochemicals/biochemical-products.html?TablePage=111283763, accessed on October 2015

Silva HD, Cerqueira MA, Vicente AA (2012) Nanoemulsions for food applications: development and characterization. Food Bioprocess Technol 5:854–867. doi:10.1007/s11947-011-0683-7

Silva NMPR, Rufino RD, Luna JM, Santos VA, Sarubbo LA (2014a) Screening of *Pseudomonas* species for biosurfactant production using low-cost substrates. Biocatal Agric Biotechnol 3:132–139. doi:10.1016/j.bcab.2013.09.005

Silva EJ, Silva NMPR, Rufino RD, Luna JM, Silva RO, Sarubbo LA (2014b) Characterization of a biosurfactant produced by *Pseudomonas cepacia* CCT6659 in the presence of industrial wastes and its application in the biodegradation of hydrophobic compounds in soil. Colloids Surf B Biointerfaces 117:36–41. doi:10.1016/j.colsurfb.2014.02.012

Simpson DR, Natraj NR, McInerney MJ, Duncan KE (2011) Biosurfactant-producing *Bacillus* are present in produced brines from Oklahoma oil reservoirs with a wide range of salinities. Appl Microbiol Biotechnol 91:1083–1093. doi:10.1007/s00253-011-3326-z

Singh BR, Dwivedi S, Al-Khedhairy AA, Musarrat J (2011) Synthesis of stable cadmium sulfide nanoparticles using surfactin produced by *Bacillus amyloliquefaciens* strain KSU-109. Colloids Surf B Biointerfaces 85:207–213. doi:10.1016/j.colsurfb.2011.02.030

Singh AM, Rautela R, Cameotra SS (2014) Substrate dependent *in vitro* antifungal activity of *Bacillus sp*. strain AR2. Microb Cell Fact 13:67. doi:10.1186/1475-2859-13-67

Sobrinho HBS, Rufino RD, Luna JM, Salgueiro AA, Campos-Takaki GM, Leite LFC, Sarubbo LA (2008) Utilization of two agroindustrial by-products for the production of a surfactant by *Candida sphaerica* UCP0995. Proc Biochem 43:912–917. doi:10.1016/j.procbio.2008.04.013

Solaiman DKY, Ashby RD, Nuñez A, Foglia TA (2004) Production of sophorolipids by *Candida bombicola* grown on soy molasses as substrate. Biotechnol Lett 26:1241–1245. doi:10.1023/B:BILE.0000036605.80577.30

Solaiman DKY, Ashby RD, Zerkowski JA, Foglia TA (2007) Simplified soy molasses-based medium for reduced-cost production of sophorolipids by *Candida bombicola*. Biotechnol Lett 29:1341–1347. doi:10.1007/s10529-007-9407-5

Sousa JR, Correia JAC, Almeida JGL, Rodrigues S, Pessoa ODL, Melo VMM (2011) Evaluation of a co-product of biodiesel production as carbon source in the production of biosurfactant by *P. aeruginosa* MSIC02. Proc Biochem 46:1831–1839. doi:10.1016/j.procbio.2011.06.016

Su W, Chen W, Lin Y (2009) Optimizing emulsan production of *A. venetianus* RAG-1 using response surface methodology. Appl Microbiol Biotechnol 84:271–279. doi:10.1007/s00253-009-1957-0

Syldatk C, Hausmann R (2010) Microbial biosurfactants. Eur J Lipid Sci Technol 112:615–616. doi:10.1002/ejlt.201000294

Tally FP, Zeckel M, Wasilewski MM, Carini C, Berman CL, Drusano GL, Oleson FB (1999) Daptomycin: a novel agent for Gram-positive infections. Expert Opin Invest Drugs 8:1223–1238. doi:10.1517/13543784.8.8.1223

Tavares LFD, Silva PM, Junqueira M, Mariano DCO, Nogueira FCS, Domont GB, Freire DMG, Neves BC (2013) Characterization of rhamnolipids produced by wild-type and engineered *Burkholderia kururiensis*. Appl Microbiol Biotechnol 97:1909–1921. doi:10.1007/s00253-012-4454-9

TechNavio (2015) Global bio-surfactants market 2015-2019, ID:3244038. http://www.researchandmarkets.com/research/bwqrsf/global

Thavasi R, Jayalakshmi S, Balasubramanian T, Banat IM (2007) Biosurfactant production by *Corynebacterium kutscheri* from waste motor lubricant oil and peanut oil cake. Lett Appl Microbiol 45:686–691. doi:10.1111/j.1472-765X.2007.02256.x

Thavasi R, Jayalakshmi S, Balasubramanian T, Banat IM (2008) Production and characterization of a glycolipid biosurfactant from *Bacillus megaterium* using economically cheaper sources. World J Microbiol Biotechnol 24:917–925. doi:10.1007/s11274-007-9609-y

Thavasi R, Jayalakshmi S, Banat IM (2011) Application of biosurfactant produced from peanut oil cake by *Lactobacillus delbrueckii* in biodegradation of crude oil. Bioresour Technol 102:3366–3372. doi:10.1016/j.biortech.2010.11.071

Thavasi R, Marchant R, Banat IM (2015) Biosurfactant applications in agriculture. In: Kosaric N, Vardar-Sukan F (eds) Biosurfactants, production and utilization-processes, technologies, and economics. CRC, Taylor & Francis, Boca Raton, FL, pp 313–325

Thompson DN, Fox SL, Bala GA (2000) Biosurfactants from potato process effluents. Appl Biochem Biotechnol 84-86:917–930. doi:10.1385/ABAB:84-86:1-9:917

Tokumoto Y, Nomura N, Uchiyama H, Imura T, Morita T, Fukuota T, Kitamoto D (2009) Structural characterization and surface-active properties of a succinoyl trehalose lipid produced by *Rhodococcus* sp. SD-74. J Oleo Sci 58:97–102. doi:10.5650/jos.58.97

Toren A, Navon-Venezia S, Ron EZ, Rosenberg E (2001) Emulsifying activities of purified alasan proteins from *Acinetobacter radioresistens* KA53. Appl Environ Microbiol 67:1102–1106. doi:10.1128/AEM.67.3.1102-1106.2001

Toribio J, Escalante AE, Soberon-Chavez G (2010) Rhamnolipids: production in bacteria other than *Pseudomonas aeruginosa*. Eur J Lipid Sci Technol 112:1082–1087. doi:10.1002/ejlt.200900256

Towner KJ (1996) Biology of *Acinetobacter* spp. In: Bergogne-Berezin E, Jily-Guillou ML, Towner KJ (eds) *Acinetobacter*—microbiology, epidemiology, infections, management. CRC, Boca Raton, FL, pp 13–36

Tran H, Ficke A, Asiimwe T, Hofte M, Raaijmakers JM (2007) Role of the cyclic lipopeptide massetolide A in biological control of *Phytophthora infestans* and in colonization of tomato plants by *Pseudomonas fluorescens*. New Phytol 175:731–742. doi:10.1111/j.1469-8137.2007.02138.x

Transparency Market Research (2012) Biosurfactants market-Global scenario, raw material and consumption trends, industry analysis, size, share and forecasts, 2011-2018. http://www.transparencymarketresearch.com/biosurfactants-market.html

Tuleva B, Christova N, Cohen R, Stoev G, Stoineva I (2008) Production and structural elucidation of trehalose tetraesters (biosurfactants) from a novel alkanothrophic *Rhodococcus wratislaviensis* strain. J Appl Microbiol 104:1703–1710. doi:10.1111/j.1365-2672.2007.03680.x

Uchida Y, Misava S, Nakahara T, Tabuchi T (1989a) Factor affecting the production of succinoyl trehalose lipids by *Rhodococcus erythropolis* SD-74 grown on n-alkanes. Agric Biol Chem 53:765–769. doi:10.1080/00021369.1989.10869386

Uchida Y, Tsuchiya R, Chino M, Hirano J, Tabuchi T (1989b) Extracellular accumulation of mono- and di- succinoyl trehalose lipids by a strain of *Rhodococcus erythropolis* grown on n-alkanes. Agric Biol Chem 53:757–763. doi:10.1080/00021369.1989.10869385

Unilever Plc. (1992) Detergent compositions. Patent: EP0499434 A1

Unilever Plc. (2015) A highly water-soluble solid laundry detergent composition that forms a clear wash liquor upon dissolution in water. Patent: ECLI:EP:BA:2015:T056213.20150624

van Bogaert INA, Soetaert W (2011) Sophorolipids. In: Soberón-Chávez G (ed) Biosurfactants—from genes to applications, Microbiology monographs, vol 20. Springer, Berlin, pp 179–210

van Haesendonck I, Vanzeveren E (2005) Rhamnolipids in bakery products. Patent: EP1560493 A1

Varnier AL, Sanchez L, Vatsa P, Boudesocque L, Garcia-Brugger A, Rabenoelina F, Sorokin A, Renault J, Kauffmann S, Pugin A, Clement C, Baillieul F, Dorey S (2009) Bacterial rhamnolipids are novel MAMPs conferring resistance to Botrytis cinerea in grapevine. Plant Cell Environ 32(2):178–193. doi:10.1111/j.1365-3040.2008.01911.x

Vasileva-Tonkova E, Sotirova E, Galabova D (2011) The effect of rhamnolipid biosurfactant produced by *Pseudomonas fluorescens* on model bacterial strains and isolates from industrial wastewater. Curr Microbiol 62(2):427–433. doi:10.1007/s00284-010-9725-z

Vedaraman N, Venkatesh N (2011) Production of surfactin by *Bacillus subtilis* MTCC 2423 from waste frying oils. Braz J Chem Eng 28:175–180. doi:10.1590/S0104-66322011000200001

Volke F, Meiche J (2005) Coupling medium for transversal ultrasonic waves. Patent: US6899677 B1

Vollenbroich D, Ozel M, Vater J, Kamp RM, Pauli G (1997) Mechanism of inactivation of enveloped viruses by the biosurfactant surfactin from *Bacillus subtilis*. Biologicals 25:289–297. doi:10.1006/biol.1997.0099

Wadekar S, Kale S, Lali A, Bhowmick D, Pratap A (2012) Sophorolipid production by *Starmerella bombicola* (ATCC 22214) from virgin and waste frying oils, and the effects of activated earth treatment of the waste oils. J Am Oil Chem Soc 89:1029–1039. doi:10.1007/s11746-011-1986-6

Wattanaphon HT, Kerdsin A, Thammacharoen C, Sangvanich P, Vangnai AS (2008) A biosurfactant from *Burkholderia cenocepacia* BSP3 and its enhancement of pesticide solubilization. J Appl Microbiol 105:416–423. doi:10.1111/j.1365-2672.2008.03755.x

Wilkinson SG, Galbraith L (1975) Studies on lipopolysaccharides from *Pseudomonas aeruginosa*. Eur J Biochem 52:331–343. doi:10.1111/j.1432-1033.1975.tb04001.x

Wittgens A, Tiso T, Arndt TT, Wenk P, Hemmerich J, Muller C, Wichmann R, Kupper B, Zwick M, Wilhelm S, Hausmann R, Syldatk C, Rosenau F, Blank LM (2011) Growth independent rhamnolipid production from glucose using the non pathogenic *Pseudomonas putida* KT2440. Microb Cell Fact 10:80. doi:10.1186/1475-2859-10-80

Worakitsiri P, Pornsunthorntawee O, Thanpitcha T, Chavadej S, Weder C, Rujiravanit R (2011) Synthesis of polyaniline nanofibers and nanotubes via rhamnolipid biosurfactant templating. Synthetic Met 161:298–306. doi:10.1016/j.synthmet.2010.11.039

Zeng G, Liu Z, Zhong H, Li J, Yuan X, Fu H, Ding Y, Wang J, Zhou M (2011) Effect of monorhamnolipid on the degradation of n-hexadecane by *Candida tropicalis* and the association with cell surface properties. Appl Microbiol Biotechnol 90:1155–1161. doi:10.1007/s00253-011-3125-6

Zeraik AN, Nitschke M (2010) Biosurfactants as agents to reduce adhesion of pathogenic bacteria to polystyrene surfaces: effect of temperature and hydrophobicity. Curr Microbiol 61(6):554–559. doi:10.1007/s00284-010-9652-z

Zhang JW, Lee SH, Gross RA, Kaplan D (1999) Surface properties of emulsan-analogs. J Chem Technol Biotechnol 74:759–765. doi:10.1002/(SICI)1097-4660(199908)74:8<759::AID-JCTB88>3.0.CO;2-T

Zhang J, Yin R, Lin X, Liu W, Chen R, Li X (2010) Interactive effect of biosurfactant and microorganism to enhance phytoremediation for removal of aged polycyclic aromatic hydrocarbons from contaminated soils. J Health Sci 56:257–266. doi:10.1248/jhs.56.257

Zhang C, Wang S, Yan Y (2011) Isomerization and biodegradation of beta-cypermethrin by *Pseudomonas aeruginosa* CH7 with biosurfactant production. Bioresour Technol 102:7139–7146. doi:10.1016/j.biortech.2011.03.086

Zhao Z, Wong JWC (2009) Biosurfactants from *Acinetobacter calcoaceticus* BU03 enhance the solubility and biodegradation. Environ Technol 30:291–299. doi:10.1080/09593330802630801

Zheng C, Luo Z, Yu L, Huang L, Bai X (2011) The utilization of lipid waste for biosurfactant production and its application in enhancing oil recovery. Petrol Sci Technol 29:282–289. doi:10.1080/10916460903117586

Zosim Z, Guntick D, Rosenberg E (1982) Properties of hydrocarbon-in-water emulsions stabilized by *Acinetobacter* RAG-1 emulsan. Biotechnol Bioeng 24:281–292. doi:10.1002/bit.260240203

# Oil Biodegradation

**Bhagwan Rekadwad and Chandrahasya Khobragade**

**Abstract** This chapter reports the sources, impacts of oil spills, and tar pollution in the marine environment. It also highlights the adverse effects of crude oil and tar pollution on the entire marine ecosystem and national economy. Effective policies measures (mechanical, biological, and societal) for recovering and managing oil spills and tar pollution in Open Oceans and on beaches, harbors, and ports are required to be in place to restore their beauty. Public awareness, government, and private participation and polices are necessary to protect, save, and preserve the pretty and precious marine/coastal biodiversity.

**Keywords** Crude oil pollution • Marine biodiversity • Oxygenase • *Pseudomonas*

## 1 Introduction

Oil and tar are disturbing the marine ecosystem. This threat is persisting for a long time. Though in recent decades, evidence has shown decreased impacts. These oil spill insistences in open ocean or near sea coast occur on a daily basis through oil exploitation process, transportation, production, filling, etc. The oil spill which occurred in the polar environments gradually cooled and resulted in the formation of ball-like structures ranging from a few centimeters to 300 cm or larger. These have higher density and sink at the bottom of the sea. It affects the deep-sea benthic meiofauna. Also, these oil balls come to seashore by sea current and giant waves in ocean (Freije 2015; Main et al. 2015). These again cause extensive damage to the coastal ecosystem. It limits the availability of various natural resources and affects the livelihood and the national economy. Therefore, it is necessary to find out the various sources of oil spills and tar ball, their impacts on the human being and society, and their control by adopting the necessary control measures. Once oil spills over an area, it is subjected to a variety of weathering processes (both abiotic and biotic). Most of the light chain hydrocarbons will be removed by environmental

B. Rekadwad (✉) • C. Khobragade
School of Life Sciences, Swami Ramanand Teerth Marathwada University, Nanded 431606, Maharashtra, India
e-mail: rekadwad@gmail.com



V.C. Kalia, P. Kumar (eds.), *Microbial Applications Vol.1*,
DOI 10.1007/978-3-319-52666-9_3

weathering process such as evaporation, dissolution (air–oil suspension), and photochemical oxidation. The heavy chain hydrocarbons are not removed by environmental weathering process. It is very difficult to remove oil slick, deeply percolated oil in sand, and oil-stained sand. Only oil-eating microorganisms have the ability to degrade heavy chain hydrocarbon.

This chapter describes the sources, impacts, and possible ways to reduce/clean the oil spills, tar, and xenobiotic pollution using eco-friendly poly-extremophilic microorganisms.

## 2 Sources and Impacts of Oil Spills Pollution

India has a coastline of 8118 km. The coastline of important states of India is coming in contact with seashore such as Gujarat, Maharashtra, Goa, Kerala, Tamil Nadu, Andhra Pradesh, Orissa, West Bengal, and Tripura. Marine ecosystems, especially in the Arabian Sea, are stressed by destructive overfishing practices, toxic pollution, and climate change. These are now threatened by additional artificial problems of oil spills and tar pollution (Figs. 1 and 2). The oil spill and tar-ball pollution have become a global problem particularly in industrialized and developing countries such as India. Every day hundreds of liters of oil contaminate the oceans and natural water bodies by curbing and hazardous activities in Open Oceans and on ports, harbors and beaches. This adversely affected the marine environment (Sukhdhane et al. 2013; Rekadwad and Khobragade 2015).

The oil reduces the penetration of light in thermocline by forming a layer on Sea surface. As a result of these, the Ocean productivity in polluted areas decreased. Ultimately, the chlorophyll production in Open Ocean is inhibited. On the other hand, the microorganisms utilize every form of available oxygen in the water. This

**Fig. 1** Dark streaks caused by tar balls lying on a south Goa beach during the last week of July 2005 (NIO, Goa, www.nio.org)

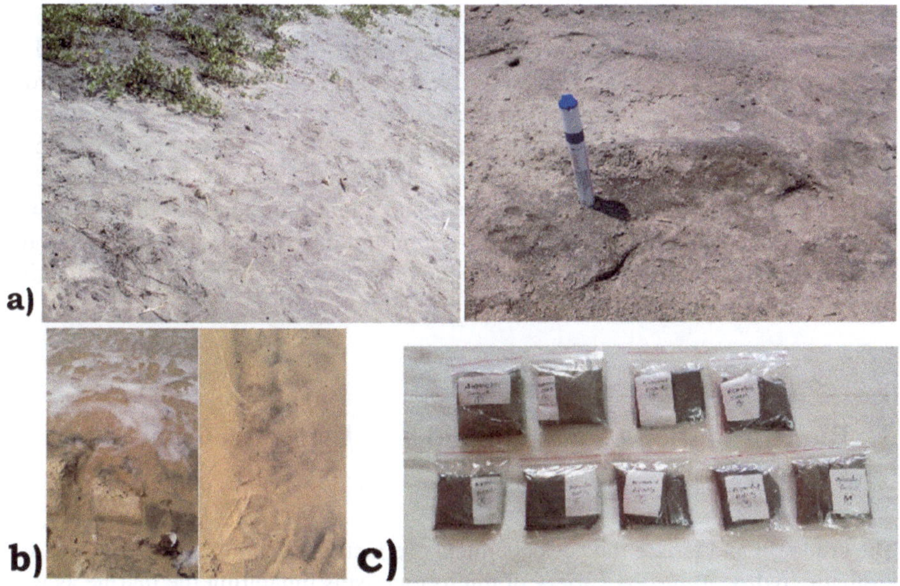


**Fig. 2** Field photographs of oil pollution in Goa: (**a**) Arambol beach, Goa; (**b**) Candolim beach, Goa; and (**c**) oil-slicked sand samples collected from Arambol beach

stalled the growth of sensitive species of planktons such as copepods, amphipods, and benthic meiofauna (Kirk et al. 2005; Shi et al. 2005; Chauhan et al. 2008; Lai et al. 2009; Phillips et al. 2009; Sun et al. 2010; Tahhan et al. 2010; Duraisamy and Latha 2011). The remainder of oil slicks, tar-ball residues, and heavy petroleum hydrocarbons formed the emulsion in water and with beach sand. The sun-heated oil spill and tar percolate deep in the sand, and sea current moves it to and fro. The oil covers Sea surface which disallows the mixing of atmospheric oxygen in seawater. It creates the anaerobic environment. The various anaerobic microorganisms include siderophore-producing bacteria flourishing under anaerobic conditions. In the produced siderophore under anaerobic conditions, the free iron present traces dissolved and it is made available for the uptake of phytoplankton (microalgae). It results in the emergence of harmful algal blooms of *Noctiluca scintillans*. It is a free-living nonparasitic species of dinoflagellate. It exhibits bioluminescence when their colonies are disturbed. *Noctiluca scintillans* and *Noctiluca*-like species form *Red tide*. This alga clogs the gills of fishes and marine organisms and their death occurs (Jonker et al. 2006; Riccardi 2010; Tao et al. 2011). The impact of oil pollution is visually seen on the salt marsh plant and mangroves. The crude oil contamination halts the growth of salt marsh plants (i.e., stunted growth of salt marsh plant), reduces the density/thickness of the stem, and lowers percentage of biomass leading to mortality of the plants (Lin and Mendelssohn 2009; Ribeiro et al. 2014). The potential fishing zones (PFZ) are coming under the influence of crude oil and tar contamination. As a result of this, a number of PFZ are reduced. The remainder of the oil slick and tar (residual crude oil) make the Sea coast dirty places. The oil contamination also affected the productivity of fishes in lagoons and ditches. These directly have an impact on the livelihood of fishermen (Kiruri et al. 2013; Rekadwad and Khobragade 2015; Warnock et al. 2015). Moreover, the coral reefs are very sensitive to changes in the surrounding environment. Coral reefs play a key role in the economy of many countries. Coral reefs have hundreds of species in a natural marine ecosystem. The demand for coral reefs is increasing steadily due to the presence of useful natural products. Coral reefs provide coastal protection and supply food and natural products useful in pharmaceutical and cosmetic industries. The major coral reef area is also under the influence of this marine pollution. It directly affects the economy by reducing foreign currency coming to the country through tourism.

## 3 Biodegradation Crude Oil and Tar Using Bacterial Cultures

The crude oil and tar are the petrochemical product that contains both aliphatic and aromatic hydrocarbons. A variety of microorganisms inhabiting the marine as well as terrestrial environment have the ability to degrade hydrocarbons under the oxic and anoxic conditions (Table 1). Halophilic facultative anaerobe *Halobacterium* produce butylated hydroxyl-toluene, 1,2-benzene-di-carboxylic acid, bis-ester, and di-butyl phthalate from these hydrocarbons. *Halobacterium* redistributes the contaminant by conversion of heavy to light hydrocarbons, which are easily degraded by other microorganisms (Hao and Lu 2009). The polycyclic aromatic hydrocarbon (PAH)-degrading bacteria have monooxygenase pathways (MOP). They have the ability to utilize and degrade crude oil, lubricating oil, grease, diesel, petrol, kerosene, naphthalene, and toluene into short-chain hydrocarbons through MOP within a short period of time (12 h) (Geetha et al. 2013). The short chain alkanes are degraded by aerobic microorganisms, mostly bacteria (*Desulfosarcina-Desulfococcus* cluster of Delta-proteobacteria) and fungi. These microorganisms inhabiting coastal and marine environments perform biodegradation through mechanisms such as the sulfate reduction and anaerobic oxidation of methane ($CH_4$), etc. They consume small alkanes in anoxic conditions. Under anoxic conditions, these bacteria perform homolytic cleavage of C–H bond present in $CH_4$, propane, n-butane, and similar types of alkanes (Musat 2015). Mixed culture of petrochemical oily sludge-degrading bacteria such as *Bacillus cereus, Bacillus cibi, Bacillus megaterium, Pseudomonas aeruginosa Polaromonas vacuolata*, and *Stenotrophomonas acidaminiphila*; *fungi such as Eupenicillium hirayamae, Aspergillus terreus, Alternaria alternata, Cladosporium sphaerospermum, Rhizopus stolonifer, Fusarium solani*, and *Paecilomyces variotii*; and yeast such as *Candida maltosa, Candida tropicalis, and Candida apicola* favors considerable reduction of aliphatic (90%) and aromatic (51%) hydrocarbons (Jeon et al. 2003; Zheng et al. 2005; Eibes et al. 2006; Wu et al. 2010; Cerqueira et al. 2011; Wiese et al. 2011; Ali et al. 2012; Ameen et al. 2014, 2015; Jagtap et al. 2014, Hasan 2014; Janani

**Table 1** Oil spills and tar ball-degrading microorganisms

| Isolate | Gene bank Accession no. | Oxic/anoxic condition | |
|---|---|---|---|
| *Acinetobacter venetianus* | DQ912805 | Oxic | Kostka et al. (2011) |
| *Acinetobacter* sp. | FJ876296 | Oxic | |
| *Alcanivorax dieselolei* | AB453732 | Anoxic | |
| *Alcanivorax* sp. | AB435642 | Oxic MPN | |
| *Bacillus* sp. | HQ588864 | Anoxic | |
| *Halomonas shengliensis* | EF121853 | Anoxic | |
| *Labrenzia* sp. | EU440961 | Oxic MPN | |
| *Marinobacter hydrocarbonoclasticus* | DQ768638 | Anoxic | |
| *Marinobacter hydrocarbonoclasticus* | DQ768638 | Anoxic | |
| *Marinobacter hydrocarbonoclasticus* | DQ768638 | Oxic MPN | |
| *Marinobacter vinifirmus* | FJ161339 | Anoxic | |
| *Marinobacter vinifirmus* | FJ161339 | Anoxic | |
| *Microbacterium schleiferi* | EU440992 | Oxic MPN | |
| *Microbulbifer* sp. | GQ334398 | Oxic MPN | |
| *Pseudidiomarina maritima* | EU600203 | Anoxic | |
| *Pseudoalteromonas* sp. | AY394863 | Oxic | |
| *Pseudomonas pachastrellae* | EU603457 | Anoxic | |
| *Pseudomonas stutzeri* | GU396288 | Anoxic | |
| *Shewanella algae* | GQ372877 | Anoxic | |
| *Vibrio alginolyticus* | GQ455008 | Anoxic | |
| *Vibrio alginolyticus* | GQ455008 | Anoxic | |
| *Vibrio hepatarius* | EU834019 | Anoxic | |
| *Vibrio hepatarius* | HM584097 | Anoxic | |
| *Vibrio* sp. | HM640395 | Oxic | |
| *Alternaria alternata* | KP033203 | Oxic | Ameen et al. (2015) |
| *Aspergillus terreus* | KP033202 | Oxic | |
| *Cladosporium sphaerospermum* | KM979605 | Oxic | |
| *Eupenicillium hirayamae* | KM979606 | Oxic | |
| *Paecilomyces variotii* | KM979604 | Oxic | |

Prathiba et al. 2014; Wang et al. 2015). The aliphatic hydrocarbons such as n-hexadecane are one of the main components of the crude oil. Its bioremediation using *Marinobacter* sp., *Mycobacterium*, *Pseudomonas*, and *Pseudomonas*-like species and filamentous fungi could be made more cost-effective by incorporating cyclodextrins (cyclic oligosaccharides). The solubility of these hydrocarbons can be increased through incorporation of suitable hydrophobic molecules into their hydrophobic sites which speed up the process of biodegradation (Li et al. 2008; Sivaraman et al. 2010; Bonin et al. 2015).

## 4 Phyto-micro-Degradation/Rhizoremediation

The salt marsh plant rhizospheres are one of the habitats for hydrocarbon-degrading microbial community. The process of removal of hydrocarbon using salt marsh plants and their rhizosphere microbial community is known as phyto-micro-degradation. The phyto-micro-degradation has the active contribution in hydrocarbon removal. This process of hydrocarbon removal is remarkable, observed in case of all plants species in salt marshes such as *Carex phacota, Juncus maritimus, Phragmites australis, Spartina patens*, and *Triglochin striata* (Kuiper et al. 2004; Ribeiro et al. 2011, 2012; Wang et al. 2011; Kurzawova et al. 2012; Khan et al. 2013). The rhizospheric microorganisms associated with legumes such as *Rhizobium* sp. and *Bradyrhizobium* sp. have mechanism to transform xenobiotic compound/polychlorinated hydrocarbons (PCB). The incorporation of alkaloids such as flavonoids, naringenin, and apigenin induces enzyme production and enhances PCB degradation by bacteria. This manipulation of rhizosphere community is known as rhizoengineering. It critically enhances the rhizoremediation of xenobiotic compound which is beyond the inherent power of plant–microbe interaction (Xu et al. 2011; Toussaint et al. 2012; Tang et al. 2013; Jha et al. 2015; Jha and Jha 2015).

## 5 Biodegradation of Crude Oil and Tar Using Immobilized Cells

The use of immobilized microbial culture (pure culture/microbial consortium) is a simple and economic way for the bioremediation of crude oil, tar, and petroleum hydrocarbons. Immobilized microbial cells have increased contact as compared to a single cell with the hydrocarbon droplets. *Pseudomonas* and *Pseudomonas*-like species produce surfactant (rhamnolipid) which enhances the dispersion of water-insoluble crude oil (*n*-alkanes) at a wide range of salinity. Like the immobilized enzymes, immobilized microbial culture has no decline in the rate. Thus, immobilized microbial pure culture and microbial consortium are effective tools in bioremediation of crude oil, tar, petrochemical wastewater, and soil and xenobiotics/recalcitrant compounds (Cunningham et al. 2004; Rahman et al. 2006; Das and Chandran 2011).

## 6 Enzyme Participating Biodegradation Pathways

The rapid and complete degradation of most of the oil pollutants by microorganism under aerobic conditions. It is an oxidative process catalyzed by enzymes such as oxygenases and peroxidases. The peripheral pathway such as TCA cycle converts pollutants into easily degradable forms. The energy required for biosyntheses is

**Table 2** Enzymes involved in biodegradation of hydrocarbons

| Enzymes | Substrates | Microorganisms | |
|---|---|---|---|
| Soluble Methane Monooxygenases | $C_1$–$C_8$ alkanes alkenes and cycloalkanes | *Methylococcus, Methylosinus, Methylocystis, Methylomonas, Methylocella* | Das and Chandran (2011), Rajasekar et al. (2012) |
| Particulate Methane Monooxygenases | $C_1$–$C_5$ (halogenated) alkanes and cycloalkanes | *Methylobacter, Methylococcus, Methylocystis* | |
| AlkB-related Alkane Hydroxylases | $C_5$–$C_{16}$ alkanes, fatty acids, alkyl benzenes, cycloalkanes, and so forth | *Pseudomonas, Burkholderia, Rhodococcus, Mycobacterium* | |
| Eukaryotic P450 | $C_{10}$–$C_{16}$ alkanes, fatty acids | *Candida maltose, Candida tropicalis, Yarrowia lipolytica* | |
| Bacterial P450 oxygenase system | $C_5$–$C_{16}$ alkanes, cycloalkanes | *Acinetobacter, Caulobacter, Mycobacterium* | |
| Dioxygenases | $C_{10}$–$C_{30}$ alkanes | *Acinetobacter sp.* | |
| Catechol 2,3-dioxygenase | C6–$C_{10}$ alkanes | *Pseudomonas putida mt-2* | Zhang and Xing (2008) |
| Cysteine dioxygenase | Cysteine | Genetically engineered *Escherichia coli* | Stipanuk et al. (2008) |
| Naphthalene dioxygenase | PAH | *Ralstonia, Sphingomonas, Burkholderia, Pseudomonas, Comamonas, Flavobacterium,* and *Bacillus* | Widada et al. (2002) |
| Manganese peroxidase | Anthracene, dibenzothiophene, and pyrene | – | Eibes et al. (2006) |

obtained through gluconeogenesis pathway. Oxygenases mediate degradation of crude oil, oil spills, and tar through various steps such as attachment of microbial cells to the substrates and production of enzymes and surfactants (Das and Chandran 2011). The crude oil, tar, chlorinated hydrocarbons, fuel additives, and their remainders in the environment-degraded by enzymes. The bacteria, archaea, and yeasts have diverse alkane-oxygenase systems involved actively in the synthesis and production of alkane-degrading enzymes such as cytochrome P-450enzymes, integral-membrane di-iron alkane hydroxylases (alkB, catechol 1,2-diooxygenase, soluble di-iron $CH_4$ monooxygenases), and membrane-bound copper $CH_4$ monooxygenases (Table 2) under oxic conditions (Saxena and Thakur 2005; Van Beilen and Funhoff 2005, 2007).

## 7 Future Perspectives

This chapter keenly explains the factors responsible for oil spills and tar-ball pollution in marine/coastal environment such as beaches and salt marshes. It also highlights the adverse effects of oil spills and tar pollution and their impacts on the pathetic condition of ports and harbors, economy, and dependence of the local population on the Oceans. It is necessary to avoid excessive port and harbor activities, unuseful and poor activities, and bad planning of state/central government in maintaining and preventing damage to the coastline. The policy measures and management are required for the complete restoration and protection of the marine as well as coastal environment and the health of Oceans and regime places especially Arabian Sea. These include the removal of tar balls and other hydrocarbons either manually or treated using modern biological techniques. Microorganisms (*Micrococcus, Rhodococcus*, and *Pseudomonas*-like species) capable of hydrocarbon biodegradation should be used to treat these polluted areas. Strict fulfillment of existing regulation and licensing by the government and use of good practices in marine/coastal environment should be adopted. There will be strict and approved protocols for oil filling and washing of ship at the designated port and harbor. The creation of public awareness and giving information to the public and private participants together through communication media (TV, press, advertisements, etc.) to make sustainable marine/coastal environment, to protect, save, and preserve the precious biodiversity of the coastal region of Goa. This will contribute to improve livelihood of people depending on Sea and enhance national economy.

**Acknowledgement** The authors wish to thank the Hon'ble Prof. Pandit B. Vidyasagar, Vice Chancellor, and Swami Ramanand Teerth Marathwada University, Nanded, for providing necessary facilities to carry out this research. BNR is also thankful to Dr. Satheesh C. Shennoi, Director, ESSO-Indian National Information Centre for Ocean Information (ESSO-INCOIS), Hyderabad, India, and Dr. N. Ramaiah, Chief Scientist, National Institute of Oceanography (NIO), Goa, India, for giving a chance to become a part of the International Training Course on "Biological Oceanographic Processes" organized by the International Training Centre for Operational Oceanography (ITCOocean), ESSO-Indian National Centre for Ocean Information Services (ESSO-INCOIS), Hyderabad, and CSIR-National Institute of Oceanography (NIO), Goa (India). The words are insufficient to thank the Dr. V. C. Kalia (Chief Scientist, CSIR—Institute of Genomics and Integrative Biology; Professor, Academy of Scientific and Innovative Research, Delhi University Campus, Delhi, India) for his continuous support and fatherlike guidance. BNR is grateful to Dr. Baban Ingole, Chief Scientist, National Institute of Oceanography (NIO), Goa, India, and Dr. G. Dharni, Scientist E, National Institute of Ocean Technology, Chennai, TN, India, for their valuable guidance. BNR is thankful to University Grants Commission for the financial support in the form of the postdoctoral fellowship (No. F./PDFSS-2013-14-ST-MAH-4350 - Website).

## References

Ali M, Khalil NM, Abd El-Ghany M (2012) Biodegradation of some polycyclic aromatic hydrocarbons by *Aspergillus terreus*. Afr J Microbiol Res 6:3783–3790. doi:10.5897/ajmr12.411

Ameen F, Moslem MA, Hadi S, Al-Sabri A (2014) Biodegradation of cellulosic materials by marine fungi isolated from South Corniche of Jeddah, Saudi Arabia. J Pure Appl Microbiol 8:3617–3626

Ameen F, Moslem M, Hadi S, Al-Sabri AE (2015) Biodegradation of diesel fuel hydrocarbons by mangrove fungi from Red Sea Coast of Saudi Arabia. Saudi J Biol Sci 23:211–218. doi:10.1016/j.sjbs.2015.04.005

Bonin P, Vieira C, Grimaud R, Militon C, Cuny P, Lima O, Guasco S, Brussaard CPD, Michotey V (2015) Substrates specialization in lipid compounds and hydrocarbons of *Marinobacter* genus. Environ Sci Pollut Res Int 22:15347–15359. doi:10.1007/s11356-014-4009-y

Cerqueira VS, Hollenbach EB, Maboni F, Vainstein MH, Camargo FAO, Peralba MDCR, Bento FM (2011) Biodegradation potential of oily sludge by pure and mixed bacterial cultures. Biores Technol 102:11003–11010. doi:10.1016/j.biortech.2011.09.074

Chauhan A, Oakeshott JG, Jain RK (2008) Bacterial metabolism of polycyclic aromatic hydrocarbons: strategies for bioremediation. Indian J Microbiol 48:95–113. doi:10.1007/s12088-008-0010-9

Cunningham CJ, Ivshina IB, Lozinsky VI, Kuyukina MS, Philp JC (2004) Bioremediation of diesel contaminated soil by microorganisms immobilised in polyvinyl alcohol. Int Biodeterior Biodegrad 54:167–174. doi:10.1016/j.ibiod.2004.03.005

Das N, Chandran P (2011) Microbial degradation of petroleum hydrocarbon contaminants: an overview. Biotechnol Res Int 2011:941810. doi:10.4061/2011/941810

Duraisamy A, Latha S (2011) Impact of pollution on marine environment—a case study of coastal Chennai. Indian J Sci Technol 4:259–262. doi:10.17485/ijst/2011/v4i3/29978

Eibes G, Cajthaml T, Moreira M, Feijoo G, Lema J (2006) Enzymatic degradation of anthracene, dibenzothiophene and pyrene by manganese peroxidase in media containing acetone. Chemosphere 64:408–414. doi:10.1016/j.chemosphere.2005.11.075

Freije AM (2015) Heavy metal, trace element and petroleum hydrocarbon pollution in the Arabian Gulf: review. J Assoc Arab University Basic Appl Sci 17:90–100. doi:10.1016/j.jaubas.2014.02.001

Geetha SJ, Joshi SJ, Kathrotiya S (2013) Isolation and characterization of hydrocarbon degrading bacterial isolate from oil contaminated sites. APCBEE Procedia 5:237–241. doi:10.1016/j.apcbee.2013.05.041

Hao R, Lu A (2009) Biodegradation of heavy oils by halophilic bacterium. Proc Natl Acad Sci USA 19:997–1001. doi:10.1016/j.pnsc.2008.11.010

Hasan I (2014) Biodegradation of kerosene by *Aspergillus niger* and *Rhizopus stolonifer*. J Appl Environ Microbiol 2:31–36. doi:10.12691/jaem-2-1-7

Jagtap SS, Woo SM, Kim TS, Dhiman SS, Kim D, Lee J (2014) Phytoremediation of diesel-contaminated soil and saccharification of the resulting biomass. Fuel 116:292–298. doi:10.1016/j.fuel.2013.08.017

Janani Prathiba G, Keerthi K, Aparna D, Sourav B, Priyadarsini RI (2014) Molecular identification of the isolated diesel degrading bacteria and optimization studies. J Biochem Technol 5:727–730 http://www.jbiochemtech.com/index.php/jbt/article/view/JBt531

Jeon CO, Park W, Padmanabhan P, DeRito C, Snape JR, Madsen EL (2003) Discovery of a bacterium, with distinctive dioxygenase, that is responsible for in situ biodegradation in contaminated sediment. PNAS 100:13591–13596. doi:10.1073/pnas.1735529100

Jha P, Jha PN (2015) Plant-microbe partnerships for enhanced biodegradation of polychlorinated biphenyls. In: Arora NK (ed) Plant microbes symbiosis: applied facets. Springer, India, p 95. doi:10.1007/978-81-322-2068-8_4

Jha P, Panwar J, Jha PN (2015) Secondary plant metabolites and root exudates: guiding tools for polychlorinated biphenyl biodegradation. Int J Environ Sci Technol 12:789–802. doi:10.1007/s13762-014-0515-1

Jonker MTO, Brils JM, Sinke AJC, Murk AJ, Koelmans AA (2006) Weathering and toxicity of marine sediments contaminated with oils and polycyclic aromatic hydrocarbons. Environ Toxicol Chem 25:1345–1353. doi:10.1897/05-296R.1

Khan S, Afzal M, Iqbal S, Khan QM (2013) Plant-bacteria partnerships for the remediation of hydrocarbon contaminated soils. Chemosphere 90:1317–1332. doi:10.1016/j.chemosphere.2012.09.045
Kirk JL, Klironomos JN, Lee H, Trevors JT (2005) The effects of perennial ryegrass and alfalfa on microbial abundance and diversity in petroleum contaminated soil. Environ Pollut 133:455–465. doi:10.1016/j.envpol.2004.06.002
Kiruri LW, Dellinger B, Lomnicki S (2013) Tar balls from deepwater horizon oil spill: environmentally persistent free radicals (epfr) formation during crude weathering. Environ Sci Technol 47:4220–4226. doi:10.1021/es305157w
Kostka JE, Prakash O, Overholt WA, Green SF, Freyer G, Canion A, Delgardio J, Norton N, Hazen TC, Huettel M (2011) Hydrocarbon-degrading bacteria and the bacterial community response in gulf of Mexico beach sands impacted by the deepwater horizon oil spill. Appl Environ Microbiol 77:7962–7974. doi:10.1128/AEM.05402-11
Kuiper I, Lagendijk EL, Bloemberg GV, Lugtenberg BJ (2004) Rhizoremediation: a beneficial plant-microbe interaction. Mol Plant Microbe Interact 17:6–15. doi:10.1094/MPMI.2004.17.1.6
Kurzawova V, Stursa P, Uhlik O, Norkova K, Strohalm M, Lipov J, Kochankova L, Mackova M (2012) Plant-microorganism interactions in bioremediation of polychlorinated biphenyl-contaminated soil. New Biotechnol 30:15–22. doi:10.1016/j.nbt.2012.06.004
Lai C, Huang YC, Wei Y, Chang J (2009) Biosurfactant-enhanced removal of total petroleum hydrocarbons from contaminated soil. J Hazard Mat. 167:609–614. doi:10.1016/j.jhazmat.2009.01.017
Li Y, Liu H, Tian Z, Zhu L, Wu Y, Tang H (2008) Diesel pollution biodegradation: synergetic effect of *Mycobacterium* and filamentous fungi. Biomed Environ Sci 21:181–187. doi:10.1016/S0895-3988(08)60026-4
Lin Q, Mendelssohn IA (2009) Potential of restoration and phytoremediation with *Juncus roemerianus* for diesel-contaminated coastal wetlands. Ecol Eng 35:85–91. doi:10.1016/j.ecoleng.2008.09.010
Main CE, Ruhl HA, Jones DOB, Yool A, Thornton B, Mayo DJ (2015) Hydrocarbon contamination affects deep-sea benthic oxygen uptake and microbial community composition. Deep-Sea Res Pt I 100:79–87. doi:10.1016/j.dsr.2014.12.008
Musat F (2015) The anaerobic degradation of gaseous, nonmethane alkanes-from in situ processes to microorganisms. Comput Struct Biotechnol J 13:222–228. doi:10.1016/j.csbj.2015.03.002
National Institute of Oceanography, Goa (2015) www.nio.org, Assessed on 14/04/2015 at 15:21:40 (GMT/UTC + 5:30).
Phillips LA, Greer CW, Farrell RE, Germida JJ (2009) Field-scale assessment of weathered hydrocarbon degradation by mixed and single plant treatments. Appl Soil Ecol 42:9–17. doi:10.1016/j.apsoil.2009.01.002
Rahman RNZA, Ghazali FM, Salleh AB, Basri M (2006) Biodegradation of hydrocarbon contamination by immobilized bacterial cells. J Microbiol 44:354–359 http://www.ncbi.nlm.nih.gov/pubmed/16820766
Rajasekar A, Maruthamuthu S, Ting YP, Balasubramanian R, Rahman PKSM (2012) Chapter 13 Bacterial degradation of petroleum hydrocarbons. In: Singh SN (ed) Microbial Degradation of Xenobiotics, Environmental Science and Engineering. Springer, Berlin Heidelberg, pp 339–369. doi:10.1007/978-3-642-23789-8_13
Rekadwad BN, Khobragade CN (2015) A case study on effects of oil spills and tar-ball pollution on beaches of Goa (India). Mar Poll Bull 100:567–570. doi:10.1016/j.marpolbul.2015.08.019
Ribeiro H, Mucha AP, Almeida CMR, Bordalo AA (2011) Hydrocarbon degradation potential of salt marsh plant-microorganisms associations. Biodegradation 22:729–739. doi:10.1007/s10532-010-9446-9
Ribeiro H, Almeida CMR, Mucha AP, Teixeira C, Bordalo AA (2012) Influence of natural rhizosediments characteristics on hydrocarbons degradation potential of microorganisms

associated to Juncus maritimus roots. Int Biodeterior Biodegrad 84:86–96. doi:10.1016/j.ibiod.2012.05.039
Ribeiro H, Almeida CMR, Mucha AP, Bordalo AA (2013) Influence of different salt marsh plants on hydrocarbon degrading microorganisms abundance throughout a phenological cycle. Int J Phytoremed 15:715–728. doi:10.1080/15226514.2012.735285
Ribeiro H, Mucha AP, Almeida CMR, Bordalo AA (2014) Potential of phytoremediation for the removal of petroleum hydrocarbons in contaminated salt marsh sediments. J Environ Manage 137:10–15. doi:10.1016/j.jenvman.2014.01.047
Riccardi N (2010) Selectivity of plankton nets over mesozooplankton taxa: implications for abundance, biomass and diversity estimation. J Limnol 69:287–296. doi:10.3274/JL10-69-2-10
Saxena P, Thakur IS (2005) Purification and characterization of catechol 1,2-dioxygenase of Pseudomonas fluorescens for degradation of 4-chlorobenzoic acid. Indian J Biotechnol 4:134–138 http://nopr.niscair.res.in/handle/123456789/5626
Shi W, Bischoff M, Turco R, Konopka A (2005) Microbial catabolic diversity in soils contaminated with hydrocarbons and heavy metals. Environ Sci Technol 39:1974–1979. doi:10.1021/es049034n
Sivaraman C, Ganguly A, Mutnuri S (2010) Biodegradation of hydrocarbons in the presence of cyclodextrins. World J Microbiol Biotechnol 26:227–232. doi:10.1007/s11274-009-0164-6
Stipanuk MH, Dominy JE Jr, Ueki I, Hirschberger LL (2008) Measurement of cysteine dioxygenase activity and protein abundance. Curr Protoc Toxicol 38:6.15.1–6.15.25. doi:10.1002/0471140856.tx0615s38
Sukhdhane KS, Priya ER, Raut SM, Jayakumar T (2013) Status of oil pollution in Indian coastal waters. Fishing Chimes 33:53–54 http://eprints.cmfri.org.in/9582/1/fc.pdf
Sun T, Cang L, Wang Q, Zhou D, Cheng J, Xu H (2010) Roles of abiotic losses, microbes, plant roots, and root exudates on phytoremediation of PAHs in a barren soil. J Hazard Mater 176:919–925. doi:10.1016/j.jhazmat.2009.11.124
Tahhan RA, Ammari TG, Goussous SJ, Al-Shdaifat HI (2010) Enhancing the biodegradation of total petroleum hydrocarbons in oily sludge by a modified bioaugmentation strategy. Int Biodeterior Biodegrad 65:130–134. doi:10.1016/j.ibiod.2010.09.007
Tang X, Qiao J, Chen C, Chen L, Yu C, Shen C, Chen Y (2013) Bacterial communities of polychlorinated biphenyls polluted soil around an E-waste recycling workshop. Soil Sed Contam 22:562–573. doi:10.1080/15320383.2013.75026
Tao Z, Bullard S, Arias C (2011) High numbers of *Vibrio vulnificus* in tar balls collected from oiled areas of the North-Central Gulf of Mexico following the 2010 BP deepwater horizon oil spill. Ecohealth 8:507–511. doi:10.1007/s10393-011-0720-z
Toussaint JP, Pham T, Barriault D, Sylvestre M (2012) Plant exudates promote PCB degradation by a *Rhodococcal rhizobacteria*. Appl Microbiol Biotechnol 95:1589–1603. doi:10.1007/s00253-011-3824-z
Van Beilen JB, Funhoff EG (2005) Expanding the alkane oxygenase toolbox: new enzymes and applications. Curr Opin Biotechnol 16:308–314. doi:10.1016/j.copbio.2005.04.005
Van Beilen JB, Funhoff EG (2007) Alkane hydroxylases involved in microbial alkane degradation. Appl Microbiol Biotechnol 74:13–21. doi:10.1007/s00253-006-0748-0
Wang Z, Liu X, Chen L, Hu X, Liu F (2011) Degradation of diesel with microorganisms in rhizosphere of *Carex phacota Spr*. Procedia Environ Sci 8:61–67. doi:10.1016/j.proenv.2011.10.012
Wang X, Wang X, Liu M, Bu Y, Zhang J, Chen J, Zhao J (2015) Adsorption-synergic biodegradation of diesel oil in synthetic seawater by acclimated strains immobilized on multifunctional materials. Mar Pollut Bull 92:195–200. doi:10.1016/j.marpolbul.2014.12.033
Warnock AM, Hagen SC, Passeri DL (2015) Marine tar residues: a review. Water Air Soil Pollut 226:68. doi:10.1007/s11270-015-2298-5

Widada J, Nojiri H, Kasuga K, Yoshida T, Habe H, Omori T (2002) Molecular detection and diversity of polycyclic aromatic hydrocarbon-degrading bacteria isolated from geographically diverse sites. Appl Microbiol Biotechnol 58:202–209. doi:10.1007/s00253-001-0880-9

Wiese J, Ohlendorf B, Blumel M, Schmaljohann R, Imhoff JF (2011) Phylogenetic identification of fungi isolated from the marine sponge *Tethya aurantium* and identification of their secondary metabolites. Mar Drugs 9:561–585. doi:10.3390/md9040561

Wu Y, Luo Z, Vrijmoed L (2010) Biodegradation of anthracene and benz[a]anthracene by two *Fusarium solani* strains isolated from mangrove sediments. Bioresour Technol 101:9666–9672. doi:10.1016/j.biortech.2010.07.049

Xu Y, Gao D, Liu J, Wang Z (2011) Plant-microbe interactions to improve crude oil degradation. Energy Procedia 5:844–848. doi:10.1016/j.egypro.2011.03.149

Zhang FSU, Xing X (2008) A simple assay of catechol 2,3-dioxygenase in whole cell catalysis of *Pseudomonas putida mt-2*. J Chem Indus Eng (China) 59:450–455 http://www.hgxb.com.cn/EN/Y2008/V59/I2/450

Zheng Y, Yanful KE, Bassi AS (2005) A review of plastic waste biodegradation. Crit Rev Biotechnol 25:243–250. doi:10.1080/07388550500346359

# Bacterial Decolourization, Degradation and Detoxification of Azo Dyes: An Eco-friendly Approach

**Shweta Agrawal, Devayani Tipre, Bhavesh Patel, and Shailesh Dave**

**Abstract** Recent decades have noticed that the wastewater loaded with dyes is one of the most challenging to treat. There has been comprehensive research on azo dye biodegradation with the help of microorganisms. It is evolving as a promising substitute to the conventional treatment with physico-chemical methods. In this chapter, the use of various bacterial cultures, either in pure or mixed form, for their capability to decolourize dye wastewaters was described. Apart from microorganisms involved, this chapter focuses on various mechanisms involved in the dye biodegradation along with the impact of several factors influencing biodegradation of dye wastewater and the role of oxidoreductases involved in bacterial dye biodegradation. The various analytical techniques used for deducing the metabolic pathway for dye remediation and ensuring its detoxification through toxicity tests are described. The chapter also highlights the huge gap between the scientific progresses in this field and lack of commercialization of research. The information gathered here will be useful for the students, teacher, industrial persons and researchers to understand the state of the art of the subject and use it for removal of dye pollution by possible green technology for the sustainable environment.

S. Agrawal
Department of Biotechnology, Sanghvi Institute of Management and Science, Indore 453331, India
e-mail: shweta.agrawal24@gmail.com

D. Tipre • S. Dave (✉)
Department of Microbiology and Biotechnology, School of Sciences, Gujarat University, Ahmedabad, Gujarat 380009, India
e-mail: devayanitipre@hotmail.com; shaileshrdave@yahoo.co.in

B. Patel
V.P. and R.P.T.P. Science College, Vallabh Vidyanagar, Gujarat 388120, India
e-mail: bhavesh1968@rediffmail.com

 
V.C. Kalia, P. Kumar (eds.), *Microbial Applications Vol.1*,
DOI 10.1007/978-3-319-52666-9_4

# 1 Introduction

The term dye is derived from old English word 'daeg' or 'daeh' meaning 'colour'. Colour can be imparted to a wide array of materials utilizing substances well known as dyes and pigments (Christie 2001). Dyes are coloured organic compounds, which strongly absorb visible light and can form physical as well as chemical bond with the textile fibre, thereby attaching firmly to it. Commercially important dyes should characteristically be colour fast and withstand rubbing and application of water (Mansoor 2008). According to the 'colour theory' proposed by O. Witt, a dye is composed of two vital components: one is the chromophores and the other one is the auxochromes. Chromophores (Gr, *Kuroma* = colour + *Phors* = carrier) have a delocalized system of electrons with conjugated double bonds and produce colour, and a molecule consisting such group is termed as a chromogen. Some groups do not produce colour themselves but are able to amplify the colour in association with chromophores, when present in a molecule. Such molecules are termed auxochromes (Gr, auxanein = to increase). –OH, –COOH, −SOH (acidic), (basic) NH, NHR and NR are some of the most effective auxochromes (Mansoor 2008).

## 1.1 Azo Dyes

Azo colourants contain an azo (−N=N−) bond as their common structural feature, which is attached on either side of the two $sp^2$ carbon atoms. The azo group is customarily diazotized to two aromatic ring systems. Majority of the commercially important azo colourants comprise of a single azo group and, hence, are denoted as monoazo dyes or pigments, but there are several dyes which contain two (diazo), three (triazo) or more such groups (Christie 2001). Dyes containing azo group are exceptionally stable and resistant to degradation by microbes, owing to their complex structure and electron-withdrawing tendency (Dave et al. 2015).

## 1.2 Present Scenario of Dye Production and Consumption in India

The dyestuff production capacity in India (Fig. 1) stood at around 200,000 tons of dye. Approximately, 700 different varieties of dyes and dye intermediates are manufactured in India (http://www.indiaenvironmentportal.org.in/content/573/united-colours-of-industry/). Currently, there are roughly 1000 small-scale units and 50 large organized units. Maharashtra and Gujarat make up 90% of dyestuff production in India because of the accessibility to the raw materials and dominance of textile industry in these states. Indian dyestuff industry meets more than 95% of

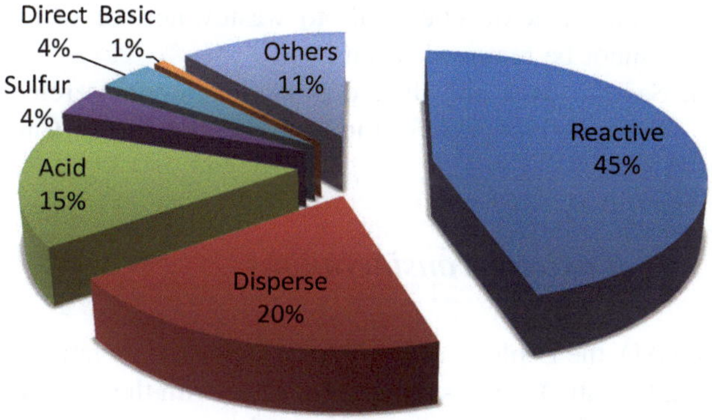


**Fig. 1** Production capacity (tonnes per annum) of different major dyes by Indian dye industry

the domestic requirement, out of which approximately 60% is consumed by the textile industry and the remaining is collectively used by leather, paper and other consumer industries. The textile industries consume approximately 80% of the total dyes produced (Mangal 2010).

## *1.3 Discharge and Toxicity of Dyestuffs*

Textile industry wastes consists of both inorganic and organic chemicals including finishing agents, sequestering and levelling agents, carriers and surfactants, etc. pH of the textile mill effluents is generally alkaline. Besides, it has high COD (3–1.4 g/L, approximately), $BOD_5$ (200–2000 mg/L, approximately), total solids (1000–10,000 mg/L, approximately), suspended solids (100–1000 mg/L, approximately), total phosphorus content (5–70 mg/L, approximately) and conductivity (1000–15,000 mS/cm) (Yonar 2011; Dave et al. 2015).

The three major problems associated with textile industry effluents are:

1. Textile industries use more than 8000 chemicals during several processes of textile manufacturing including dyeing and printing (Kant 2012). Approximately, 10–25% of the dyes are lost during textile dyeing and 2–20% is directly discharged as wastewater, without any treatment in diverse environmental bodies (Carmen and Daniela 2012). Several dyes are visible in water at an extremely low (1 mg/L) concentration. The average dye concentrations in the textile industry wastewaters have been found to be 300 mg/L. The release of dyes in the effluent may in consequence cause an ecotoxic hazard, potential threat of bioaccumulation as well as aesthetic problems (Agrawal et al. 2014a)
2. A textile industry of average size, with a daily manufacture of nearly 8000 kg of fabric, consumes approximately 1.6 million litres of water daily, 16% of which is used in dyeing and 8% in printing; 17–20% of the water pollution caused by the industries is contributed by the textile dyeing and finishing treatment given to fabric, as is estimated by the World Bank. Textile dyeing solely contributes to

some 72 toxic chemicals to wastewater, of which as many as 30 chemicals cannot be removed (Kant 2012).
3. Several dyes and their degradation products released to the environment are toxic, carcinogenic or mutagenic to life forms (Carmen and Daniela 2012).

## 1.4 Toxicity Considerations

ETAD, the Ecological and Toxicological Association of the Dyestuffs Manufacturing Industry), was established in 1974 with the objectives of reducing environmental hazards, safeguarding users and consumers and cooperation with government and public concerns with respect to the toxicological effect of their products (Chequer et al. 2011). Toxic effects of dyes may be categorized broadly into two ways. Acute toxicity denotes the consequences of short-term exposure to a substance, for example, in a single oral administration, while chronic toxicity denotes the consequences of regular exposure over a longer period of time (Christie 2001).

Dyestuffs usually exhibit low acute toxicity. $LD_{50}$ values higher than 2000 mg/kg had been reported by ETAD for 4000 different dyes tested. Additionally, more than 90% of the dyes tested showed $LD_{50}$ values beyond 2000 mg/kg, proving that exposure to azo dyes need not necessarily cause acute toxicity; however, in the context of systemic bioavailability, concern arises upon inhalation and skin contact of the azo dyes, owing to the likely formation of carcinogenic aromatic amines (Chequer et al. 2011). Various toxicological studies have reported that a lethal concentration value ($LC_{50}$) of more than 1 mg/L, for fishes, for the tested 98% of dyes, and 59% of the tested dyes have an $LC_{50}$ value above 100 mg/L (Carmen and Daniela 2012).

Chronic and occupational hazards of dyes, particularly of azo dyes, have been investigated several times. The workers involved in dye manufacturing have showed increased rates in bladder cancer since 1895. Azo dyes can also be absorbed upon exposure to skin occurring either as an occupational hazard or during application as cosmetic products (Chequer et al. 2011). Evidently, some reactive dyes cause occupational asthma, allergic conjunctivitis, rhinitis, contact dermatitis or other allergic reactions in textile industry employees (Hunger 2003). In the same way, if dye-laden waste enters the water supply, probably by contamination of the groundwater, the common public may therefore be exposed to the azo dyes by the oral route (Chequer et al. 2011).

Evidently, reduction of azo dyes in addition leads to generation of aromatic amines, and a number of them are reported to be mutagenic and carcinogenic. In mammals, water-soluble azo dyes are predominantly metabolized through azo bond cleavage by azoreductase enzyme of the hepatic and extrahepatic tissue or by the microflora of the intestine in the body. The aromatic amines show acute toxicity by causing bladder cancer (Hunger 2003).

## 2 Dye Removal Techniques

Currently, majority of the practices under use for the dye wastewater treatment are chemical, physical or physico-chemical processes, which are generally expensive and of limited applicability (Jadhav and Phugare 2012). Along with the limitations associated with each technique, several factors such as the type of the dye, effluent composition, dosage and cost of the chemicals required for wastewater treatment, environmental fate and costs of treatment process and handling of generated waste products influence the technical and economic viability of each single technique (Van Der Zee 2002).

### *2.1 Physico-chemical Techniques*

Physical techniques involve use of strictly physical phenomenon, without any gross chemical or biological changes to improve or treat the effluent. On the contrary, chemical processes use chemical reactions to improve water quality (http://envirosystems.net/effluent-treatment-plant/). Table 1 illustrates a detailed account of various physico-chemical techniques.

### *2.2 Biological Methods*

The process in which biological systems are used to remove pollutants is known as bioremediation (Ali 2010). To date, many microorganisms including bacteria, yeast, fungi, actinomycetes and algae either in the form of pure or mixed cultures of microorganisms or their enzymes have been reported for azo dye degradation. The remediation of synthetic dyes from industrial discharges using microorganisms is beneficial as the process is comparatively economical, low operational costs in addition to mineralization of the pollutant so that the products are non-toxic (Jadhav and Phugare 2012).

Microbial decolourization of dyes may occur in two ways: either through biosorption, which is adsorption of the pollutant on the growing/living or dead microbial biomass, or through biodegradation by microbial metabolism. Biosorption of dyes is not a suitable solution to the problem because the dyes are not degraded but on the contrary gets entrapped into the biosorbent. The discharge of the bacterial biomass laden with adsorbed dyes is also a huge problem. Biodegradation is therefore preferred and seems to be promising (Ali 2010).

**Table 1** Different physico-chemical techniques used for the dye decolourization and biodegradation

| Process | Treatment stage | Methodology | Advantages | Disadvantages | References |
|---|---|---|---|---|---|
| Fenton's reagent | Pre/Main treatment) | Oxidation mainly using $H_2O_2$-Fe(II) | Effective decolourization of soluble and insoluble dyes<br>No alteration in volume<br>Simple equipment and easy implementation<br>No energy input is necessary to activate hydrogen peroxide | Sludge generation and its disposal<br>Expensive long reaction time | Lang (2009), Yonar (2011), Carmen and Daniela (2012), Zaharia et al. (2012) |
| Precipitation Coagulation/ flocculation | Pre/Main treatment | Use of coagulants, viz. lime, organic polymer, magnesium, iron and aluminium salts | Short detention time<br>Economic Relatively good removal efficiencies<br>This method was used to partly remove COD and colour from raw wastewater | Sludge production and its disposal<br>Agglomerates separation<br>and treatment<br>Selected operating condition | Lang (2009), Carmen and Daniela (2012), Jadhav and Phugare (2012), Zaharia et al. (2012) |
| Sonication | Pre-treatment | Use of varying frequency of ultrasound waves | Complete decolourization and increase in rate of decolourization under spent dye bath conditions<br>Simple to use<br>Very effective in integrated system | Decolourization efficiency decreased with an increased dye concentration<br>Comparatively new method and in anticipation of full-fledged application | Lang (2009), Carmen and Daniela (2012), Jadhav and Phugare (2012), Zaharia et al. (2012) |
| Ion exchange | Main treatment | Retention on ion exchange resins | Removal of soluble dyes.<br>Adsorbent is not lost at regeneration<br>Reclamation of solvent after use | Specific application; not applicable to all dyes<br>Expensive | Lang (2009), Carmen and Daniela (2012), Zaharia et al. (2012) |

| | | | | | |
|---|---|---|---|---|---|
| Ozonation and ultrasound enhanced ozonation | Main treatment | Varying ozone, ultrasound and ultrasound enhanced ozone operational conditions. First-order rate constant increased to 200% using ultrasonic power inputs compared to ozonation alone | Effective azo dye removal<br>Applied in gaseous state<br>No alteration of volume | Generation of highly oxidizing reactive radicals<br>Ozone generation process coupled with its very short half-life (20 min) | Lang (2009), Carmen and Daniela (2012), Jadhav and Phugare (2012), Zaharia et al. (2012) |
| Adsorption | Pre/Post-treatment | Adsorption of dyes using various adsorbent, viz. activated carbon, cellulosic bacteria, yeast, fungal chitin biomass, soil material, pulp mill waste, etc. | Equilibrium capacity varies with dyes and adsorbents<br>Economically attractive Good removal efficiency | Requires long retention times<br>High cost of organic solvents<br>Huge quantities are required | Carmen and Daniela (2012), Jadhav and Phugare (2012), Zaharia et al. (2012) |
| Electrolysis | Pre treatment | Electric current is applied to the wastewater using electrodes. | Breakdown compounds are non-hazardous<br>No additional chemicals required | Cost intensive<br>Requires the high amount of energy | Lang (2009), Carmen and Daniela (2012), Jadhav and Phugare (2012), Zaharia et al. (2012) |
| Photocataysis | Post treatment | Oxidation using mainly $H_2O_2$-UV | No sludge production<br>Process carried out at ambient conditions<br>Inputs are nontoxic and inexpensive<br>Complete mineralization with shorter detention times | Formation of by-products<br>Effective for small amount of coloured compounds<br>Expensive<br>Mineralization with short detention | Lang (2009), Carmen and Daniela (2012), Zaharia et al. (2012) |
| Membrane filtration | Main treatment | Nanofiltration and reverse osmosis using membranes with a molecular weight cut-off below 10,000 Da | Low spatial requirements<br>Removes all dyes<br>Reuse of chemicals and water<br>Chemicals and water can be easily recovered and reused | Flux decline<br>Membrane fouling, necessitating frequent cleaning and replacement<br>Concentrated sludge formation<br>High capital costs<br>The process fails to separate dissolved solids | Lang (2009), Carmen and Daniela (2012), Jadhav and Phugare, (2012), Zaharia et al. (2012) |

### 2.3 Dye Decolourization Using Bacteria

Some examples of the biodegradation of textile dyes using pure and bacterial consortium/mixed cultures are given in Tables 2 and 3, respectively. Omnipotence, rapid growth, facultative nature and extraordinary adaptability are the qualities that make bacterial community suitable for the bioremediation (Gomare et al. 2009). Additionally, bacterial treatment system is eco-friendly and economical, produces less sludge and can lead to a higher degree of degradation and mineralization as well as it does apply to a wide spectrum of azo dyes (Saratale et al. 2011).

Reportedly, efforts for isolation of azo dye-degrading bacteria started in the late 1970s where species of *Bacillus* and *Aeromonas* were used. Several bacterial genera and species adapted for decolourization of dyes, either as pure cultures or as consortia, have been described to date (Saratale et al. 2011).

Reportedly, azo dyes cannot be degraded completely by individual bacterial strains, in general, and also the dye degradation products generated are usually aromatic amines that are often carcinogenic and need additional treatments. On the contrary, mixed/consortial cultures are especially useful in such situations, as different strains may attack the different bonds in the azo dye molecule or may use the dye intermediates generated by the symbiotic strains for additional degradation, and this synergistic metabolism of the microbial culture may lead to better degradation and mineralization. Moreover, isolation of a pure culture from dye-laden effluent samples followed by long-term adaptation processes is obligatory for efficient decolourization and biodegradation of azo dyes and is a time-consuming and tedious process. In spite of this, if pure culture is added it is not possible to sterilize huge volume of dye containing wastewater for treatment, and even if it is sterilized, it is not possible to run the treatment under aseptic conditions (Saratale et al. 2011).

## 3 Mechanism of Dye Biodegradation

### 3.1 Anaerobic Reduction of Azo Dye

Several bacteria, when encounter anaerobic conditions, degrade azo dyes by reducing its highly electrophilic azo bond with the help of non-specific enzymes. Anaerobic reduction of azo dyes leads to reductive cleavage of azo linkage, consequently leading to the generation of aromatic amines (Santos et al. 2007), which are usually colourless, and therefore, reduction of azo dyes is also denoted as dye decolourization. Anaerobic reduction of azo dyes includes different mechanisms, which may be broadly categorized into direct and indirect mechanisms (Fig. 2a, b). Direct enzymatic reduction of azo dyes and indirect reduction of azo dyes are catalysed by enzymatically generated/regenerated redox mediating compounds.

**Table 2** Decolourization of azo dyes by isolated bacterial strains

| Bacterium | Dye | Concentration (mg/L) | Conditions (pH, temperature, agitation) | Time (h) | Decolourization (%) | References |
|---|---|---|---|---|---|---|
| *Acinetobacter calcoaceticus* NCIM-2890 | Direct Brown MR | 50 | 7.0,<br>30°C,<br>Static | 48 | 91.3 | Ghodake et al. (2009) |
| *Aeromonas hydrophila* | Reactive Black 5<br>Reactive Red 141<br>Reactive Red 198<br>Reactive Blue 171<br>Direct Yellow86 Reac-tive Green 19 | 300 | 7.5,<br>30°C, agitation (125 rpm), | – | 80.9<br>66.5<br>60.2<br>36.0<br>30.9<br>23.2 | Hsueh et al. (2009) |
| *Bacillus* sp. VUS | Navy Blue 2GL | 50 | 7.0,<br>40°C,<br>Static | 18 | 94 | Dawkar et al. (2009) |
| *Citrobacter* sp. | Azo and triphenylmeth-ane<br>dyes | 5 mM | 7–9,<br>35–40°C, static | 1 | 100 | An et al. (2002) |
| *Bacillus* sp. strain AK1 | Metanil Yellow<br>Amaranth<br>Congo Red<br>Fast Red<br>Ponceau S<br>Reactive Black | 200 | 7.2<br>37°C | 27 | 99<br>97<br>80<br>62<br>46<br>31 | Anjaneya et al. (2011) |
| *Brevibacillus laterosporus* MTCC 2298 | Golden Yellow HER; | 50 | 7.0,<br>30°C,<br>Static | 48 | 87 | Gomare et al. (2009) |
| *Citrobacter* sp. CK3 | Reactive Red 180; | 200 | 7.0,<br>32°C, anaerobic | 36 | 96 | Wang et al. (2009) |
| *Escherichia coli* JM109 | Direct Blue 71 | 150 | 9.0,<br>30°C, anaerobic | 12 | 100 | Jin et al. (2009) |

(continued)

**Table 2** (continued)

| Bacterium | Dye | Concentration (mg/L) | Conditions (pH, temperature, agitation) | Time (h) | Decolourization (%) | References |
|---|---|---|---|---|---|---|
| *Lysinibacillus* sp. strain AK2 | Metanil Yellow<br>Amaranth<br>Congo Red<br>Fast Red<br>Ponceau S<br>Reactive Black | 200 | 7.2<br>37°C | 12 | 100<br>99<br>98<br>71<br>58<br>43 | Anjaneya et al. (2011) |
| *Pseudomonas aeruginosa* NGKCTS | Reactive Red BS | 300 | 7.0<br>31°C<br>static | 5.5 | 91 | Sheth and Dave (2009) |
| *Providencia* sp . SRS82 | Acid Black 210 | 100 | 7.0<br>32°C<br>static | 90 min | 100 | Agrawal et al. (2014b) |
| *Alcaligenes faecalis* PMS-1 | Reactive Orange 13 | 400 | 7.0<br>32°C<br>static | 16 | 100 | Shah et al. (2012) |
| *Micrococcus luteus* | Direct Orange 16 | 100 | 8.0<br>37°C<br>static | 6 | 96 | Singh et al. (2015) |

**Table 3** Decolourization of dyes by bacterial consortium

| Name of strain constituting the consortium | Name of dye | mg/L | Condition (pH, Temp., agitation) | Time (h) | Decolorization (%) | References |
|---|---|---|---|---|---|---|
| Bacterial consortium (*Proteus vulgaris* NCIM-2027 *and Micrococcus glutamicus* NCIM-2168) | Green HE4BD<br>Golden Yellow HE4R<br>Violet 5R<br>Orange 3R<br>Red ME4BL<br>Red M2BN | 50 | 6.6<br>37°C<br>Static | 24 | 100<br>90.6<br>88.0<br>87.9<br>70.3<br>61.7 | Saratale et al. (2010) |
| NAR (*Citrobacter freundii* A1, *Enterococcus casseliflavus* C1, and *Enterobacter cloacae* L17) | Amaranth | 100 | Not<br>45°C<br>Static | 0.5 | 100 | Chan et al. (2012) |
| Two isolated bacterial strains (BF1, BF2) and *Pseudomonas putida* (*MTCC1194*) | Chrysoidine<br>Procion Brilliant Red<br>Procion Navy Blue,<br>Procion Green, and Direct Blue<br>Supranol Red | 50 | 9.0<br>28°C<br>120 | 240<br>216 | >90<br>>90<br>60<br>80<br>80<br>80 | Senan and Abraham (2004) |
| Consortium with eight different bacteria | Congo Red | 100 | 7.5<br>30°C<br>Static | 12 | 100 | Lade et al. (2015) |
| Consortium DAS (3 different *Pseudomonas* species) | Reactive Orange 16 | 100 | 7.0<br>30<br>Static | 48 | 100 | Jadhav et al. (2010) |
| Consortium SDS (*Providencia* sp. SDS and *Pseudomonas aeruginosa* BCH | Red HE3B<br>Red HE7B<br>Scarlet GDR<br>Brown 3 REL<br>Green HE4BD<br>Direct red 5B<br>Ramzol black 5B | 50 | –<br>–<br>Static | 1<br>1<br>3<br>7<br>2<br>1.5<br>1 | 100<br>97<br>98<br>99<br>100<br>98<br>100 | Phugare et al. (2011) |

(continued)

**Table 3** (continued)

| Name of strain constituting the consortium | Name of dye | mg/L | Condition (pH, Temp., agitation) | Time (h) | Decolorization (%) | References |
|---|---|---|---|---|---|---|
| Consortium DMC (*Pseudomonas aeruginosa* PAO1, *Stenotrophomonasmaltophila*, and *Proteus mirabilis*) | Direct Black 22<br>Reactive Navy Blue<br>HER<br>Procion Red H7B<br>Reactive Black B | 100 | 7.0<br>45<br>Static | 12 | 91<br>94<br>97<br>90 | Mohana et al. (2008) |
| Consortium PMB11 (*Bacillus odysseyi* SUK3, *Morganella morganii* SUK5, and *Proteus* sp. SUK7) | Reactive Blue 59 Navy<br>Blue HE2R Red HE3B<br>Red HE7B<br>Red HE8B<br>Red 6BI<br>Red BLI<br>Orange 3RLI<br>Green HE4B<br>Green HE4BD 97 | 50 | –<br>30<br>Static | 3<br>6<br>12<br>12<br>12<br>12<br>10<br>8<br>24<br>24 | 92<br>91<br>99<br>87<br>87<br>86<br>83<br>85<br>81<br>83 | Patil et al. (2008) |
| Consortium GR (*Proteus vulgaris* NCIM-2027 and *Micrococcus glutamicus* NCIM-2168) | Scarlet R | 50 | –<br>37<br>Static | 3 | 100 | Saratale et al. (2009) |

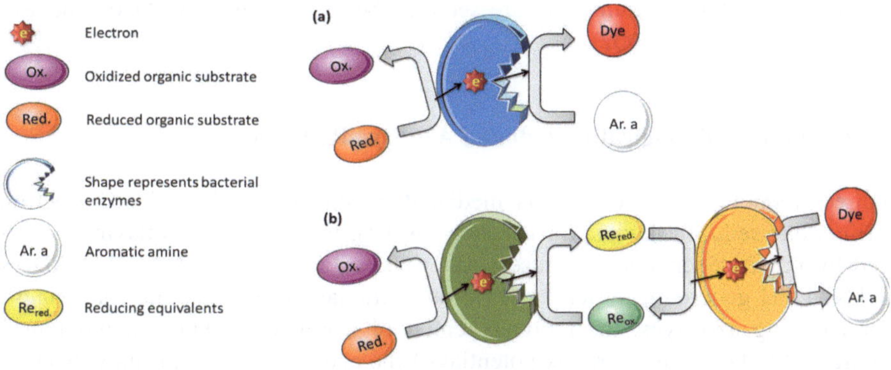


**Fig. 2** Mechanisms of (**a**) direct enzymatic reduction of azo dye and (**b**) indirect biologically mediated azo dye reduction

Besides this, azo dyes can also be solely reduced chemically by biogenic bulk reductants like sulphide (Van Der Zee 2002).

Anaerobic reduction is usually non-specific, as various groups of azo dyes are decolourized by an assorted group of bacteria. Under static condition, depletion of oxygen is effortlessly achieved, consequently permitting obligate and facultative anaerobic bacteria azo dye reduction. Thus, anaerobic decolourization of azo dyes is a casual process, where the dye molecule probably serves as an electron acceptor for the electrons facilitated through the carriers of the electron transport chain. Contrastingly, dye decolourization may be accredited to non-specific reactions occurring extracellularly amongst reduced compounds formed by the anaerobic bacteria (Van Der Zee 2002).

### 3.1.1 Direct Enzymatic Azo Dye Reduction

Oxidation of organic substrates leads to generation of reducing equivalents which are then transferred, by the enzymes, to the azo dyes. The enzymatic reduction of azo dyes is usually categorized into different categories: one carried out by specialized enzymes termed as azoreductases, present in the bacteria utilizing azo dyes as sole source of carbon and energy, and second carried out by non-specific enzymes catalysing the reduction of a wide variety of electron-withdrawing substrates along with azo dyes (Santos et al. 2007). Specialized azo dye reducing enzymes, called as 'azoreductases', are reported with certain aerobic and facultative anaerobic bacteria that could utilize simple azo molecules as sole source of carbon and energy. Non-specific enzymes catalyse reduction of azo dyes and have also been isolated from aerobically grown bacteria like *Escherichia coli, Shigella dysenteriae* and *Bacillus* sp. These enzymes were characterized as flavoproteins. Presence of specific azoreductases in bacteria grown anaerobically is seldom reported. The azo bond reducing enzymes were reported to be dispersed in the bacterial cytoplasm devoid of any association with the cell membranes or other organized cellular structures and are secreted before acting as an 'azoreductase' in vivo (Van Der Zee 2002).

### 3.1.2 Indirect Biologically Mediated Azo Dye Reduction

Indirect reduction of azo dyes is mediated by enzymatically reduced electron carriers like reduced flavins, for instance $FADH_2$, $FMNH_2$ and riboflavins, generated by flavin-dependent reductases (anaerobic azoreductases); NADH, NADPH and NADPH-generating system; and various artificial redox mediating compounds. For reducing azo dyes, the redox potential of a redox mediator should range, theoretically, between the redox potential of the azo dye and the primary electron donor. As estimated, the value of redox potential for the redox mediators catalysing azo dye reduction should lie between –430 and –100 mV (Van Der Zee 2002). Reductive decolourization of the azo dyes facilitated by the reducing equivalents occurs in two steps: a non-specific reduction of redox mediator by an enzyme is the initial step which is followed by a chemical reoxidation of the redox mediator by the azo dyes (Santos et al. 2007).

### 3.1.3 Site of the Reaction

Dye oxidation and the catalytic reduction of the electron carrier can occur intracellularly as well as extracellularly. Reducing cofactors like $FMNH_2$, $FADH_2$, NADH and NADPH, along with the dye-degrading enzymes are found in the cytoplasm. Intact cells require a membrane transport system for the reduction of azo dyes by these redox mediators. Additionally, as FAD and FMN cannot freely pass through the bacterial cell walls, this offers grave obstacle, particularly for the azo dyes comprising highly polar groups (Van Der Zee 2002; Solis et al. 2012). Cell lysis releases these cofactors in the extracellular environment, and therefore, cell extracts or starving or lysed cells exhibit higher reduction rates of azo dyes as compared to intact or resting cells. In addition to this, the absence of a direct relationship between the structure of a dye (size, molecular weight and degree of sulphonation) and the dye reduction rate suggests that the mechanisms of intracellular reduction of azo dyes are insignificant. Presumably, anaerobic biological reduction of azo dye ensues outside the cells, catalysed directly by reduced electron carriers regenerated with the help of these periplasmic enzymes (Van Der Zee 2002).

## *3.2 Aerobic Oxidation of Dyes*

Even though azo dyes are usually resistant to bacterial degradation in aerobic conditions, few selected aerobic bacterial strains with specialized azo dye-degrading enzymes were noticed to degrade azo dyes under absolutely aerobic conditions (Ola et al. 2010). Additionally, it is presumed that the initial step in the biodegradation of azo dyes is their reduction to the corresponding aromatic amines, a reaction catalysed by the enzyme azoreductase. The aromatic amines generated are then further mineralized aerobically (Sarayu and Sandhya 2010).

Sulphonated aromatic amines, except simple sulphonated amino benzene and amino naphthalene compounds, are specifically challenging to degrade because of the hydrophilic nature of their sulphonate group, which hampers membrane transport. Yet another additional transformation that the aromatic amines, substituted with ortho-hydroxy group, may undergo upon contact with oxygen is autoxidation. Several aromatic amines, like substituted anilines, amino benzidines and naphthyl amines, initially get oxidized to oligomers and in due course to dark-coloured polymers with reduced solubility and, therefore, are effortlessly removed from the aqueous phase (Van Der Zee 2002).

## *3.3 Combined Anaerobic–Aerobic Degradation of Azo Dyes*

It is generally presumed that most azo dyes can be reductively decolourized under anaerobic conditions, while the biotransformation products produced during dye biodegradation, being insusceptible to anaerobic degradation, are readily metabolized under aerobic conditions. Consequently, anaerobic degradation followed by aerobic treatment is usually recommended for treating the dye containing effluent from textile processing and dye production industries. Two distinct strategies are usually detected: integrated treatment in a single reactor and sequential treatment in distinct reactors.

### 3.3.1 Sequential Anaerobic–Aerobic Treatment of Azo Dyes

It involves sequential treatment of dye-laden effluent under anaerobic conditions and aerobic conditions. Various investigators have studied this technique over the past few years and have pointed out that the anaerobic phase of the sequential system shows noticeable colour, COD and organic matter removal. Furthermore, it increases the biodegradability of azo dyes for subsequent aerobic treatment. However, evidence on complete biodegradation of azo dyes is not clearly available (Rai et al. 2005).

### 3.3.2 Integrated Anaerobic–Aerobic Treatment

The ability of aerobic as well as anaerobic microorganisms to coexist symbiotically in a biofilm underlies the success of this technique. Providing oxygen to an oxygen-tolerant anaerobic consortium or, instead, exposing a biofilm to low concentration of oxygen together with a co-substrate are some of the ways to generate an integrated anaerobic–aerobic system. The integrated systems although showed transient accumulation of aromatic amines, generated from the bioreduction of the azo dyes. These compounds were ultimately mineralized, probably with the help of facultative aerobic bacteria already present in the anaerobic sludge or by subsequent addition of the aerobic enrichment culture to the anaerobic sludge (Rai et al. 2005). Although, an increase in colour during aerobic treatment has also been observed, in some studies, indicating autoxidation of aromatic amines (Van der Zee and Villaverde 2005). Koupaie et al. (2013) operated two fixed-bed sequencing batch biofilm integrated anaerobic–aerobic reactors (FB-SBBR) to estimate azo dye Acid Red 18 decolourization and biodegradation. FB-SBBR1 and FB-SBBR2 were packed with volcanic pumice stones and a type of plastic material composed of polyethylene, respectively. They observed that decolourization of the dye in both the reactors followed first-order kinetics in correlation to dye concentration. 1-naphthylamine-4-sulfonate was produced as main sulphonated aromatic amine during the anaerobic phase and more than 63.7% and 71.3% of which was successfully removed during the aerobic reaction phase of both the reactors.

As reported recently, the sequential anaerobic–aerobic system proved advantageous as compared to integrated system. Primarily, rapid azo dye reduction was detected under sequential anaerobic–aerobic states due to utilization of the co-substrate only for providing electrons for reduction of azo dyes; however, in an integrated system, the co-substrate is partly used in creating anaerobic micro niches and partly for azo dye reduction. Additionally, aerobic degradation of the co-substrate under integrated systems generally increases the unavailability of either oxygen or co-substrate, which ultimately is unfavourable for decolourization and degradation of azo dyes. These unfavourable effects certainly do not arise in the sequential anaerobic–aerobic conditions. Thus, a dynamic balance between co-substrate and oxygen is badly required, which was difficult to attain in the integrated anaerobic–aerobic systems (Rai et al. 2005).

The performance of sequential as well as integrated anaerobic–aerobic processes for the bioremediation azo dye Reactive Black 5-laden synthetic wastewaters was studied and compared. Both the processes showed colour removal to a significant extent during the anaerobic phase, while partial mineralization of the anaerobic dye metabolites, approximately to a similar degree, was achieved later during the aerobic stage. The major proportion of the COD was removed during the anaerobic stage for the sequential processes and during the aerobic stage in integrated process. Reportedly, both the processes lead to similar extent of overall COD removal (Bonakdarpour et al. 2011).

## 4 Factors Affecting Biodegradation of Dyes

Microbial metabolism is significantly affected by changes in abiotic parameters, such as pH, temperature, availability of oxygen, metal ions, salts, etc.; as a result, microbial dye degradation activities are also affected (Ali 2010). Hence, while analysing the ability of various microorganisms pertaining to degradation of dyes reported to be as xenobiotic compounds, the effects of such parameters are considered. Optimization of these abiotic conditions will substantially facilitate advancements of industrial scale processes as well as bioreactors for bioremediation. A number of these components influencing the biodegradation of azo dyes are therefore reviewed here.

### *4.1 pH*

The initial pH values of the textile industry dye wastewater vary widely. Enhanced decolourization and biodegradation activities are generally shown by fungi and yeasts at acidic or neutral pH (3.0–5.0) and by bacteria at neutral or basic pH, and the rate of dye decolourization tends to decrease under strongly acidic/alkaline pH conditions. Probably, pH affects solubilization and transportation of dye molecules through the plasma membrane as well as the growth of the microorganism, substrate-limiting stage for the dye decolourization (Agrawal et al. 2014b). Thus, organisms showing considerable dye degradation over a wide range of pH need to be preferred for actual waste treatment.

### *4.2 Temperature*

Biodegradation activities associated with microorganisms are unquestionably altered significantly because of changes in temperature. The rate of reaction usually increases to two times with every 10°C increase in temperature, nevertheless, this relationship loses linearity beyond optimum temperature. The degradation activity of the microorganisms decreases beyond the optimum temperature probably, because of the reduced growth and reduced reproduction rate in conjunction with deactivation of enzymes accountable for dye degradation. Optimum growth and degradation temperatures vary amongst different microorganisms, with most of them growing at 25–35 °C (Ali 2010). The reduction in dye decolourization rate at elevated temperatures can be accredited to the denaturation of the enzymes and ultimately the loss of viability of the cells (Anjaneya et al. 2011).

### *4.3 Initial Dye Concentration*

The time required for decolourization of the azo dye is directly proportional to the initial azo dye concentration in the system (Sheth and Dave 2009). The rate of dye decolourization usually slows down with escalating concentration of the dye. This decline in dye decolourization rate with escalation in initial concentration of dye is caused due to the toxic nature of the dyes and its metabolites towards the multiplying microorganisms at elevated dye concentrations. Contrastingly, it has also been reported that the dye reduction rate was independent of dye concentration exhibiting compatibility with a reduction mechanism that is non-enzymatic and is regulated by processes that are independent of the dye concentration (Dave et al. 2015).

## *4.4 Dye Structure*

Decolourization and biodegradation of azo dyes is extensively affected by the diversity in their structure, together with changes in their chemical structures like isomerism or the existence of diverse functional groups. Dyes having simple molecular structure and low molecular weights reportedly have higher rates of degradation as compared to those dyes having high molecular weight. Also, the degradation rate is faster with respect to monoazo dyes than with diazo or triazo dyes. Azo dyes possessing amino or hydroxyl groups are expected to degrade early as compared to those having methyl, sulpho, methoxy or nitro groups (Saratale et al. 2011).

Azo dyes are reduced to their anionic form by a rapid single-electron transfer reaction, leading to a subsequent second slower electron transfer reaction to generate a stable dianion. Therefore, functional groups with higher electron density might not favour dianion formation, leading to absence or reduced capability for decolourization. Additionally, permeation of azo dyes through the bacterial cell membrane is the rate-limiting step for the decolourization of sulphonated azo dyes by bacteria. Moreover, the electron-dense hydrogen bonds, present in the proximity of the azo bond, enhance the azo-hydrazone tautomerism of hydroxy azo compounds and therefore have a noteworthy impact on the azo dye reduction rate (Saratale et al. 2011).

## *4.5 Carbon and Nitrogen Content in the Medium*

Azo dyes do not act as carbon sources, and therefore, the bioremediation of fabric dyes without supplementation of carbon or nitrogen sources is challenging. Decolourization of azo dyes by mixed or pure cultures usually necessitates complex organic substrates, for instance yeast extract, peptone or possibly a mixture of complex natural and organic substrates and carbohydrates. Reportedly, reducing equivalents from different carbon sources are transmitted to the dyes during azo dye decolourization by azo bond reduction. However, addition of carbon sources proved to be ineffective to enhance dye decolourization, almost certainly, because of the inclination of the cells for assimilation of the surplus carbon sources over utilization of the dye compound and also carbon source decolourization (Saratale et al. 2011). By far the most readily utilizable carbon sources for almost all of the microorganisms can be glucose. On the other hand, glucose seems to be a costly carbon source, and therefore, other inexpensive carbon sources like starch, molasses and fructose are used in dye wastewater treatment decolourization (Kaushik and Malik 2009). Usually, the addition of the organic nitrogen sources like peptone, urea, yeast and beef extract, etc., can regenerate NADH, which serves as an electron donor for the azo dye reduction by microbes and hence effective decolourization has been witnessed. In making the approach economically viable and essentially applicable, certain researchers have utilized lignocellulosic agricultural waste as a possible supplement for successful decolourization (Saratale et al. 2011).

### 4.6 Agitation and/or Aeration

Contradictory results have been reported regarding the impact of shaking or agitation on microbial decolourization of textile dyes. It has been observed that azo dyes are usually recalcitrant under aerobic conditions, and therefore for efficient dye degradation, aeration along with agitation, which enhances the oxygen concentration in solution, ought to be avoided (Ali 2010). Bacterial azo dye reduction is strongly inhibited in the presence of oxygen, probably either due to the direct inhibition of the enzyme azoreductase or due to the favoured reduction of oxygen instead of the azo compounds (Pearce et al. 2003).

Bacteria generally degrade azo dyes to colourless toxic aromatic amines, under anaerobic environment, some of which will be readily metabolized under aerobic conditions. Apart from a few, the aromatic amines produced from azo dye decolourization are typically resistant to biodegradation under anaerobic conditions. Hence, anaerobic degradation of azo dyes is usually more suitable than aerobic biodegradation, but the intermediate metabolites (carcinogenic aromatic amines) generated after anaerobic degradation have to be degraded aerobically; therefore, a sequential anaerobic and aerobic treatment of textile industry wastewater containing aromatic compounds is essential before they are discarded to the environment (Rai et al. 2005).

### 4.7 Salts

Effluent from textile processing and dye producing industries usually comprises of 15–20% of salts along with azo dye residues. As a result, the characterization and development of halotolerant bacterial strains capable of degrading azo dyes may ease the establishment of biotreatment processes for the remediation of high salt containing azo dye wastewaters from textile industry (Khalid et al. 2008).

The dye decolourization rates initially increase with an escalation in salt concentration approximately up to 3–5 g/L of salt concentration, above which decolourization rates started decreasing. This can be accredited to the variation of enzyme activity and molecular transportation across the plasma membrane of microorganisms. With elevated salt concentrations, the cell has increased osmotic pressure which in turn affects the membrane properties. The decrease in the substrate transportation reduces, generally, the decolourization rate of dyes. The presence of lower concentrations of salt might also stimulate the growth of some bacterial strains from non-marine environments; however, higher salt concentrations had lethal effects on their growth (Meng et al. 2012).

### *4.8 Electron Donor*

The presence of an electron donor is a precondition for reduction of azo dyes. Hypothetically, only four reducing equivalents are required per azo linkage, indicating that the fewer amounts of electron donors are required. However, competition for electron donors by other reactions leads to an increase in the amount of the reducing equivalents required. Azo dyes in addition to other organic constituents of textile dyeing wastewater lack electron donors and therefore are not enough as substrate to support the growth of anaerobic bacteria. Therefore, the reductive cleavage of azo bond is apparently induced by supplementation of electron donors, for instance glucose, acetate or formate ions. In addition to this, the azo dye reduction possibly will also be kinetically benefitted by a higher electron donor concentration. Hence, it is crucial to deduce the physiological electron donors for every dye bioremediation process, because the electron donors not only stimulate the mechanism of azo dye reduction but additionally induce the particular cluster of enzymes accountable for the particular reduction process. It has also been noted that the presence of certain electron donors inhibits the process of electron transport probably because of the competition for electrons from the electron donors (Saratale et al. 2011).

## 5 Bacterial Enzyme Systems for Dye Biodegradation

Enzymatic reactions have proved their efficiency in textile processing for many years and now are progressively gaining importance as biocatalysts in textile waste processing. Bilirubin oxidase, laccase and azoreductase are some enzymes responsible for the biodegradation of dyes using non-lignolytic fungi. Likewise, various bacterial enzymes, viz. veratryl alcohol oxidase, tyrosinase, NADH-dependent dichlorophenolindophenol (NADH-DCIP) reductase, azoreductase and laccase, are also reported for dye decolourization. It was generally observed that because of the susceptibility of the enzyme to inactivation because of the presence of different chemicals, enzymatic treatment will probably be effective in the fields having a maximum concentration of dye and minimum concentration of interfering substances (Saratale et al. 2011).

## *5.1 Laccase*

Laccases, alternatively known as benzenediol oxygen oxidoreductases and with EC number EC 1.10.3.2, are principally polyphenol oxidases (PPO), belonging to a small category of enzymes called blue oxidase. Using molecular oxygen as the electron acceptor, they catalyse cross-linking of monomers, ring cleavage of aromatic compounds, degradation of polymers and oxidation of several substituted phenolic compounds (Sharma et al. 2007; Kunamneni et al. 2008). Genes encoding laccases have been reported in various gram-positive as well as gram-negative bacteria, which also includes extremophiles, e.g. in *Oceanobacillus iheyensis, Aquifex aeolicus* and *Pyrobaculum aerophilum* (Ferrer et al. 2010). Besides this, bacterial laccases overcome the drawbacks of in-process applications and instability of fungal laccases. They are extremely active and stable at high temperatures and pH values (Sharma et al. 2007).

## *5.2 Lignin Peroxidase*

Lignin peroxidases (abbreviated as LiP and recognized with the EC number, E.C. 1.11.1.14), the enzyme first reported in the year 1983, are also referred to as ligninase or diaryl propane oxygenase. This enzyme is a member of the category oxidoreductases, having hydrogen peroxide as an electron acceptor (peroxidases), and is broadly categorized as ligninases (Husain 2010). LiP catalyses a variety of reactions, including hydroxylation, ortho-demethylation, benzylic oxidation, cleavage of bond between two carbon atoms, phenol dimerization, hydroxylation, oxidation of non-phenolic aromatic lignin moieties and oxidation of azo dyes and pesticides. Bacterial lignin peroxidases have been utilized for dye bioremediation through *Brevibacillus laterosporus* and *Acinetobacter calcoaceticus* (Gomare et al. 2008; Ghodake et al. 2009). Innumerable investigations have revealed a noteworthy induction in the cellular activity of lignin peroxidase in a diverse variety of bacteria during decolourization of azo dyes, symptomatic of their probable role in decolourization of dyes.

## *5.3 Azoreductase*

Azobenzene reductase also referred to as azoreductase (EC 1.7.1.6) (Maier et al. 2004) is an enzyme, which is a member of the family of oxidoreductases,

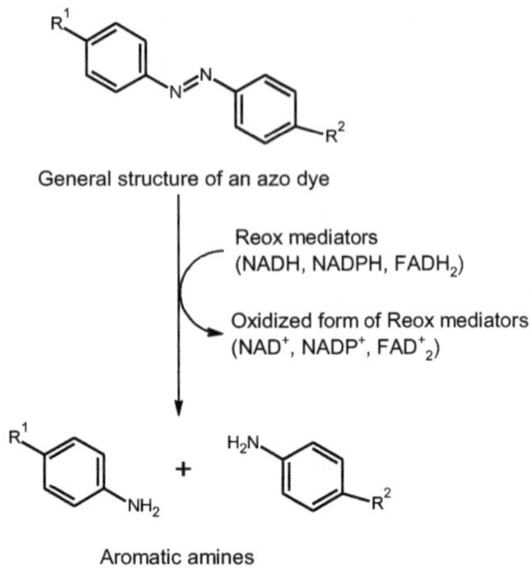


**Fig. 3** Reaction catalysed by bacterial azoreductases

specifically catalysing oxidation–reduction reactions only with the help of reducing equivalents FADH, NADH and NADPH as acceptor. Several azo bond reducing enzymes have been recognized as FMN-independent reductases, FMN-dependent reductases, NADH-dependent reductases, NADH-DCIP reductases and NADPH-dependent reductases (Saratale et al. 2011).

Azoreductases are present in bacteria, algae and yeast. Activity of the enzyme azoreductases has been detected in a number of bacteria, such as *Pseudomonas luteola*, *Rhodococcus* sp., *Xenophilus azovorans* KF46F, *Shigella dysenteriae* Type I, *Klebsiella pneumoniae* RS-13, *Clostridium perfringens*, *Bacillus laterosporus* RRK1 (Sandhya et al. 2008), *Caulobacter subvibrioides* C7-D (Mazumder et al. 1999), *Bacillus badius* (Misal et al. 2011), *Staphylococcus aureus* ATCC 25923 (Chen et al. 2005), *Pigmentiphaga kullae* K24 (Chen et al. 2010) and *Rhodobacter sphaeroides* AS1.1737 (Bin et al. 2004). It is localized either to the interior or to the exterior of the bacterial plasma membrane (Saratale et al. 2011). Azoreductases in bacteria can be broadly categorized into two different types: anaerobic and aerobic azoreductases (Solís et al. 2012). Azoreductase as well as the flavin reductase is the crucial enzyme accountable for the azo dye decolourization; conversely, they lead to generation of toxic aromatic amines upon reductive hydrolysis of azo linkage (Saratale et al. 2012; Telke et al. 2015). The reaction catalysed by the enzyme azoreductase is shown in Fig. 3.

## *5.4 Polyphenol Oxidase or Tyrosinases*

Polyphenol oxidase (PPO) enzymes also referred as tyrosinases (EC1.14.18.1) are copper-dependent enzymes. They catalyse the hydroxylation at ortho position of monophenols to o-diphenols (often known as monophenolase) and further catalyse the oxidation of o-diphenols to produce o-quinones (also known as o-diphenolase), in the presence of molecular oxygen. The enzyme remains in an inactive deoxy state and undergoes a conformational change to the active oxy state upon binding of a new molecule of oxygen. The enzyme also serves as an indicator of the oxidative enzymes accountable for azo dye degradation (Saratale et al. 2011). Activity of the enzyme polyphenol oxidases has been detected in a wide range of bacteria, viz. *Streptomyces glaucescens*, *S. antibioticus*, *Bacillus licheniformis*, *B. sphaericus* and *B. natto* (Telke et al. 2015).

### *5.5 NADH-DCIP Reductase*

This enzyme with EC number (EC 1.6.99.3) catalyses reduction of 2,6-dichlorophenol (DCIP) to its leuco form, with the help of NADH as an electron donor. The enzyme is usually known as diaphorase (Nishiya and Yamamoto 2007). NADH-DCIP reductases are considered as marker enzymes for the mixed function oxidase system of bacteria and fungi and are responsible for azo bond reduction and detoxification of xenobiotic compounds. DCIP is converted from blue colour to colourless, upon reduction (Saratale et al. 2011).

The substantial induction of non-specific reductase was observed during the biodegradation of the dye Malachite green and the non-specific reductase was termed as MG reductase. It reduced the Malachite Green to Leucomalachite Green with the help of NADH as an electron donor (Parshetti et al. 2006).

## 6 Characterization of Dye Metabolites

With the aim of understanding the environmental fate of the dyes and to reveal the possible mechanism of dye decolourization, study of the metabolic products of azo dyes is crucial. Several analytical techniques are utilized to deduce the degradation products produced upon biodegradation of azo dyes and some of them are described in the following section.

### *6.1 UV–Visible Spectroscopy*

UV–vis spectroscopy (UV–vis) is the principal technique used to ascertain whether dye decolourization is due to absorption or biodegradation (Wilson and Walker 2010). The characteristic peaks in the absorption spectrum of a molecule usually decrease in comparison to each other along with deep colouration of bacterial cells because of dye removal due to adsorption. Contrastingly, in case the dye remediation is due to biodegradation, the main absorbance peaks in the visible region will either completely disappear or some new peaks towards the UV region will appear as well as the cells retain their original colour (Wijetunga et al. 2007). The true colour of the mixture of dyes and industrial effluents, independent of the hue, is measured with the help of ADMI 3WL, the tri-stimulus filter method of the American Dye Manufacturers Institute (Saratale et al. 2011).

The UV–Vis spectra of the Navy blue 2GL, a textile azo dye, upon treatment by *Bacillus* sp. VUS showed decolourization and reduction in the concentration of the dye over time without any shift in $\lambda_{max}$ until complete decolourization of the medium (Dawkar et al. 2009). Similarly, noteworthy changes appearing in both the ultraviolet and visible spectra of the dye Acid Black 210, upon treatment by the bacterium *Providencia* sp. SRS83, point out that molecular structure of AB210 changed manifestly upon decolourization undoubtedly owing to the biodegradation (Agrawal et al. 2014b).

## 6.2 Fourier Transform Infrared Spectroscopy

Fourier transform infrared (FTIR) spectroscopy involves absorption of infrared light, including the wavelength range of about 700–25,000 nm, by a molecule resulting in transition to higher levels of vibration. An infrared spectrum arises because of absorption of incident light of a specific wavelength by a molecule which in turn will then disappear from the transmitted light. The observed spectrum will demonstrate an absorption band. The bonds between atoms can be contemplated as flexible springs, and bond vibrations can therefore be either stretching or bending (deformation) actions.

A comprehensive analysis of the structure of a molecule is possible, because the wavenumber associated with a specific functional group differs marginally, due to the influence of the molecular environment. For instance, it is possible to differentiate between C–H vibrations in methylene (–CH–) and methyl groups ($–CH_3$). FTIR is increasingly used for the analysis of peptides and proteins (Wilson and Walker 2010). The FTIR spectrum facilitates estimation of type as well as strength of interactions amongst different functional groups of the azo dye molecule upon bacterial treatment, and consequently it serves as a precious analytical tool (Saratale et al. 2011).

FTIR spectrum of the azo dye and its metabolites formed upon biodegradation, upon comparison, clearly reveals dye degradation by bacterial species. The FTIR spectra of dye metabolites indicated disappearance of specific peaks, in the range of 1575.0–1630.0 $cm^{-1}$ for azo compounds, thereby confirming breaking of azo bonds present in the dye (Lade et al. 2015). Rest of the peaks indicate the bending and/or stretching vibrations of different groups of the dye and hence support dye degradation (Agrawal et al. 2014b)

## 6.3 High-Performance Liquid Chromatography

A chromatogram is a pictorial record of the detector response as a function of retention time or elution volume, consisting of a series of peaks or bands, ideally symmetrical in shape, demonstrating the elution of individual analytes. This time is characteristic of the analyte and helps in its identification (Wilson and Walker 2010). High-performance liquid chromatography (HPLC) technique can be utilized for confirming dye degradation, through the advent of novel peaks in the chromatogram with altered retention times in comparison to the original dye. These metabolites generated upon dye biodegradation are represented by the formation of new peaks.

## 6.4 Mass Spectrometry

Mass spectrometry (MS) assists in obtaining the mass spectra of the mixture of compounds, offering an extremely effective qualitative tool for the analysis of the dye biodegradation. The technique is known as liquid chromatography–mass spectrometry (LC–MS) when a liquid is used as mobile phase and gas chromatography–mass spectrometry (GC–MS) when the mobile phase employed is a gas. Both the techniques have proved to be of help for the estimation of molecular weights as well as the structural details of metabolites of dye biodegradation formed after bacterial treatment and can help in elucidating the microbial pathways of azo dye degradation (Saratale et al. 2011). GCMS analysis was utilized to deduce the metabolites generated upon biodegradation of triazo dye Acid Black 210 with the help of *Providencia* sp. SRS83; the retention time and *m/z* values helped in predicting the metabolites generated and probable pathway (Fig. 4) followed by the organism for dye biodegradation (Agrawal et al. 2014b). The dye metabolites generated during anaerobic phase are further mineralized during the aerobic phase, which was deduced by Chan et al. (2012). The dye degradation metabolites, their molecular weights generated from LCMS/MS and the biodegradation pathways can be correlated and hypothesized with reference to the metabolic pathways illustrated on the MetaCyc database at http://metacyc.org/ and KEGG Pathway database at http://www.genome.jp/kegg/pathway.html. The hypothetical pathway for the mineralization of an azo dye is represented in Fig. 5.

## 6.5 COD

Additionally, the degree of azo dye decolourization in addition to degradation can be estimated by calculating the percentage of degradation through the measurement

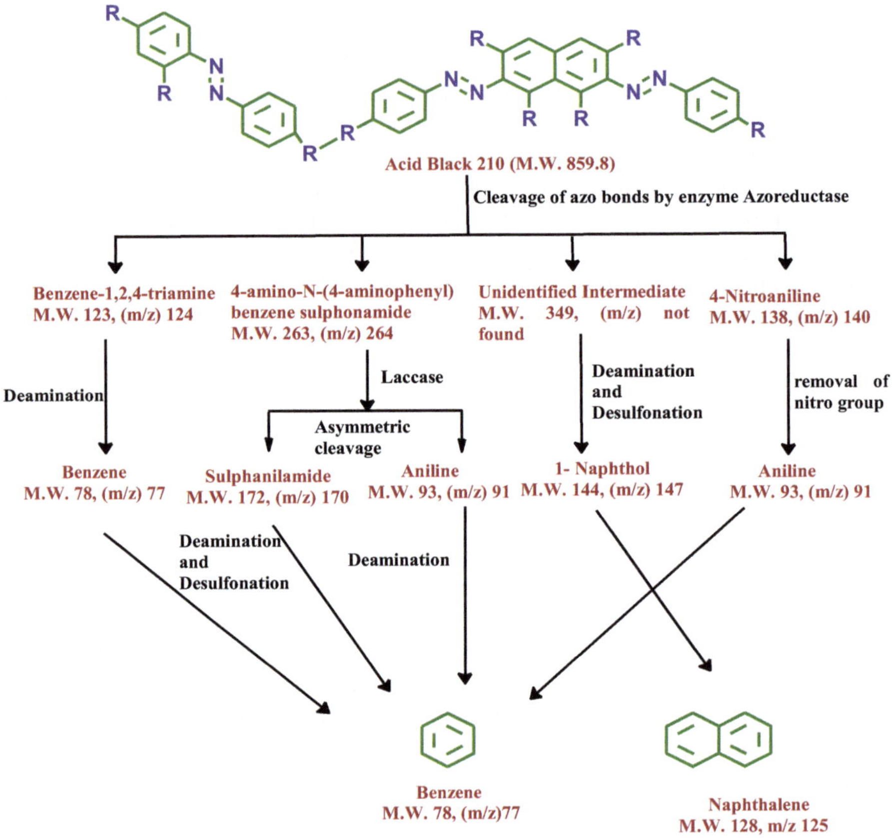


**Fig. 4** Proposed pathway for the biodegradation of a model azo dye by *Providencia* sp. SRS82

of the decrease in the ratio of both the chemical oxygen demand (COD) and biochemical oxygen demand (BOD), before and after the treatment.

Analysis and interpretation of dye biodegradation by sophisticated techniques mentioned above are not much helpful for dye containing effluent, as with individual dye biodegradation by bacteria, because composition of industrial wastewater is very complex and practically unknown. So, parameters such as COD, BOD, ADMI values and total organic carbon are estimated to estimate the degree of mineralization and biodegradation of dye-laden effluent. A reduction in values of COD, ADMI and TOC values indicates mineralization of complex dye molecule (APHA 1998).

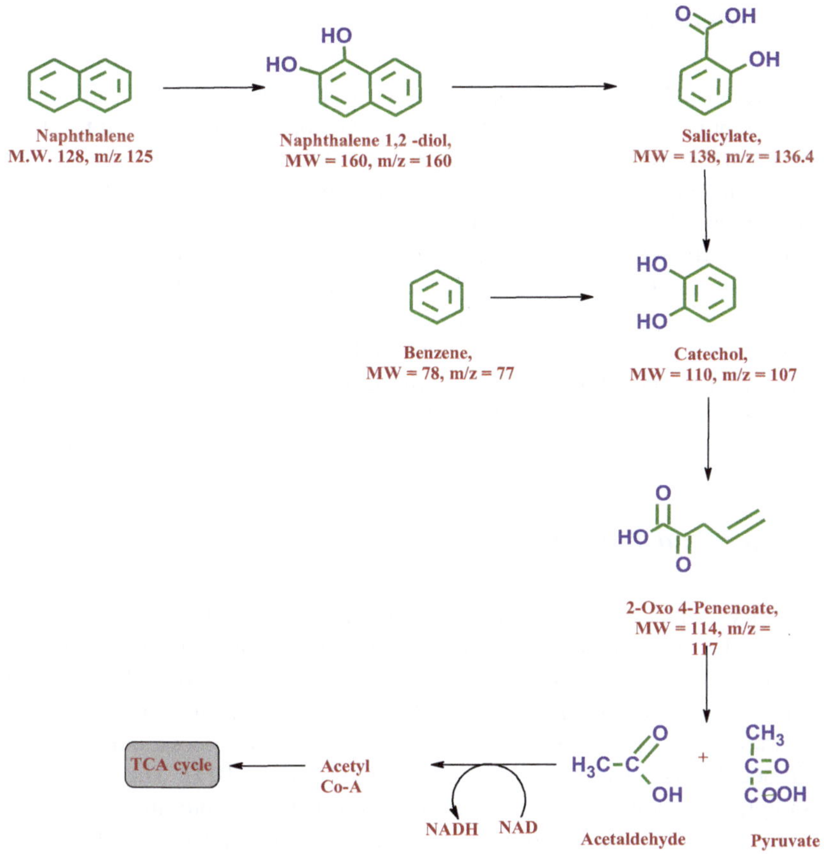


**Fig. 5** Proposed pathway for the aerobic biodegradation of dye metabolites

## 7 Toxicity Study

The release of textile industry and dyestuff manufacturing industry effluent may lead to unusual colouration of the surface water which causes the ultimate ecological distress pertaining to water quality as well as directly disturbs the aquatic biota. The synthetic dyestuffs and their degradation products released by the textile industry are mutagenic and carcinogenic. Certainly, lack of adequate statistics related to toxicity of dyes with reference to cell populations makes bioremediation unpredictable and undependable for on-site processes. Thus, it is very essential to find out phytotoxicity and microbial toxicity of the treated dyeing effluent.

## 7.1 Phytotoxicity

Phytotoxicity studies involve treatment of the seeds of model plants with the parent dye and its biodegradation products at a specific concentration. The impact of this treatment on the seed germination percentage and length of the plumule and radicle can be estimated and can then be compared with the control (treated with water). Various groups of researchers have treated the seeds of model plants with water, the dye and its extracted metabolite separately and subsequently compared percentage of seed germination along with the lengths of plumule in addition to the radical. Reportedly, seed germination (%), shoot length and root length for the tested plants were less upon dye treatments as compared to those treated with metabolites obtained after its decolourization confirming detoxification of the dye molecule upon microbial treatment (Saratale et al. 2011).

## 7.2 Microbial Toxicity

The toxicity of the azo dyes and their metabolites can be analysed by evaluating the toxicity of the dye sample before and after bacterial treatment towards the known microbial cultures. This is a sensitive method producing reproducible and comparable results. Furthermore, if the biodecolourizer itself is the test organism, then the results indicate the performance of the dye degradation process and suitability of the bio decolourizer for this process too (Saratale et al. 2011). The toxicity of the dye sample and its metabolites can also be compared by counting the changes in microbial cells count per millilitre as compared to that in nutrient broth (Ali 2010).

Correspondingly, some researchers have investigated the toxic effect of a variety of dyes and their biodegradation products generated upon bacterial treatment on plate assay by measuring their zone of inhibition at equal concentrations and used common laboratory microorganisms along with cellulose, nitrogen and phosphate solubilizing bacteria, owing to their significance in agriculture. The outcomes suggest that upon treatment with bacteria the extract of dye metabolites proved to be less toxic than the dye itself, indicating azo dye detoxification (Saratale et al. 2011).

## 7.3 Other Toxicity Tests

Higher plants are considered as outstanding genetic models to estimate mutagenicity of toxic compounds; therefore, cytotoxicity and genotoxicity tests conducted upon *Allium cepa* (onion) are nowadays generally studied. *A. cepa* species also have the advantage of being cheap and simplicity in handling. Mitotic index (MI) and chromosomal aberrations in the cells of roots of *A. cepa* are utilized to detect genotoxicity. MI of sample-treated dye is usually higher as compared to those treated with biodegraded metabolites. MI, cell death assay and certain nuclear abnormalities are evaluated to estimate cytotoxicity of dyes. Analysis of the increase or decrease in MI provides a useful insight for the estimation of the cytotoxicity of a test compound. Reportedly, total numbers of cells with chromosomal aberrations are considerably higher in dye-treated samples as compared to those treated with water and dye metabolites, indicating genotoxicty of the dyes (Dave et al. 2015). The different forms of chromosomal aberrations induced under the influence of dyes are laggards, chromosomal breakages and anaphase bridges and micronuclei formation. Additionally, the frequencies of total alterations (TA) were also detected to be higher in dye-treated samples (Jadhav et al. 2011)

Environmental pollutants can lead to induction of the oxidative stress which can subsequently bring about lipid peroxidation and/or protein oxidation or even damages to DNA in plant and mammalian cells. Antioxidants or scavenging enzymes, for instance superoxide dismutase, ascorbate peroxidase, catalase, glutathione reductase and peroxiredoxins, are present in the plant systems as their defence against such oxidative stress. Consequently, to analyse the toxicity status of dye and its degraded metabolites, various researchers assayed the status of antioxidant enzymes, protein oxidation and lipid peroxidation. Reportedly, the activities of these enzymes in *A. cepa* root cells were elevated in dye-treated samples as compared to those samples that were exposed to dye metabolites which showed virtually similar levels of these with respect to control set, clearly indicating the relatively less toxic nature of the dye metabolites generated upon its treatment with bacteria (Jadhav et al. 2011).

The Ames test or the *Salmonella* mutagenicity assay is extensively utilized to deduce mutagenicty and carcinogenicity of azo dyes and their metabolites. Similarly, toxicity of the certain dyes has been studied using bioluminescent marine bacterium *Vibrio fischeri. Pseudokirchneriella subcapitata, a green unicellular alga,* has also been used as test organism to analyse the toxicity of the azo dyes (Dave et al. 2015).

## 8 Research to Implementation

Earlier reviews on biodegradation of dyes have shown that a lot of literature is already available on the usage of different bacteria for dye removal; however, detailed economic and market studies are yet not available. However, imposition of strict regulations regarding pollution has pressurized dye manufacturing and dye using industries towards usage of advanced and technologically sophisticated methods for effluent treatment. Industrialists are still reluctant to use the microbial or bacterial reactors. The challenge with bioremediation of dyes is to implement this process to an industrial scale.

The overwhelming publications in the arena of the remediation of azo dyes from wastewaters have been exploring the different applicable facets of the microbial techniques, with simultaneous hunt for novel microorganisms giving efficient dye degradation along with an insight into the elucidation of the key biochemical and biophysical processes essential to azo dye decolourization and degradation.

However, decrease in the dye-laden waste and water recycling along with development of effective, accessible, cost-effective and eco-friendly treatment processes are the need of the hour. Therefore, the future research activities should emphasize on the following facets:

1. Application of molecular biology techniques to acquire knowledge regarding the dye biodegradation pathway, which in turn can be used to develop genetically modified bacterial strains with enhanced biodegradation capacities along with tolerance to alkaline and high-temperature conditions might increase the economic feasibility and viability of the process.
2. Advancements in techniques for enzyme immobilization or development of microbial fuels cells, which can produce electricity during the bioremediation process, may help in increasing the efficiency of the process.
3. On the basis of successful laboratory results, efforts should now be directed towards application and scale-up of bacterial decolourization techniques on real industrial effluents. It is relatively easy to demonstrate it at lab scale, but it is a herculean task to demonstrate it at a pilot scale, but to actually scale it up to a large scale requires considerable financial and technological effort and therefore multidisciplinary researches should be conducted in this field.
4. Besides this, multifaceted researches can also be directed towards process design leading to reduction in use of water, reduction in discharge of toxic chemicals, and use of dyes with modified structures that can be degraded easily.

# 9 Problem Delineated

The microbial degradation of azo dyes has drawn significant attention, as a cost-effective process for dye bioremediation. Pure, mixed or consortial bacterial cultures can degrade a variety of azo dyes, discharged in textile industry effluent, under aerobic, microaerophilic or anaerobic conditions. The initial step in the dyes biodegradation is usually the reduction of the azo dye to its corresponding aromatic amines that are then degraded aerobically. The effectiveness of the process for the degradation of azo dyes is determined by the type of dye in addition to its structure and the physico-chemical parameters of the effluent, such as temperature, pH and salt concentration, availability of nutrient sources, oxygen and electron donors. The dye biodegradation process is mediated by different oxidoreductive enzymes. A variety of sophisticated analytical techniques help in ascertaining the degradation of dye as well as in elucidating the probable pathway followed by the bacterium for dye degradation. The detoxification of the dye after bacterial treatment is confirmed by less toxicity of dye metabolites to tested microbes, plants, cells and enzymes. The development of innovative processes, such as advanced oxidation process and microbial fuel cells combined with recombinant microbiological strains and the addition of novel efforts and methodologies in this area, will play a crucial role in increasing efficiency of dye bioremediation and therefore consequently environment protection.

## References


Agrawal S, Tipre D, Dave SR (2014a) Isolation, characterization and study of microorganisms capable of decolourizing triazo dye Acid Black 210. Ind J Environ Protect 34:540–546

Agrawal S, Tipre D, Patel B, Dave S (2014b) Optimization of triazo Acid Black 210 dye degradation by *Providencia* sp. SRS82 and elucidation of degradation pathway. Proc Biochem 49:110–119. doi:10.1016/j.procbio.2013.10.006

Ali H (2010) Biodegradation of synthetic dyes—A review. Water Air Soil Pollut 213:251–273. doi:10.1007/s11270-010-0382-4

An SY, Min S, Cha IH, Choi Y, Cho YS, Kim CH, Lee YC (2002) Decolourization of triphenylmethane and azo dyes by *Citrobacter* sp. Biotechnol Lett 24:1037–1040. doi:10.1023/A:1015610018103

Anjaneya O, Souche YS, Santoshkumar M, Karegoudar TB (2011) Decolourization of sulfonated azo dye Metanil Yellow by newly isolated bacterial strains: *Bacillus* sp . strain AK1 and *Lysinibacillus* sp. strain AK2. J Hazard Mater 190:351–358. doi:10.1016/j.jhazmat.2011.03.044

APHA, Standard Methods for the Examination of Water and Wastewater (1998) APHA-AWWA-WEF, 20th edn. Washington DC, USA. Method 2120 E. isbn: 0-875530788

Bin Y, Jiti Z, Jing W, Cuihong D, Hongman H, Zhiyong S, Yongming B (2004) Expression and characteristics of the gene encoding azoreductase from *Rhodobacter sphaeroides* AS1. FEMS Microbiol Lett 236:129–136. doi:10.1016/j.femsle.2004.05.034

Bonakdarpour B, Vyrides I, Stuckey DC (2011) Comparison of the performance of one stage and two stage sequential anaerobic-aerobic biological processes for the treatment of reactive-azo-dye-containing synthetic wastewaters. Int Biodeter Biodegr 65:591–599. doi:10.1016/j.ibiod.2011.03.002

Carmen Z, Daniela S (2012) Textile organic dyes—characteristics, polluting effects and separation/elimination procedures from industrial effluents—A critical overview. In: Puzyn T, Mostrag-Szlichtyng A (eds) Organic pollutants ten years after the Stockholm convention—environmental and analytical update. InTech, Croatia, pp 55–86. isbn: 978-953-307-917-2

Chan GF, Rashid ANN, Chua LS, Nasiri R, Roslan M, Ikubar MRM (2012) Communal microaerophilic–aerobic biodegradation of Amaranth by novel NAR-2 bacterial consortium. Bioresour Technol 105:48–59. doi:10.1016/j.biortech.2011.11.094

Chen H, Feng J, Kweon O, Xu H, Cerniglia CE (2010) Identification and molecular characterization of a novel flavin-free NADPH preferred azoreductase encoded by azoB in *Pigmentiphaga kullae* K24. BMC Biochem 11:13–22. doi:10.1186/1471-2091-11-13

Chen H, Hopper SL, Cerniglia CE (2005) Biochemical and molecular characterization of an azoreductase from *Staphylococcus aureus*, a tetrameric NADPH-dependent flavoprotein. Microbiology 151:1433–1441. doi:10.1099/mic.0.27805-0

Chequer FMD, Dorta DJ, De Oliveira DP (2011) Azo dyes and their metabolites: does the discharge of the azo dye into water bodies represent human and ecological risks. In: Hauser PJ (ed) Advances in treating textile effluent. InTech, Croatia EU, pp 28–48, isbn: 978-953-307-704-8

Christie RM (2001) Colour chemistry, vol 27. The Royal Soiety of Chemistry, Cambridge, UK. doi:10.1039/9781847550590

Dave SR, Patel TL, Tipre DR (2015) Bacterial degradation of azo dye containing wastes. In: Singh SN (ed) Microbial degradation of synthetic dyes in waste water. Springer, Cham, Switzerland, pp 57–83. doi:10.1007/978-3-319-10942-8_3

Dawkar VV, Jadhav UU, Ghodake GS, Govindwar SP (2009) Effect of inducers on the decolourization and biodegradation of textile azo dye Navy blue 2GL by *Bacillus* sp. VUS. Biodegradation 20:777–787. doi:10.1007/s10532-009-9266-y

Ferrer M, Beloqui A, Golyshin PN (2010) Screening metagenomic libraries for laccase activities. In: Wolfgang RS, Rolf D (eds) Metagenomics: methods and protocols. Humana Press, London, pp 189–202. doi:10.1007/978-1-60761-823-2_13

Ghodake G, Jadhav S, Dawkar V, Govindwar S (2009) Biodegradation of diazo dye Direct brown MR by *Acinetobacter calcoaceticus* NCIM 2890. Int Biodeter Biodegr 63:433–439. doi:10.1016/j.ibiod.2008.12.002

Gomare SS, Jadhav JP, Govindwar SP (2008) Degradation of sulfonated azo dyes by the purified lignin peroxidase from *Brevibacillus laterosporus* MTCC 2298. Beiotechnol Bioprocess Eng 13:1–8. doi:10.1007/s12257-008-0008-5

Gomare SS, Tamboli DP, Kagalkar AN, Govindwar SP (2009) Eco-friendly biodegradation of a reactive textile dye Golden Yellow HER by *Brevibacillus laterosporus* MTCC 2298. Int Biodeter Biodegr 63:582–586. doi:10.1016/j.ibiod.2009.03.005

Hsueh CC, Chen BY, Yen CY (2009) Understanding effects of chemical structure on azo dye decolourization characteristics by *Aeromonas hydrophila*. J Hazard Mater 167:995–1001. doi:10.1016/j.jhazmat.2009.01.077

Hunger K (ed) (2003) Industrial dyes chemistry, properties, applications. Wiley-VCH Verlag GmbH and Co. KGaA, Weinheim. isbn: 978-3-527-30426-4.

Husain Q (2010) Peroxidase mediated decolourization and remediation of wastewater containing industrial dyes: a review. Rev Environ Sci Biotechnol 9:117–140. doi:10.1007/s11157-009-9184-9

Jadhav JP, Kalyani DC, Telke AA, Phugare SS, Govindwar SP (2010) Evaluation of the efficacy of a bacterial consortium for the removal of color, reduction of heavy metals, and toxicity from textile dye effluent. Bioresour Technol 101:165–173. doi:10.1016/j.biortech.2009.08.027

Jadhav JP, Phugare SS (2012) Textile dyes: general information and environmental aspects. In: Nemr A El (ed) Non-conventional textile waste water treatment. Nova Science, New York, pp 1–345. isbn: 978-1-62100-228-4.

Jadhav SB, Phugare SS, Patil PS, Jadhav JP (2011) Biochemical degradation pathway of textile dye Remazol red and subsequent toxicological evaluation by cytotoxicity, genotoxicity and oxidative stress studies. Int Biodeter Biodegr 65:733–743. doi:10.1016/j.ibiod.2011.04.003

Jin R, Yang H, Zhang A, Wang J, Liu G (2009) Bioaugmentation on decolourization of C. I. Direct Blue 71 by using genetically engineered strain *Escherichia coli* JM109 (pGEX-AZR). J Hazard Mater 163:1123–1128. doi:10.1016/j.jhazmat.2008.07.067

Kant R (2012) Textile dyeing industry an environmental hazard. Nat Sci 4:22–26. doi:10.4236/ns.2012.41004

Kaushik P, Malik A (2009) Fungal dye decolourization: Recent advances and future potential. Environ Int 35:127–141. doi:10.1016/j.envint.2008.05.010

Khalid A, Arshad M, Crowley DE (2008) Decolourization of azo dyes by *Shewanella* sp. under saline conditions. Appl Microbiol Biotechnol 79:1053–1059. doi:10.1007/s00253-008-1498-y

Koupaie EH, Moghaddam MRA, Hashemi SH (2013) Evaluation of integrated anaerobic/aerobic fixed-bed sequencing batch biofilm reactor for decolorization and biodegradation of azo dye Acid Red 18 : Comparison of using two types of packing media. Bioresour Technol 127:415–421. doi:10.1016/j.biortech.2012.10.003

Kunamneni A, Plou FJ, Ballesteros A, Alcalde M (2008) Laccases and their applications: a patent review. Recent Pat Biotechnol 2:10–24. doi:10.2174/187220808783330965

Lade H, Govindwar S, Paul D (2015) Mineralization and detoxification of the carcinogenic azo dye Congo Red and real textile effluent by a polyurethane foam immobilized microbial consortium in an upflow column bioreactor. Int J Environ Public Health 12:6894–6918. doi:10.3390/ijerph120606894

Lang AR (ed.) (2009) Dyes and pigments: new research. Nova Science, New York. isbn: 978-1-60876-195-1

Maier J, Kandelbauer A, Erlancher A, Cavaco-Paulo A, Gubits GM (2004) A new alkali thermostable azoreductase from *Bacillus* sp. strain SF. Appl Environ Microbiol 70:837–844. doi:10.1128/AEM.70.2.837-844.2004

Mangal VP (2010) The future of Indian dyes and dye intermediates. Textile Rev www.fibre2fashion.com, accessed on June 6, 2014

Mansoor I (2008) Textile dyes. Rahber Publishers, Karachi

Mazumder R, Logan JR, Mikell JAT, Hooper SW (1999) Characteristics and purification of an oxygen insensitive azoreductase from *Caulobacter subvibrioides* strain C7-D. J Ind Microbiol Biotechnol 23:76–483

Meng X, Liu G, Zhou J, Shiang FQ, Wang G (2012) Azo dye decolourization by *Shewanella aquimarina* under saline conditions. Bioresour Technol 114:95–101. doi:10.1016/j.biortech.2012.03.003

Misal SA, Lingojwar DP, Shinde RM, Gawai KR (2011) Purification and characterization of azoreductase from alkaliphilic strain *Bacillus badius*. Proc Biochem 46:1264–1269. doi:10.1016/j.procbio.2011.02.013

Mohana S, Shrivastava S, Divecha J, Madamwar D (2008) Response surface methodology for optimization of medium for decolourization of textile dye Direct Black 22 by a novel bacterial consortium. Biresour Technol 99:562–569. doi:10.1016/j.biortech.2006.12.033

Nishiya Y, Yamamoto Y (2007) Characterization of a NADH: dichloroindophenol oxidoreductase from *Bacillus subtilis*. Biosci Biotechnol Biochem 71:611–614. doi:10.1271/bbb.60548

Ola IO, Akintokun AK, Akpan I, Omomowo IO, Areo VO (2010) Aerobic decolourization of two reactive azo dyes under varying carbon and nitrogen source by *Bacillus cereus*. Afr J Biotechnol 9:672–677

Parshetti G, Kalme S, Saratale G, Govindwar S (2006) Biodegradation of Malachite Green by *Kocuriarosea* MTCC 1532. Acta Chim Solv 53:492–498

Patil PS, Shedbalkar UU, Kalyani DC, Jadhav JP (2008) Biodegradation of Reactive Blue 59 by isolated bacterial consortium PMB11. J Ind Microbiol Biotechnol 35:1181–1190. doi:10.1007/s10295-008-0398-6

Pearce CI, Lloyd JR, Guthrie JT (2003) The removal of colour from textile wastewater using whole bacterial cells: a review. Dye Pigment 58:179–196. doi:10.1016/S0143-7208(03)00064-0

Phugare SS, Kalyani DC, Patil AV, Jadhav JP (2011) Textile dye degradation by bacterial consortium and subsequent toxicological analysis of dye and dye metabolites using cytotoxicity, genotoxicity and oxidative stress studies. J Hazard Mater 186:713–723. doi:10.1016/j.jhazmat.2010.11.049

Rai HS, Bhattacharyya MS, Singh J, Bansal TK, Vats P, Banerjee UC (2005) Removal of dyes from the effluent of textile and dyestuff manufacturing industry: a review of emerging techniques with reference to biologocal treatment. Crit Rev Environ Sci Technol 35:219–238. doi:10.1080/10643380590917932

Sandhya S, Sarayu K, Uma B, Swaminathan K (2008) Decolorizing kinetics of a recombinant *Escherichia coli* SS125 strain harboring azoreductase gene from *Bacillus latrosporus* RRK1. Bioresour Technol 99:2187–2191. doi:10.1016/j.biortech.2007.05.027

Santos B, Cervantes FJ, Lier JB (2007) Review paper on current technologies for decolourisation of textile wastewaters: perspectives for anaerobic biotechnology. Bioresour Technol 98:2369–2385

Saratale RG, Gandhi SS, Purankar MV, Kurade MB, Govindwar SP, Oh SE, Saratale GD (2012) Decolorization and detoxification of sulfonated azo dye C.I. Remazol Red and textile effluent by isolated *Lysinibacillus* sp. RGS. J Biosci Bioeng 115:658–667. doi:10.1016/j.jbiosc.2012.12.009

Saratale RG, Saratale GD, Chang JS, Govindwar SP (2010) Decolourization and biodegradation of reactive dyes and dye wastewater by a developed bacterial consortium. Biodegradation 21:999–1015. doi:10.1007/s10532-010-9360-1

Saratale RG, Saratale GD, Chang JS, Govindwar SP (2011) Bacterial decolourization and degradation of azo dyes: a review. J Taiwan Inst Chem Eng 42:138–157. doi:10.1016/j.jtice.2010.06.006

Saratale RG, Saratale GD, Kalyani DC, Chang JS, Govindwar SP (2009) Enhanced decolourization and biodegradation of textile azo dye Scarlet R by using developed microbial consortium-GR. Bioresour Technol 100:2493–2500. doi:10.1016/j.biortech.2008.12.013

Sarayu K, Sandhya S (2010) Aerobic biodegradation pathway for Remazol Orange by *Pseudomonas aeruginosa*. Appl Biochem Biotechnol 160:1241–1253. doi:10.1007/s12010-009-8592-1

Senan RC, Abraham TE (2004) Bioremediation of textile azo dyes by aerobic bacterial consortium aerobic degradation of selected azo dyes by bacterial consortium. Biodegradation 15:275–280

Shah PD, Dave SR, Rao MS (2012) Enzymatic degradation of textile dye Reactive Orange 13 by newly isolated bacterial strain *Alcaligenes faecalis* PMS-1. Int Biodeter Biodegr 69:41–50. doi:10.1016/j.ibiod.2012.01.002

Sharma P, Goel R, Capalash N (2007) Bacterial laccases. World J Microbiol Biotechnol 23:823–832. doi:10.1007/s11274-006-9305-3

Sheth NT, Dave SR (2009) Optimisation for enhanced decolourization and degradation of reactive Red BS C.I. 111 by *Pseudomonas aeruginosa* NGKCTS. Biodegradation 20:827–836. doi:10.1007/s10532-009-9270-2

Singh S, Chatterji S, Nandini PT, Prasad ASA, Rao KVB (2015) Biodegradation of azo dye Direct Orange 16 by *Micrococcus luteus* SSN2. Int J Environ Sci Technol 12:2161–2168. doi:10.1007/s13762-014-0588-x

Solis M, Solis A, Perez HI, Manjarrez N, Flores M (2012) Microbial decolouration of azo dyes: a review. Proc Biochem 47:1723–1748. doi:10.1016/j.procbio.2012.08.014

Telke AA, Kadam AA, Govindwar SP (2015) Bacterial enzymes and their role in decolourization of azo dyes. In: Singh SN (ed) Microbial degradation of synthetic dyes in waste water. Springer, Cham, Switzerland, pp 57–83. doi:10.1007/978-3-319-10942-8_7

Van Der Zee FP (2002) Anaerobic azo dye reduction. Ph.D. Thesis, Wageningen University, Wageningen, Netherlands.

Van Der Zee FP, Villaverde S (2005) Combined anaerobic-aerobic treatment of azo dyes—a short review of bioreactor studies. Water Res 39:1425–1440. doi:10.1016/j.watres.2005.03.007

Wang H, Qiang J, Wei X, Tian Y, Jing X, Ling T (2009) Bacterial decolorization and degradation of the reactive dye Reactive Red. Int Biodeter Biodegr 63:395–399. doi:10.1016/j.ibiod.2008.11.006

Wijetunga S, Xiufen L, Wenquan R, Chen J (2007) Removal mechanisms of acid dyes of different chemical groups under anaerobic mixed culture. Ruhuna J Sci 2:96–110

Wilson K, Walker JM (2010) Principles and techniques of biochemistry and molecular biology. 7th Ed England: Cambridge University Press. 10: 0521731674.

Yonar T (2011) Decolorisation of textile dyeing effluents using advanced oxidation processes. In: Hauser PJ (ed) Advances in treating textile effluent. InTech, Croatia EU, pp 28–48. isbn: 978-953-307-704-8

Zaharia C, Suteu D, Muresan A (2012) Option and solutions for textile effluent decolourization using some specific physico-chemical treatment steps. Environ Eng Manag J 11:493–509

http://envirosystems.net/effluent-treatment-plant/, Accessed on October 20, 2015

http://metacyc.org/

http://planningcommission.gov.in/aboutus/committee/wrkgrp12/wg_chem0203.pdf, Accessed on November 23, 2015.

http://www.genome.jp/kegg/pathway.html

http://www.indiaenvironmentportal.org.in/content/573/united-colours-of-industry/, Accessed on June 18, 2014.

# Biological Significance of Degradation of Polyhydroxyalkanoates

**Subhasree Ray and Vipin Chandra Kalia**

**Abstract** Polyhydroxyalkanoates (PHAs) are biodegradable polymers produced by microbes under nutrient-deprived conditions. These polymers act as food reserves of the organism to survive under adverse environmental conditions. These molecules show high similarity with petroleum-based plastics. Hence, these are promoted as biodegradative alternatives to plastics. PHAs are degraded by the depolymerase enzyme primarily to generate energy for microbial growth. Extracellular degradation occurs on the PHA released from the lysed cell. Various factors are known to influence the degradative process such as humidity, temperature, monomeric composition, etc. The PHA degradation accompanies a decline in the polymer molecular weight and an increase in its crystallinity. Microbes involved in the PHA degradation contribute toward maintenance of ecosystem through the carbon cycle. The by-products of PHA degradation process can be subjected to different biological applications, especially in the energy and medical fields.

**Keywords** Intracellular depolymerase • PHA degradation • Hydroxy acids • Monomers

## 1 Introduction

Polyhydroxyalkanoates (PHAs) represent an interesting alternative to petroleum-derived plastics (Kalia et al. 2003; Reddy et al. 2003; Kumar et al. 2013, 2015a, 2016). Bacterial system can accumulate PHAs—high-molecular-weight biopolymers as storage compounds of carbon and reducing equivalents (Seebach and Fritz 1999). PHAs are also known to be osmotic neutral reservoirs of carbon and energy source. These inclusion bodies may contribute to more than 90% of the cell mass. These molecules can be biochemically synthesized from renewable resources (e.g.,

S. Ray (✉) • V.C. Kalia
Microbial Biotechnology and Genomics, CSIR-Institute of Genomics and Integrative Biology (IGIB), Delhi University Campus, Mall Road, Delhi 110007, India

Academy of Scientific & Innovative Research (AcSIR), New Delhi, India
e-mail: subhasreeray141@gmail.com; vckalia@igib.res.in

 
V.C. Kalia, P. Kumar (eds.), *Microbial Applications Vol.1*,
DOI 10.1007/978-3-319-52666-9_5

biowaste, crude glycerol, etc.) and degrade completely in natural conditions (Jendrossek 2009; Kumar et al. 2009, 2015a, b, c, 2016; Patel et al. 2015a, b, 2016; Kalia et al. 2016). More than 90 bacterial genera are known to incorporate around 150 different hydroxy acid (HA) monomers to the PHA polymers. These polymers have been categorized as short-chain length (scl, $C_3$–$C_5$), medium-chain length (mcl, $C_6$–$C_{14}$), and long-chain length (lcl, $>C_{14}$) PHAs depending on their HA monomers. The PHA biosynthesis is controlled by *phaCAB* operon which governs a three-step conversion process of acetyl-CoA > acetoacetyl-CoA > 3-hydroxybutyryl-CoA. The latter is used by a broad substrate-specific key enzyme—PHA synthase—to produce PHAs (Singh et al. 2009, 2013, 2015; Kumar et al. 2014; Ray and Kalia 2016). All the proteins involved in PHA synthesis such as synthase, phasins, regulatory proteins, and those responsible for its degradation depolymerase are found to be integrated into a dense phospholipid layer, a structure known as "carbonosomes" (Jendrossek 2009; García-Hidalgo et al. 2013).

PHAs are quite attractive and have diverse biotechnological applications. PHAs act as a raw material for a variety of bioactive molecules such as vitamins, pheromones, fatty acids, probiotics, 3-hydroxyalkanoates, etc. However, their most important characteristic is their biodegradable nature. Most studies focus on PHA synthesis; however, it is equally interesting to learn how they are metabolized. Like all biological materials, it is quite easy to envisage the potential factors, which may be instrumental in their degradation, and the most relevant are the enzymes and the physicochemical composition of the substrate (Figs. 1 and 2).

## 2 Biodegradation of PHAs

The PHA biodegradation involves depolymerizing enzymes that convert them to their monomeric constituents, which undergo mineralization. The biological depolymerization process involves both extracellular and intracellular depolymerase enzymes. Intracellular degradation of PHAs involves hydrolysis of the stored material for carbon and energy generation. Extracellular degradation occurs on the polymer released from the cell by lysis. This degradation process depends upon the physicochemical composition of the polymer: stereoregularity, composition, crystallinity, and accessibility (Jendrossek 2001; Jendrossek and Handrick 2002). Under anaerobic conditions, the end products generally include $CO_2$, $H_2O$, $CH_4$/$H_2S$, and microbial biomass. In contrast, under aerobic conditions primary end products include $CO_2$, $H_2O$, and microbial biomass.

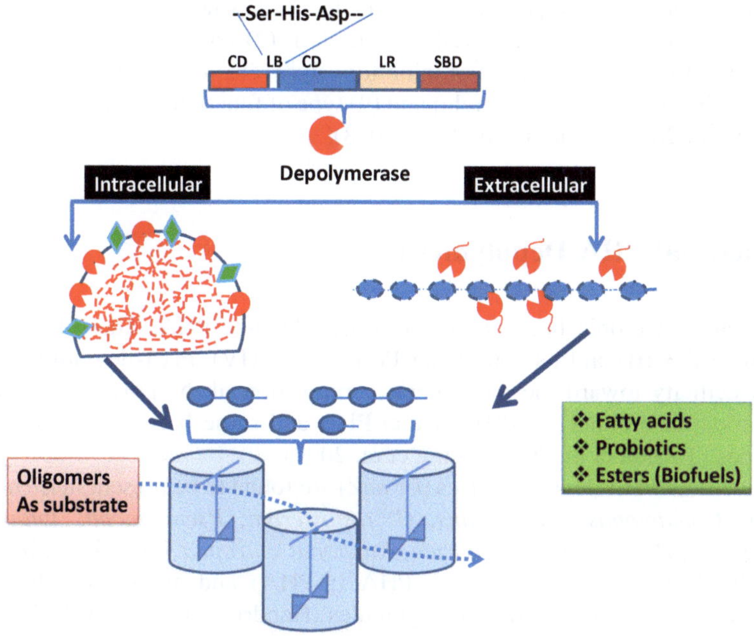


**Fig. 1** PHA depolymerase activities and potential applications

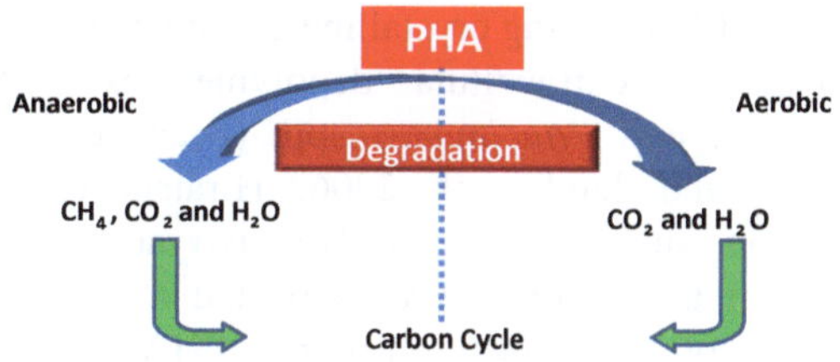


**Fig. 2** General mechanism of biodegradation

## 3 PHA Depolymerase

The PHA-depolymerizing enzymes (*phaZ*) have gained significant attention in the last few decades (Abe et al. 2005). Depolymerases acting on scl-PHAs do not act on mcl-PHAs. Both bacteria and fungi possess these depolymerases, with higher (R) stereospecificity. Depending upon the substrate specificity, extracellular PHA depolymerases are classified into (1) scl-PHA depolymerase and (2) mcl-PHA depolymerase (Jendrossek 1998). *Bacillus megaterium* produces intracellular n-PHB depolymerase, capable of exhibiting extracellular PHB depolymerase activity as well. This unique property makes it more feasible for biodegradation (Cai et al. 2009). Standard methods for evaluating biodegradability have been established (Kalia et al. 2000).

PHA degradation competency is found to be dependent upon the following variables: (1) the environment (water, soil, etc.); (2) temperature, (3) shape, and texture of PHA; (4) the presence of inhibitors, dyes, etc.; (5) mobility; (6) tacticity; (7) crystalline; (8) molecular weight; and (9) type of functional groups present in its structure (Gu 2000; Artham and Doble 2008).

## 4 Bacterial PHA Depolymerases

Aerobic and anaerobic bacteria are known to degrade PHA. These bacteria can metabolize P(3HB) and its copolymer P(3HB-co-3 HV). PHA-degrading bacteria show specificity toward the monomeric composition of the polymer. In general, they can degrade either scl-PHA or mcl-PHA and some bacteria can metabolize both types (Takeda et al. 2000; Wang et al. 2009).

Intracellular PHA depolymerases (*i-PhaZ*) are found to be present in *Alcaligenes faecalis*, *Comamonas acidovorans*, *C. testosteroni*, *Pseudomonas fluorescens*, *P. lemoignei*, *Rhodospirillum rubrum*, *Ralstonia eutropha* H16, *R. pickettii*, etc. The *i-PhaZ* are specific for native PHA (n-PHA) and are usually located as transmembrane proteins on the PHA granules (Handrick et al. 2004; Uchino et al. 2007; Gebauer and Jendrossek 2006). Based on the physical state and location of the polymer, the PHA degradation process could either occur in amorphous state or denatured state. Apparently, the n-PHA gets metabolized by *i-PhaZ* releasing HA monomers, whereas the d-PHA being crystalline gets converted into HA monomers or oligomers by respective extracellular depolymerases (*e-PhaZ*) recorded in *Alcaligenes*, *Acidovorax*, and *Paucimonas* spp. (Braaz et al. 2003; Sugiyama et al. 2004; Gebauer and Jendrossek 2006; Hiraishi et al. 2010). Several PHA-degrading bacteria, such as *Stenotrophomonas* and *Pseudomonas* (among Gram-negative bacteria) and *Streptomyces* spp. and *Rhodococcus equi* (among Gram-positive bacteria), predominantly act on mcl-PHA. The genus *Rhodococcus* has gained importance due to its ability to degrade several hydrophobic substances which include scl-PHA, petroleum hydrocarbon, benzene ring, and PCB (Kobayashi et al. 2004; Kim et al. 2007; Ihssen et al. 2009) (Table 1).

Basically PHA depolymerase enzyme consists of three functional domains (amino acids): (a) a catalytic domain (320–420), (b) a linker region (50–100), and (c) a substrate-binding domain (40–60). The catalytic domain constitutes of a lipase box since they contain lipase-like catalytic triad that includes specific amino acid residues, i.e., serine (S), aspartic acid (D), and histidine (H), along with a pentapeptide signature sequence Gly-Xaa1-Ser-Xaa2-Gly, which is present in serine hydrolase (Jendrossek and Handrick 2002). Serine hydrolase side chain behaves as a nucleophile, which contains oxygen that attacks the ester bond. Here, the imidazole ring of the histidine residue enhances the rate of acidity and reactivity of the oxygen atom, while the carboxylic group of aspartate stabilizes the imidazole ring. Here, the hydroxyl group of the serine hydrolase side chain plays an important role in depolymerization (Jendrossek and Pfeiffer 2014)

**Table 1** Biological applications of PHA deplolymerases

| Origin | Bio-products | Applications | References |
|---|---|---|---|
| *Pseudomonas* sp. | R-hydroxyalkanoic acids | Helps in establishing PHA producers in soil and rhizosphere, and improves metabolism | Eugino et al. (2010) |
| *Pseudomonas fluroscens* | | Drug delivery, protein microarray, protein purification, antibody immobilization in clinical diagnostics | Ihssen et al. (2009) |
| *Pseudomonas putida* CA-3 | Monomers (3-hydroxydecanoic acids; R10) | Anti-proliferative activity | O'Connor et al. (2013) |
| *P. fluoroscens* GK13; *P. putida* KT2442 | 3-Hydroxyalkanoic acids | Potential to inhibit *Staphylococcus aureus* growth (Antibacterial property) | Martinez et al. (2014) |
| | PHACOS | Bactericidal to Gram-positive and Gram-negative bacteria | |
| *Escherichia coli* | | Immobilized cell factories for biocatalysis and bio-transformation, Chaperone protein levels. | Wang et al. (2009) |

(Table 2). PhaZ7 of *P. lemoignei* has considerable similarity to *B. subtilis* lipase—LipA. It has an additional domain. This lid-like domain reveals the presence of many hydrophobic amino acid residues, which includes Tyr105 (Tseng et al. 2006; Shah et al. 2007; Kim et al. 2007; Hermawan and Jendrossek 2010; Eugino et al. 2007, 2010; O'Conner et al. 2013) (Table 2). $PhaZ1_{Rru}$ has a similar sequence with the extracellular *PhaZ* of *Acidovorax* sp. having type II catalytic domain (lipase box) at N-terminus. In *R. rubrum*, it seems that it has an additional nonsoluble PHB depolymerase, comparatively less active than $PhaZ1_{Rru}$ (Kobayashi et al. 2003; Abe et al. 2005) (Table 2).

## 5 Fungal PHA Depolymerases

The process to degrade PHA is not limited to bacteria; many fungi and yeast play a major role in degrading P-3(PHB) and its copolymer P(3HB-co-3 HV) (Kim and Rhee 2003). Fungi are known to be potential candidates for PHA degradation process due to their higher rate of surface growth and high depolymerase activity (Table 2). Fungal *e-PhaZ* is specific for denatured PHA (d-PHA) and is found to be secreted by *Ascomycetes*, *Basidiomycetes*, *Deuteromycetes*, *Mastigomycetes*, *Myxomycetes*, *Zygomycetes*, etc. (Kim et al. 2000; Sang et al. 2002). *Penicillium* spp., *Aspergillus* spp., and *Variovorax* spp. are known to degrade PHB and its copolymer P(3HB-co-3 HV) by extracellular PHB depolymerase (Kim et al. 2000; Han and Kim 2002; Jendrossek and Handrick 2002; Nadhman et al. 2012; Seo et al. 2012) (Table 2). PHB depolymerase of *P. funiculosum* is a trimer having a M.Wt. of

**Table 2** Biochemical characteristics of PHA depolymerases

| Wild type strains | Depolymerase gene and source | MWt (KDa); aa | Active site | Type | pH; °C | References |
|---|---|---|---|---|---|---|
| *Bacteria* | | | | | | |
| *Bacillus thuringiensis* ATCC35646 | *phaZ; Escherichia coli* JM109 | 33; 300 | $GWS_{102}MG$ | IC-MCL | 8;– | Tseng et al. (2006) |
| *Bacillus sp. AF3* | | 37 | | | 7;37 | Shah et al. (2007) |
| *Ralstonia eutropha* H16 | *phaZd1 ; phaZd2* *E. coli* JM109 S17-1 | 39.2;362 | $GMS_{190}AG$ | IC | 8.5;– | (Jendrossek and Pfeiffer 2014) |
| | | 38.4;365 | $GMS_{193}AG$ | IC | 8.5;– | |
| *R. eutropha* | *PhaZ5;PhaZ7* *Pseudomonas lemoignei* | | | IC, EC | 8.5;– | (Jendrossek and Pfeiffer 2014) |
| *R. eutropha* H16 | *PhaZa1 E. coli* | | | IC-MCL | 7;– | Uchino et al. (2008) |
| *R. eutropha* H16 | $PhaZ_{Reu}$ *E.coli* | 78; 1400 | | IC-MCL | 8;– | Sugiyama et al. (2004) |
| *Ralstonia picketti* T1 | | | | EC-SCL | 8;– | |
| *Wautersia eutropha* H16 | $PhaZd_{weu}$ HH16DZD1 | 39 | $S^{190}D^{266}H^{330}$ | IC-MCL | 8.5–9;– 20–30 | Abe et al. (2005) |
| | *E. coli* (PE3ReZd1) | | | | | |
| *Pseudomonas putida* KT2442 | *E. coli* DH5ὰ M15; S17-1; PpAZ1; PpAZ2; Ppaz3 | 31 | $S^{102}H^{248}A^{221}$ | IC-MCL | 8.8;– | De Eugenio et al. (2007) |
| *Acidovorax* sp.SA1 | | | | EC-SCL | | Kobayashi et al. (2003) |
| *Paucimonas lemoignei* | *phaZ5 B. subtilis* WB800 Plasmid Pwb980 | 42.2 | | EC-SCL | 8;– | Braaz et al. (2003) |
| *Streptomyces ascomycinius* | $PhaZ_{sa}$*(fkbU)* *E. coli; Rhodococcus sp.* T104 | 48.4 | Ser $_{131}$-Asp$_{209}$-His$_{269}$ | EC | 6;45 | Hidalgo et al. (2013) |

| | | | | | | |
|---|---|---|---|---|---|---|
| *Rhodobacter spheroides* (ATCC17023) | *PhaZ$_{Rsh}$* (*ZP_00006106*) *E. coli* BLR(DE3)-PlYsS | 46 | $C_{178}A_{351}H_{384}$ | IC-MCL | 8;– | Kobayashi et al. (2004) |
| Bacterial strain HSJCM10698 | *Type 2 dPHB* *E. coli* JM109 PUC19 | 46 | D-x-D | EC | 8;– | Takeda et al. (2000) |
| *Thermus thermophilus* HB8 | | 42 | S-D-(E)-H | EC | 8;– | Uchino et al. (2007) |
| *Rhodospirillum rubrum S1* | ApdA | 17.5 | | IC-MCL | 1–12;– | Handrick et al. (2004) |
| | *PhaZ1 R. rubrum SmRif* | 37 | $S_{42}Asp_{138}His_{178}$ | IC- SCL | 8;– | Abe et al. (2005) |
| *Fungi* | | | | | | |
| *Penicillium simplicissimum* LAR14 | *dPHB* | 36 | | EC | 5;– | (Han and Kim 2002) |
| *Bdellovibrio bacteriovorus* HD100 | *PhaZ$_{BD}$* | 30 | S-H-D | EC-MCL | 10;4–45 | Martinez et al. (2012) |
| *Paucimonas lemoignei* | *PhaZ7* | | | EC-SCL | 9;– | Hermawan and Jendrossek (2010) |
| *Variovorax* sp. DSH1 | | 26.5 | | EC-MCL | | Seo et al. (2012) |
| *Aspergillus* sp.NA-25 | | 57 | | EC | 7;45 | Nadhman et al. (2012) |

*mcl* Medium chain length, *scl* Short chain length, *IC* Intra cellular, *EC* Extra cellular

33 kDa. It contains a catalytic triad which has residues Ser39, Asp121, and His155, respectively.

### *5.1 Biological Significance of PHA Depolymerases*

The by-products generated by the action of PHA depolymerase have found important biological significances as biofuels, fuel additives, and probiotics and in the pharmaceutical industry. (Magdouli et al. 2015).

### *5.2 Biofuels*

The major contributions of these HAs are toward biofuels, where they can be used directly to enhance the fuel efficiency. Besides, the esters of HAs (HAMEs) and (3HBME) with a combustion heat of 20–30 kJ/g were reported to be as good as biodiesel in terms of their combustion energy (Gao et al. 2011; Magdouli et al. 2015). Combustion heats were reduced when these biofuels were added to n-butanol and n-propanol (Chen 2011).

## 6 Medical Applications

### *6.1 Biocontrol Agents*

The usage of antibiotics in animal husbandry has been increasing at an alarming rate. However, it is difficult to envisage the complete elimination of antibiotic from the system. Bacterial infections could be controlled by using short-chain fatty acids (SCFAs), as bacteriostatic agents. These bioactive molecules act by reducing the expression of virulence genes. Thus SCFAs can be exploited as biocontrol agents in animal husbandry. PHA degraded within the gastrointestinal tract has the potential to act as biocontrol agents. scl-PHAs and mcl-PHAs have been proposed to be the bioactive molecules for controlling bacterial infections (Defoirdt et al. 2009).

### *6.2 Antibacterials*

The degradative product of PHAs can be used as building blocks of various chiral compounds and polyesters. 3-(HAs) can be used for the production of macrolides like elaiophylidene, pyrenophorin, grahamimycin A1, and colletodiol. The bioconversion of 3HA to dioxanone amulets is used in the synthesis of different hydroxycarboxylic acids such as 2-alkylated 3HB and β-lactones. During adiposis treatment, 3HB is employed as an orally administered drug (Chen and Wu 2005). Pure monomeric units, 3-hydroxybutyric acid, can be used as a precursor of carbapenem and related antibiotics (Shivakumar et al. 2011). The HAs are employed for preparing biodegradable, implantable rods, which can be used to deliver antibiotics (sulperazone and duocid) in chronic osteomyelitis therapy (Chen and Wu 2005). *Pseudomonas fluorescens* GK13 harbors gene for PhaZGK13 depolymerase to depolymerize PHAs to monomers (HAs). The hydrolyzed products of polymer have greater potential to reduce bacterial infection, specifically those caused by *Staphylococcus aureus* (Martinez et al. 2014).

## 6.3 Drug Carriers

Dendrimers, a novel class of polymers, have been synthesized using 3HB monomers. These dendrimers possess biodegradability, monodispersity, and surface functional moieties, which enable them as promising drug carriers (Chen and Wu 2005). Monomers such as 3HB and 4HB prove effective in preparing novel β- and γ-peptides, which have improved resistance against peptidases and thus live longer in mammalian serum, making it suitable for cargo-drug delivery. 3HB suppresses glycolysis during hemorrhagic shock. Monomers are also used in the synthesis of fragrance (S-citronellol) and sex hormones. Also they possess antibacterial, antiproliferative, and hemolytic properties (Philip et al. 2007). The PHAs may be used to encapsulate cells.

## 6.4 Medical Devices

The HAs can be helpful in making various medical devices due to their biocompatible nature as they possess diverse specifications, mainly degradation rate and mechanical properties. The devices are sutured fastener (reattach tissue to the bone); meniscus repair device (repair of meniscus lesions); rivets and tacks (reattachment of soft tissues); staples and screws (fixation of soft tissues); surgical mesh, repair patch, and adhesion barriers (general surgery); cardiovascular patch (vascular patch grafting); orthopedic pins (bone and soft tissue fixations); stents; articular cartilage repair; etc. (Williams and Martin 2005; Valappil et al. 2006).

## 6.5 Anti-Osteoporosis Effect

A high concentration of 3HB induces ketoacidosis in humans (Tokiwa and Calabia 2007). 3HB oligomers are found to be a good energy substrate for injured patients as they undergo rapid diffusion in peripheral tissues. Also, it plays a major role in preventing brain damage by enhancing cardiac efficiency. It can regenerate mitochondrial energy in the heart. It has been shown that 3HB has potential to cure Parkinson's and Alzheimer's diseases by reducing the death rate of the human neuronal cells (Kashiwaya et al. 2000). (R)-3HB has a positive impact on osteoblasts growth and anti-osteoporosis activity. It enhances calcium deposition and serum alkaline phosphatase activity. It reduces the serum osteocalcin level and also prevents ovariectomization (bone mineral density reduction) (Zhao et al. 2007; Ren et al. 2010; Chen 2011).

## 6.6 Memory Enhancer

A unique property of the HA monomers and their oligomers is their ability to stimulate the $Ca^{2+}$ channels and act as memory enhancers (Cheng et al. 2006; Xiao et al. 2007; Zou et al. 2009; Chen 2011; Magdouli et al. 2015).

## 6.7 Energy Factories

Depolymerase degrades PHA to produce oligomers and monomers. They are further utilized as a part of carbon assimilation machinery. So this hydrolytic enzyme provides the degradative by-product as C source to the predator bacterium in case of the intracellular degradation system. Very likely this enzyme serves as an energy reservoir factory for cells (Martinez et al. 2012).

## 6.8 Bio-Indicator

HAs are used for sensing the pollution and can be utilized as bio-indicators (Foster et al. 2001).

## 7 Conclusion

In summary, PHA depolymerase-producing organisms contribute toward ecosystem maintenance (through carbon cycle) and waste utilization and also provide building blocks for a plethora of industrially important substances. Thus, gaining the knowledge about the mechanism of PHA depolymerase from diverse microorganisms may enable us to exploit it for efficient bioconversion of PHAs.

## 8 Opinion

PHA depolymerase degrades PHA and its copolymers producing oligomers and monomers as their bio-product. (R)-3HB is a known chiral active compound. But the major concern with this polymer is their stinky smell during their production process. Gram-negative bacteria produce PHB along with lipopolysaccharides from their cell wall as a secretory material. Due to endotoxin, PHAs are not favorable for various applications such as medical and other industrial applications. So there is time to look forward for production of (R)-3HB through PHA depolymerization process, since chemical hydrolysis causes breakage of chemical bonds and lowers its quality. Employing depolymerase enzymes to degrade PHA industrial wastes will allow its recycling, leading to the production of value-added by-products—oligomers and monomers. These monomers have the potential to be helpful in PHA polymerization process.

**Acknowledgments** The authors wish to thank the Director of CSIR-Institute of Genomics and Integrative Biology, New Delhi, India, ESC0108 (WUM) and BSC0111 (INDEPTH), for providing the necessary funds, facilities, and moral support. Authors are also thankful to Academy of Scientific and Innovative Research (AcSIR), New Delhi.

## References

Abe T, Kobayashi T, Saito T (2005) Properties of a novel intracellular poly (3-hydroxybutyrate) depolymerase with high specific activity (*PhaZd*) in *Wautersia eutropha* H16. J Bacteriol 187:6982–6990. doi:10.1128/JB.187.20.6982-6990.2005

Artham T, Doble M (2008) Biodegradation of aliphatic and aromatic polycarbonates. Macromol Biosci 8:14–24. doi:10.1002/mabi.200700106

Braaz R, Handrick R, Jendrossek D (2003) Identification and characterisation of the catalytic triad of the alkaliphilic thermotolerant PHA depolymerase *PhaZ7* of *Paucimonas lemoignei*. FEMS Microbiol Lett 224:107–112. doi:10.1016/S0378-1097(03)00425-7

Cai L, Yuan MQ, Liu F, Jian J, Chen GQ (2009) Enhanced production of medium-chain-length polyhydroxyalkanoates (PHA) by PHA depolymerase knockout mutant of *Pseudomonas putida* KT2442. Bioresour Technol 100:2265–2270. doi:10.1016/j.biortech.2008.11.020

Chen GQ (2011) Biofunctionalization of polymers and their applications. In: Nyanhongo GS, Steiner W, Gübitz G (eds) Biofunctionalization of polymers and their applications. Springer, Berlin, pp 29–45. doi:10.1007/10_2010_89

Chen GQ, Wu Q (2005) The application of polyhydroxyalkanoates as tissue engineering materials. Biomaterials 26:6565–6578. doi:10.1016/j.biomaterials.2005.04.036

Cheng S, Chen GQ, Leski M, Zou B, Wang Y, Wu Q (2006) The effect of D, L-β-hydroxybutyric acid on cell death and proliferation in L929 cells. Biomaterials 27:3758–3765. doi:10.1016/j.biomaterials.2006.02.046

De Eugenio LI, Garcia P, Luengo JM, Sanz JM, San Roman J, Garcia JL, Prieto MA (2007) Biochemical evidence that *phaZ* gene encodes a specific intracellular medium chain length polyhydroxyalkanoate depolymerase in *Pseudomonas putida* KT2442 characterization of a paradigmatic enzyme. J Biol Chem 282:4951–4962. doi:10.1074/jbc.M608119200

De Eugenio LI, Escapa IF, Morales V, Dinjaski N, Galan B, Garcia JL, Prieto MA (2010) The turnover of medium-chain-length polyhydroxyalkanoates in *Pseudomonas putida* KT2442 and the fundamental role of *PhaZ* depolymerase for the metabolic balance. Environ Microbiol 12:207–221. doi:10.1111/j.1462-2920.2009.02061.x

Defoirdt T, Boon N, Sorgeloos P, Verstraete W, Bossier P (2009) Short-chain fatty acids and poly-β-hydroxyalkanoates: (New) Biocontrol agents for a sustainable animal production. Biotechnol Adv 27:680–685. doi:10.1016/j.biotechadv.2009.04.026

Foster LJR, Saufi A, Holden PJ (2001) Environmental concentrations of polyhydroxyalkanoates and their potential as bioindicators of pollution. Biotechnol Lett 23:893–898. doi:10.1023/A:1010528229685

Gao X, Chen JC, Wu Q, Chen GQ (2011) Polyhydroxyalkanoates as a source of chemicals, polymers, and biofuels. Curr Opin Biotechnol 22:768–774. doi:10.1128/AEM.01184-06

García-Hidalgo J, Hormigo D, Arroyo M, de la Mata I (2013) Novel extracellular PHB depolymerase from *Streptomyces ascomycinicus*: PHB copolymers degradation in acidic conditions. PLoS One 8:e71699. doi:10.1371/journal.pone.0071699

Gebauer B, Jendrossek D (2006) Assay of poly (3-hydroxybutyrate) depolymerase activity and product determination. Appl Environ Microbiol 72:6094–6100. doi:10.1128/AEM.01184-06

Gu JD (2000) Microbiological deterioration and degradation of synthetic polymeric materials: recent research advances. Int Biodeterior Biodegrad 52:69–91. doi:10.1016/S0964-8305(02)00177-

Han JS, Kim MN (2002) Purification and characterization of extracellular poly (3-hydroxybutyrate) depolymerase from *Penicillium simplicissimum* LAR13. J Microbiol 40:20–25

Handrick R, Reinhardt S, Schultheiss D, Reichart T, Schüler D, Jendrossek V, Jendrossek D (2004) Unraveling the function of the *Rhodospirillum rubrum* activator of polyhydroxybutyrate (PHB) degradation: the activator is a PHB-granule-bound protein (phasin). J Bacteriol 186:2466–2475. doi:10.1128/JB.186.8.2466-2475.2004

Hermawan S, Jendrossek D (2010) Tyrosine 105 of *Paucimonas lemoignei* PHB depolymerase PhaZ7 is essential for polymer binding. Polym Degrad Stab 95:1429–1435. doi:10.1016/j.polymdegradstab.2010.01.002

Hiraishi T, Komiya N, Maeda M (2010) Y443F mutation in the substrate-binding domain of extracellular PHB depolymerase enhances its PHB adsorption and disruption abilities. Polym Degrad Stab 95:1370–1374. doi:10.1016/j.polymdegradstab.2010.01.022

Ihssen J, Magnani D, Thoony-Meyer L, Ren Q (2009) Use of extracellular medium chain length polyhydroxyalkanoate depolymerase for targeted binding of proteins to artificial poly [(3-hydroxyoctanoate)-co-(3-hydroxyhexanoate)] granules. Biomacromolecules 10:1854–1864. doi:10.1021/bm9002859

Jendrossek D (1998) Microbial degradation of polyesters: a review on extracellular poly (hydroxyalkanoic acid) depolymerases. Polym Degrad Stab 59:317–325. doi:10.1016/S0141-3910(97)00190-0

Jendrossek D (2001) Microbial degradation of polyesters. In: Babel W, Steinbüchel A (eds) Biopolyesters. Springer, Berlin, pp 293–325. doi:10.1007/3-540-40021-4_10

Jendrossek D (2009) Polyhydroxyalkanoate granules are complex subcellular organelles (carbonosomes). J Bacteriol 191(10):3195–3202

Jendrossek D, Handrick R (2002) Microbial degradation of polyhydroxyalkanoates. Ann Rev Microbiol 56:403–432. doi:10.1146/annurev.micro.56.012302.160838

Jendrossek D, Pfeiffer D (2014) New insights in the formation of polyhydroxyalkanoate granules (carbonosomes) and novel functions of poly (3-hydroxybutyrate). Environ Microbiol 16:2357–2373. doi:10.1111/1462-2920.12356

Kalia VC, Raizada N, Sonakya V (2000) Bioplastics. J Sci Ind Res 59:433–445

Kalia VC, Chauhan A, Bhattacharyya G, Rashmi (2003) Genomic databases yield novel bioplastic producers. Nat Biotechnol 21:845–846. doi: 10.1038/nbt0803-845

Kalia VC, Prakash J, Koul S (2016) Biorefinery for glycerol rich biodiesel industry waste. Indian J Microbiol 56:113–125. doi:10.1007/s12088-016-0583-7

Kashiwaya Y, Takeshima T, Mori N, Nakashima K, Clarke K, Veech RL (2000) d-β-Hydroxybutyrate protects neurons in models of Alzheimer's and Parkinson's disease. Proc Natl Acad Sci USA 97:5440–5444. doi:10.1073/pnas.97.10.5440

Kim DY, Rhee YH (2003) Biodegradation of microbial and synthetic polyesters by fungi. Appl Microbiol Biotechnol 61:300–308. doi:10.1007/s00253-002-1205-3

Kim H, Ju HS, Kim J (2000) Characterization of an extracellular poly (3-hydroxy-5-phenylvalerate) depolymerase from *Xanthomonas* sp. JS02. Appl Microbiol Biotechnol 53:323–327. doi:10.1007/s002530050028

Kim DY, Kim HW, Chung MG, Rhee YH (2007) Biosynthesis, modification, and biodegradation of bacterial medium-chain-length polyhydroxyalkanoates. J Microbiol 45:87–97

Kobayashi T, Shiraki M, Abe T, Sugiyama A, Saito T (2003) Purification and properties of an intracellular 3-hydroxybutyrate-oligomer hydrolase (*PhaZ2*) in *Ralstonia eutropha* H16 and its identification as a novel intracellular poly (3-hydroxybutyrate) depolymerase. J Bacteriol 185:3485–3490. doi:10.1128/JB.185.12.3485-3490.2003

Kobayashi T, Nishikori K, Saito T (2004) Properties of an intracellular poly (3-hydroxybutyrate) depolymerase (PhaZ1) from *Rhodobacter sphaeroides*. Curr Microbiol 49:199–202. doi:10.1007/s00284-004-4339-y

Kumar T, Singh M, Purohit HJ, Kalia VC (2009) Potential of *Bacillus* sp. to produce polyhydroxybutyrate from biowaste. J Appl Microbiol 106:2017–2023. doi:10.1111/j.1365-2672.2009.04160.x

Kumar P, Patel SKS, Lee JK, Kalia VC (2013) Extending the limits of *Bacillus* for novel biotechnological applications. Biotechnol Adv 31:1543–1561. doi:10.1016/j.biotechadv.2013.08.007

Kumar P, Singh M, Mehariya S, Patel SKS, Lee JK, Kalia VC (2014) Ecobiotechnological approach for exploiting the abilities of *Bacillus* to produce co-polymer of polyhydroxyalkanoate. Indian J Microbiol 54:1–7. doi:10.1007/s12088-014-0457-9

Kumar P, Mehariya S, Ray S, Mishra A, Kalia VC (2015a) Biodiesel industry waste: a potential source of bioenergy and biopolymers. Indian J Microbiol 55:1–7. doi:10.1007/s12088-014-0509-1

Kumar P, Mehariya S, Ray S, Mishra A, Kalia VC (2015b) Biotechnology in aid of biodiesel industry effluent (glycerol): biofuels and bioplastics. In: Kalia VC (ed) Microbial factories. Springer, New Delhi, pp 105–119. doi:10.1007/978-81-322-2598-0

Kumar P, Ray S, Patel SKS, Lee JK, Kalia VC (2015c) Bioconversion of crude glycerol to polyhydroxyalkanoate by *Bacillus thuringiensis* under non-limiting nitrogen conditions. Int J Biol Macromol 78:9–16. doi:10.1016/j.ijbiomac.2015.03.046

Kumar P, Ray S, Kalia VC (2016) Production of co-polymers of polyhydroxyalkanoates by regulating the hydrolysis of biowastes. Bioresour Technol 200:413–419. doi:10.1016/j.biortech.2015.10.045

Magdouli S, Brar SK, Blais JF, Tyagi RD (2015) How to direct the fatty acid biosynthesis towards polyhydroxyalkanoates production? Biomass Bioenergy 74:268–279. doi:10.1016/j.biombioe.2014.12.017

Martínez V, de la Peña F, García-Hidalgo J, de la Mata I, García JL, Prieto MA (2012) Identification and biochemical evidence of a medium-chain-length polyhydroxyalkanoate depolymerase in the *Bdellovibrio bacteriovorus* predatory hydrolytic arsenal. Appl Environ Microbiol 78:6017–6026. doi:10.1128/AEM.01099-12

Martínez V, Dinjaski N, de Eugenio LI, de la Peña F, Prieto MA (2014) Cell system engineering to produce extracellular polyhydroxyalkanoate depolymerase with targeted applications. Int J Biol Macromol 71:28–33. doi:10.1016/j.ijbiomac.2014.04.013

Nadhman A, Hasan F, Shah Z, Hameed A, Shah AA (2012) Production of poly (3-hydroxybutyrate-co-3-hydroxyvalerate) depolymerase from Aspergillus sp. NA-25. Appl Biochem Microbiol 48:482–487. doi:10.1134/S0003683812050080

O'Connor S, Szwej E, Nikodinovic-Runic J, O'Connor A, Byrne AT, Devocelle M, O'Donovan N, Gallagher WM, Babu R, Kenny ST, Zinn M (2013) The anti-cancer activity of a cationic antimicrobial peptide derived from monomers of polyhydroxyalkanoate. Biomaterials 34:2710–2718. doi:10.1016/j.biomaterials.2012.12.032

Patel SKS, Kumar P, Singh S, Lee JK, Kalia VC (2015a) Integrative approach for hydrogen and polyhydroxybutyrate production. In: Kalia VC (ed) Microbial factories waste treatment. Springer, New Delhi, pp 73–85. doi:10.1007/978-81-322-2598-0_5

Patel SKS, Kumar P, Singh S, Lee JK, Kalia VC (2015b) Integrative approach to produce hydrogen and polyhydroxybutyrate from biowaste using defined bacterial cultures. Bioresour Technol 176:136–141. doi:10.1016/j.biortech.2014.11.029

Patel SKS, Lee JK, Kalia VC (2016) Integrative approach for producing hydrogen and polyhydroxyalkanoate from mixed wastes of biological origin. Indian J Microbiol 56:293–300. doi:10.1007/s12088-016-0595-3

Philip S, Keshavarz T, Roy I (2007) Polyhydroxyalkanoates: biodegradable polymers with a range of applications. J Chem Technol Biotechnol 82:233–247. doi:10.1002/jctb.1667

Ray S, Kalia VC (2016) Microbial co-metabolism and Polyhydroxyalkanoate co-polymers. Indian J Microbiol 55:1–9. doi:10.1007/s12088-016-0622-4

Reddy CSK, Ghai R, Rashmi KVC (2003) Polyhydroxyalkanoates: an overview. Bioresour Technol 87:137–146. doi:10.1016/S0960-8524(02)00212-2

Ren Q, de Roo G, Witholt B, Zinn M, Thony-Meyer L (2010) Influence of growth stage on activities of polyhydroxyalkanoate (PHA) polymerase and PHA depolymerase in Pseudomonas putida U. BMC Microbial 10:1. doi:10.1186/1471-2180-10-254

Sang BI, Hori K, Tanji Y, Unno H (2002) Fungal contribution to in situ biodegradation of poly (3-hydroxybutyrate-co-3-hydroxyvalerate) film in soil. Appl Microbiol Biotechnol 58:241–247. doi:10.1007/s00253-001-0884-5

Seebach D, Fritz MG (1999) Detection, synthesis, structure, and function of oligo (3-hydroxyalkanoates): contributions by synthetic organic chemists. Int J Biol Macromol 25:217–236

Seo BS, Kim DY, Ni YY, Son KH, Park HY, Rhee YH (2012) Non-ionic polysorbate surfactants: alternative inducers of medium-chain-length poly (3-hydroxyalkanoates) (MCL-PHAs) for production of extracellular MCL-PHA depolymerases. Bioresour Technol 121:47–53. doi:10.1016/j.biortech.2012.06.118

Shah AA, Hasan F, Hameed A, Ahmed S (2007) Isolation and characterization of poly (3-hydroxybutyrate-co-3-hydroxyvalerate) degrading bacteria and purification of PHBV depolymerase from newly isolated *Bacillus* sp. AF3. Int Biodeter Biodegrad 60:109–115. doi:10.1016/j.ibiod.2007.01.004

Shivakumar S, Jagadish SJ, Zatakia H, Dutta J (2011) Purification, characterization and kinetic studies of a novel poly (β) hydroxybutyrate (PHB) depolymerase *PhaZ* from *Penicillium citrinum* S2. Appl Biochem Biotechnol 164:1225–1236. doi:10.1007/s12010-011-9208-0

Singh M, Patel SKS, Kalia VC (2009) *Bacillus subtilis* as potential producer for polyhydroxyalkanoates. Microb Cell Fact 8:38. doi:10.1186/1475-2859-8-38

Singh M, Kumar P, Patel SKS, Kalia VC (2013) Production of polyhydroxyalkanoate co-polymer by *Bacillus thuringiensis*. Indian J Microbiol 53:77–83. doi:10.1007/s12088-012-0294-7

Singh M, Kumar P, Ray S, Kalia VC (2015) Challenges and opportunities for the customizing polyhydroxyalkanoates. Indian J Microbiol 55:235–249. doi:10.1007/s12088-015-0528-6

Sugiyama A, Kobayashi T, Shiraki M, Saito T (2004) Roles of poly (3-hydroxybutyrate) depolymerase and 3HB-oligomer hydrolase in bacterial PHB metabolism. Curr Microbiol 48:424–427. doi:10.1007/s00284-003-4227-x

Takeda M, Kitashima K, Adachi K, Hanaoka Y, Suzuki I, Koizumi JI (2000) Cloning and expression of the gene encoding thermostable poly (3-hydroxybutyrate) depolymerase. J Biosci Bioeng 90:416–421. doi:10.1016/S1389-1723(01)80011-6

Tokiwa Y, Calabia BP (2007) Biodegradability and biodegradation of polyesters. J Polym Environ 15:259–267. doi:10.1007/s10924-007-0066-3

Tseng CL, Chen HJ, Shaw GC (2006) Identification and characterization of the *Bacillus thuringiensis phaZ* gene, encoding new intracellular poly-3-hydroxybutyrate depolymerase. J Bacteriol 188:7592–7599. doi:10.1128/JB.00729-06

Uchino K, Saito T, Gebauer B, Jendrossek D (2007) Isolated poly (3-hydroxybutyrate) (PHB) granules are complex bacterial organelles catalyzing formation of PHB from acetyl coenzyme A (CoA) and degradation of PHB to acetyl-CoA. J Bacteriol 189:8250–8256. doi:10.1128/JB.00752-07

Uchino K, Saito T, Jendrossek D (2008) Poly (3-hydroxybutyrate) (PHB) depolymerase PhaZa1 is involved in mobilization of accumulated PHB in *Ralstonia eutropha* H16. App Environ Microbiol 74:1058–1063. doi:10.1128/AEM.02342-07

Valappil SP, Misra SK, Boccaccini AR, Roy I (2006) Biomedical applications of polyhydroxyalkanoates, an overview of animal testing and *in vivo* responses. Expert Rev Med Devices 3:853–868. doi:10.1586/17434440.3.6.853

Wang Q, Yu H, Xia Y, Kang Z, Qi Q (2009) Complete PHB mobilization in *Escherichia coli* enhances the stress tolerance: a potential biotechnological application. Microb Cell Fact 8:1. doi:10.1186/1475-2859-8-47

Williams SF, Martin DP (2005) Applications of polyhydroxyalkanoates (PHA) in medicine and pharmacy. Biopolym Online. doi:10.1002/3527600035.bpol4004

Xiao XQ, Zhao Y, Chen GQ (2007) The effect of 3-hydroxybutyrate and its derivatives on the growth of glial cells. Biomaterials 28:3608–3616. doi:10.1016/j.biomaterials.2007.04.046

Zhao YH, Li HM, Qin LF, Wang HH, Chen GQ (2007) Disruption of the polyhydroxyalkanoate synthase gene in *Aeromonas hydrophila* reduces its survival ability under stress conditions. FEMS Microbiol Lett 276:34–41. doi:10.1111/j.1574-6968.2007.00904.x

Zou XH, Li HM, Wang S, Leski M, Yao YC, Yang XD, Huang QJ, Chen GQ (2009) The effect of 3-hydroxybutyrate methyl ester on learning and memory in mice. Biomaterials 30:1532–1541. doi:10.1016/j.biomaterials.2008.12.012

# Microbial Biofouling: A Possible Solution to Treat Harmful Microorganisms in Ship Ballast Water

**Bhagwan Rekadwad and Chandrahasya Khobragade**

**Abstract** Ships carry huge quantities of seawater in a specially designed ballast water tank required to sink in the sea. While loading cargoes, stored ballast water is discharged from ballast water tanks and vice versa. This compulsory exchange of ballast water is taking place for the past many decades. The movement of ballast water results in extensive transport and exchange of billions tons of national water in a global ocean along with many animals, plants, and microorganisms (especially pathogenic microorganisms such as bacteria, fungi, and pathogenic yeast). Ballast water exchange and transport results in invasion of unnecessary species into new environment. They accommodate into a new environment and utilize every available natural resource and become dominant. The complete eradication and removal of dominating pathogenic microorganisms is becoming a close to impossible task. These pathogenic species can be removed or their growth can be suppressed using halophilic antibiotic producers, halophilic bacterial bionts from marine invertebrates, and the use of mobile genome (host-specific bacteriophages/viruses). Only such natural methods will eradicate the pathogenic microorganisms from ballast water tank and will clean it. Moreover, the use of antipathogenic microorganisms checks microbial biofouling/invasion of new species through eco-friendly means.

**Keywords** Ballast water • Biofouling • Microbial invasion • National water • Open ocean

## 1 Introduction

Biofouling is defined as the unwanted and undesirable accumulation of a biotic deposit on man-made surfaces such as on the ship external jacket, in ballast water tanks, and in pipes and pipe-like structures. Biofilm-forming microorganisms such as pathogenic and nonpathogenic bacteria, fungi, and algae and biofouling by organisms like hydroids, barnacles, tubeworms, and bivalves on submerged

B. Rekadwad (✉) • C. Khobragade
School of Life Sciences, Swami Ramanand Teerth Marathwada University, Nanded 431606, India
e-mail: rekadwad@gmail.com



V.C. Kalia, P. Kumar (eds.), *Microbial Applications Vol.1*,
DOI 10.1007/978-3-319-52666-9_6

surfaces and in ship ballast tank pose a serious problem (Oliver 2010; Flemming and IWW 2011). The biofilm-coated submerged surfaces, pipes, and pipe-like parts of a ship in an aquatic milieu have harmful bacteria, microalgae, and pathogenic protozoan species called "disease hive" which can be removed using physical and chemical means. But one can't deny the risk of aquatic microbial invasions in ship ballast water tanks in the form of pathogenic biofilms (Drake et al. 2005). World common harbors are the places where ship has to be ballasted/de-ballasted. Through this routine and compulsory process, almost all ships acquired and contain most similar population of dominated and detrimental biofilm at the base, submerged in ballast tanks, or as deposited cover or sediments in a corner and pipes. This indicates that one ship is sufficient to seed other waters (Baier et al. 2014). Since many decades (nineteenth century), ship has used and exchanged ballast water at the common harbors and during succession. Ship safety and stability is the only reason for the exchange of ballast water at the harbors and in the Open Ocean. The causes of exchange may be one of the following: ballast water required for propulsion of ships, to acquire weight of lost and/or consumed fuel or water, and to sink in the ocean. On the other hand, from the same time period ship already exchanged a huge quantity of National water in shipping routes through their ballast water tanks. This activity of ballast water exchange results in biological invasions of many aquatic biota and pathogenic microbiota, which feeds on aquatic organisms or causes harm to human populations. These include toxic dinoflagellates, *Vibrio cholerae*, and numerous heterotrophic aerobic bacteria (HPC), etc. (Drake et al. 2007). This chapter describes a possible solution to treat and eradicate harmful microorganisms in the ship ballast water tank.

## 2 Halophilic Antibiotic Producers

Polyextremophilic microbes thrive under new extremes such as high/low temperature, high pressure/salt concentration, low quantity of nutrients, oxygen level below required level, and presence of adverse chemicals, etc., and have to compete with other microorganisms and enemies for their survival (van der Wielen et al. 2005; DasSarma and DasSarma 2012). These microbes secrete antibiotics and antibiotic-like compounds having inhibition potential which is necessary to maintain their constant growth and to fight infections (Kalia 2014a). Halophilic organisms have the potential to produce a variety of antibiotics (Oren 2010; Ghosh et al. 2010; Kalia 2013) and antimicrobial pigments (Venil et al. 2013). The antibiotics and pigments produced by a certain group of halophilic microorganisms such as *Haloarcula, Halobacterium, Halococcus, Haloferax, Halogeometricum, Halorubrum*, and *Haloterrigena* species have potential to kill or inhibit the different pathogenic bacteria (*Bacillus subtilis, Escherichia coli, Vibrio cholerae, Klebsiella pneumoniae, Pseudomonas aeruginosa, Staphylococcus aureus, Salmonella typhi*), fungi (*Aspergillus niger, Penicillium species*), *and* Yeast (*Candia albicans).* Three identified strains of halophilic microorganism possess pVC1, pVC2, pVC3, and megaplasmids (Schwartz 2009). Other plasmids that have plasmids include *C. israelensis* (pH 11, 48 kb size), *Halomonas subglaciescola* (pHS1—70 kb size) (Huma et al. 2011), and *H. elongate* (possesses pMH1 plasmid) (Todkar et al. 2012). The organic extracts of some marine bacteria (Selvin et al. 2009; Aguila-Ramírez et al. 2014; Kalia and Kumar 2015) associated with the sponge *Aplysina gerardogreeni* belong to the genera *Bacillus, Micrococcus, Paracoccus, Pseudobacter, Pseudovibrio, Psychrobacter, Staphylococcus,* and *Terribacillus* showing antifouling (AF) activity (Plouguerné et al. 2008; Chambers et al. 2011; Thabard et al. 2009; Plouguerné et al. 2010; Kumar et al. 2015). These diverse microbes have roles in ballast water treatment.

## 3 Use of Halophilic Bacterial Bionts from Marine Invertebrates for Ballast Water Treatment

The organisms in marine ecosystem including invertebrates such as bryozoans (moss animals) (Selvin et al. 2009), soft corals, sea fans and sea whips (coelenterates), nudibranchs and sea hares (mollusks), tunicates, echinoderms (starfish, sea cucumbers), and sponges (Selvin et al. 2009) are the biotechnologically as well as pharmacological important target groups having bacterial species (bacterial bionts). These produced compounds have structurally unique and effective secondary metabolites (Kijjoa and Sawangwong 2004). These natural products comprise ranges of chemical classes such as alkaloids, acetogenins, peptides, polyketides, and terpenes having pronounced pharmacological potential (Rodrigues et al. 2004; Wang et al. 2010).

In the limelight of available reports, bacteria isolated from bivalves, corals, and marine sponges of the Indian Ocean region have produced effective antibacterial compounds. It is reported that 100 heterotrophic halophilic bacteria species are isolated from nine sponges, five corals, and one bivalve species (Gnanambal et al. 2005; Anand et al. 2006; Chandran et al. 2009; Donio et al. 2013).

The crude broth (centrifuged broth supernatant) of the 46 isolated bacteria showed inhibitory activity (Kalia et al. 2011, 2014a, b; Kalia 2013) against pathogenic bacteria (*Aerobactor aerogenes, E. coli, Staphylococcus citreus, Proteus vulgaris, S. typhi,* and *Serratio marcescens*). Furthermore, obtained results have been confirmed in the presence of bioactive compounds by partially purified extracts (ethyl acetate extract) of crude supernatant. Surprisingly, highest numbers of bacterial bionts having pharmacologically important products (Tokunaga et al. 2008, 2010) show inhibitory activity against harmful human pathogens. Here, we mention that the bacterial bionts isolated are the richest producers of bioactive compounds. The diversity of marine organisms and the highly competitive, environmental habitats in which access to space and nutrients is limited are responsible for this stunning variety (Velho Pereira and Furtado 2012; Kalia 2014a, b, 2015; Kumar et al. 2013, 2015) and capability to flourish; said adverse habitats can be used to treat ballast water by incorporating them into the ballast water tank. These microorganisms may be used for scaling up from benchtop to pilot plant scales for the production of antibiotics (Rezanka et al. 2010).

## 4 Use of Mobile Genome (Host-Specific Bacteriophages/Virus) to Control Microbial Biofouling

"The host specific mobile genome showing host killing activity is known as Viruses". Viruses are the submicroscopic organisms that specifically attack only their host and timely kill them.

Viruses occupy the twilight zone that separates the "living" from the "non-living." They do not have a cellular organization and contain DNA/RNA either in single-stranded or double-stranded form but never both. They use the host cellular mode of replication system and hijack host synthesis machinery for their multiplication. They utilize host metabolic processes, product for production of new progeny virions either through lytic or lysogenic cycle. They can be classified based on their host such as animal viruses, plant viruses, bacteria viruses, archaeal viruses, algal viruses, and fungal viruses (Rohwer and Breitbart 2005; La Scola et al. 2008).

Ballast water is the source of many invasive animal, plant, algae, and pathogenic bacteria and fungi. These include bacteria (*A. aerogenes, B. subtilis, E. coli, K. pneumoniae, P. vulgaris, P. aeruginosa, S. aureus, S. citreus, S. marcescens, S. typhi, V. cholera*), fungi (*A. niger*, *Penicillium* species), and yeast (*C. albicans*). Each of these microorganisms is capable of infecting ballast water tanks and causing serious infections in human beings in the place where de-ballasted. Likely, each of these pathogenic organisms has their natural enemies in the form of bacteriophages and viruses (Table 1).

Effective control measures should be adopted to treat ballast water. The mobile genomes play a very important role to treat ballast water pathogens. The concentrated and active viruses can be used to treat ballast water. For this the colder water-soluble gelatin will be used to prepare Gelatin virus balls (GVB) and Liquid virus filled and sealed gelatin capsules (LVFSGC) (Yu et al. 2008). After loading of ballast water into ballast water tank, GVB and LVFSGC can be released into the ballast water tank (s) and will be slowly dispersed. The movement of ship causes uniform mixing and liberation and dispersion of viruses from GVB and LVFSGC will attach to their host. Ultimately, the disease-causing agent may be killed and the pathogenic infection will be removed from the ballast water tank without any harm to the environment (Freshney 2010; Atlas 2010; Bosch et al. 2015). All these processes should be done before being the ballast water discharged. Hence, the GVB and LVFSGC will be used for the release of viruses which systematically kill disease-causing agents.

**Table 1** Bacteriophages and viruses of pathogenic bacteria inhabit ballast water

| Microorganism | Viruses/Bacteriophages | References |
|---|---|---|
| *Staphylococcus citreus* | CoNS phages | Deghorain and Van Melderen (2012), Xia and Wolz (2014) |
| *Staphylococcus aureus* | φ11, φ80, and φ80α of serogroup B, φ81 | Mašlaňová et al. (2013) |
| *Proteus vulgaris* | Phage 4a, 4b, 34a, 26a,36b, 21b, 21c, Fr2, Fr5 | Schmidt and Jeffries (1974) |
| *Serratio marcescens* | KSP20, KSP90, KSP100, P2-like phage, T4-type phage, and phiEco32 phage, wIF3, SM701 | Petty et al. (2006), Yu et al. (2008), and Matsushita et al. (2009) |
| *Salmonella typhi* | E1 ($n = 8$), UVS ($n = 5$), E1, *A* phage, biotype I, øSH19, Vi01-like phage family, Vi-phage-type, DT7a | Lalko and Gunnel (1967), NIID (1998), Goode et al. (2003), Atterbury et al. (2007), Trung et al. (2007), De Lappe et al. (2009), Pickard et al. (2010), Wall et al. (2010), Hooton et al. (2011), Lettini et al. (2014), Srirangara et al. (2015) |
| *Aerobacter aerogenes* | phi-mp | Souza et al. (1972) |
| *Escherichia coli* | T4 and T4-like phages | Brüssow (2005) |
| *Vibrio cholerae* | Vibriophage | Jensen et al. (2006), Bosch et al. (2015) |
| *Klebsiella pneumoniae* | Phage SS, *Myoviridae* phage (KP15 and KP27), *Siphoviridae* phage (KP16 and KP36), *Podoviridae phage,* T7-like phages (KP32) | Chhibber et al. (2008), Kęsik-Szeloch et al. (2013) |
| *Pseudomonas aeruginosa* | phi6 and phiKMV | Bosch et al. (2015) |
| *Bacillus subtilis* | SP5, SP6, SP7, SP8, SP9, SP13, SP3, SP10, PBS1, SP alpha, SP beta | Brodetsky and Romig (1965), Krasowska et al. (2015) |
| *Ostreococcus* | Prasinoviruses, OtV-2 | Weynberg et al. (2011), Clerissi et al. (2012) |
| *Aspergillus niger* | Mycovirus, CSP, Virus-like particle (hepatitis B) | Plüddemann and Van Xyl (2003), Refos et al. (2013) |
| *Penicillium* sp. | PsV-S, PsV-F, AfV-S, AfV-F | Lemke and Nash (1974), Border et al. (1972) |
| *Candida albicans* | Hybrid phage displaying the Sap epitope VKYTS | Yang et al. (2007) |

# 5 Perspectives

The USA and other countries including India, China, and Africa having Ocean regime proposed and adopted strict International rules to control or to reduce risks of microbial invasions. The current ballast water treatment measures do not have effective control over the ship ballast water microbial invasions. Moreover, additional care must be taken when using microbial anti-biofoulers to treat harmful infections/microorganisms occupying space in the ballast water tank. It is an acceptable method because the use of halophilic antibiotics and pigment producers, halophilic bacterial bionts, and mobile genome (s) (GVB and LVFSGC) could be the most cost-effective, eco-friendly, and alternative biotherapy that definitely achieves an acceptable hygiene (cleanliness).

**Acknowledgments** The authors wish to thank the Registrar of Swami Ramanand Teerth Marathwada University, Nanded, for providing all available necessary facilities. BNR is thankful to University Grants Commission, New Delhi, India, for granting Postdoctoral Fellowship. BNR is grateful to Dr. G. Dharni, Scientist E, National Institute of Ocean Technology, Chennai (TN), India, for his valuable guidance. BNR is also thankful to Dr. Satheesh C. Shennoi, Director, ESSO-Indian National Information Centre for Ocean Information (ESSO-INCOIS), Hyderabad (India), and Dr. N. Ramaiah, Chief Scientist, National Institute of Oceanography (NIO), Goa, India, for giving a chance to become a part of the International Training Course on "Biological Oceanographic Processes" organized by the International Training Centre for Operational Oceanography (ITCOocean), INCOIS, Hyderabad, and CSIR-NIO, Goa (India).

## References

Aguila-Ramírez RN, Hernández-Guerrero CJ, González-Acosta B, Id-Daoud G, Hewitt S, Pope J, Hellio C (2014) Antifouling activity of symbiotic bacteria from sponge *Aplysina gerardogreeni*. Int Biodeterior Biodegrad 90:64–70. doi:10.1016/j.ibiod.2014.02.003

Anand TP, Bhat AW, Shouche YS, Roy U, Siddharth J, Sarma SP (2006) Antimicrobial activity of marine bacteria associated with sponges from the waters off the coast of South East India. Microbiol Res 161:252–262. doi:10.1016/j.micres.2005.09.002

Atlas RM (2010) Handbook of microbiological media, 4th edn. University of Louisville, CRC Press, Florence, KY

Atterbury RJ, Van Bergen MA, Ortiz F, Lovell MA, Harris JA, De Boer A, Wagenaar JA, Allen VM, Barrow PA (2007) Bacteriophage therapy to reduce *salmonella* colonization of broiler chickens. Appl Environ Microbiol 73:4543–4549. doi:10.1128/AEM.00049-07

Baier RE, Lundquist DC, Forsberg RL, Meyer AE (2014) Suppressing biodiversity in the world's waterbodies: ballast biofilms are the dental plaque of the oceans. J Biodivers Biopros Dev 1:1000109. doi:10.4172/2376-0214.1000109

Border DJ, Buck KW, Chain EB, Kempson-Jones GF, Lhoas P, Ratti G (1972) Viruses of *Penicillium* and *Aspergillus* species. Biochem J 127:4P–6P

Bosch TCG, Grasis JA, Lachnit T (2015) Microbial ecology in *Hydra*: why viruses matter. J Microbiol 53:193–200. doi:10.1007/s12275-015-4695-2

Brodetsky AR, Romig WR (1965) Characterization of *Bacillus subtilis* bacteriophages. J Bacteriol 90:1655–1663

Brüssow H (2005) Phage therapy: the *Escherichia coli* experience. Microbiol 151:2133–2140. doi:10.1099/mic.0.27849-0

Chambers L, Hellio C, Stokes K, Dennington S, Goodes L, Wood R, Walsh F (2011) Investigation of *Chondus crispus* as a potential source of new antifouling agents. Int Biodeterior Biodegrad 65:939–946. doi:10.1016/j.ibiod.2011.07.002

Chandran B, Rameshwaran G, Ravichandran S (2009) Antimicrobial activity from the gill extraction of *Perna viridis* (Linnaeus, 1758). Glob J Biotech Biochem 4:88–92

Chhibber S, Kaur S, Kumari S (2008) Therapeutic potential of bacteriophage in treating *Klebsiella pneumoniae* B5055-mediated lobar pneumonia in mice. J Med Microbiol 57:1508–1513. doi:10.1099/jmm.0.2008/002873-0

Clerissi C, Desdevises Y, Grimsley N (2012) Prasinoviruses of the marine green alga *Ostreococcus tauri* are mainly species specific. J Virol 86:4611–4619. doi:10.1128/JVI.07221-11

DasSarma S, DasSarma P (2012) Halophiles. In: eLS. Wiley, Chichester. doi:10.1002/9780470015902.a0000394.pub3

De Lappe N, Doran G, O'Connor J, O'Hare C, Cormican M (2009) Characterization of bacteriophages used in the *Salmonella enterica* serovar *Enteritidis* phage-typing scheme. J Med Microbiol 58:86–93. doi:10.1099/jmm.0.000034-0

Deghorain M, Van Melderen L (2012) The Staphylococci phages family: an overview. Viruses 4:3316–3335. doi:10.3390/v4123316

Donio MBS, Ronica FA, Viji VT, Velmurugan S, Jenifer JSCS, Michaelbabu M, Dhar P, Citarasu T (2013) *Halomonas* sp. BS4, A biosurfactant producing halophilic bacterium isolated from solar salt works in India and their biomedical importance. SpringerPlus 2:149. doi:10.1186/2193-1801-2-149

Drake LA, Meyer AE, Forsberg RL, Baier RE, Doblin MA, Heinemann S, Johnson WP, KochM RPA, Dobbs FC (2005) Potential invasion of microorganisms and pathogens via "interior hull fouling": biofilms inside ballast-water tanks. Biol Invasions 7:969–982. doi:10.1007/s10530-004-3001-8

Drake LA, Doblin MA, Dobbs FC (2007) Potential microbial bioinvasions via ships' ballast water, sediment, and biofilm. Mar Pollut Bull 55:333–341. doi:10.1016/j.marpolbul.2006.11.007

Flemming HC, IWW (2011) Microbial biofouling: unsolved problems, insufficient approaches, and possible solutions. In: Flemming HC, Wingender J, Szewzyk U (eds) Biofilm highlights. Springer series on biofilms 5. Springer, Berlin. doi:10.1007/978-3-642-19940-0_5

Freshney RI (2010) Culture of animal cells: a manual of basic technique and specialized application, 6th edn. Wiley, Hoboken, NJ. doi:10.1002/9780470649367

Ghosh R, Chattopadhay PK, Chattopadhay B, Pal D (2010) Antibiotic resistance profiles of halophilic microorganisms isolated form tannery effluents. Indian J Biotechnol 9:80–86

Gnanambal KM, Chellaram C, Patterson J (2005) Isolation of antagonistic bacteria from the surface of the Gorgorian corals at Tuticorin, South East coast of India. Indian J Mar Sci 34:316–319

Goode D, Allen VM, Barrow PA (2003) Reduction of experimental *Salmonella* and *Campylobacter* contamination of chicken skin by application of lytic bacteriophages. Appl Environ Microbiol 69:5032–5036. doi:10.1128/AEM.69.8.5032-5036.2003

Hooton SPT, Timms AR, Rowsell J, Wilson R, Connert IF (2011) *Salmonella typhimurium*-specific bacteriophage FSH19 and the origins of species specificity in the Vi01-like phage family. Virol J 8:498. doi:10.1186/1743-422X-8-498

Huma N, Shankar P, Kushwah J, Bhushan A, Joshi J, Mukherjee T, Raju SC, Purohit HJ, Kalia VC (2011) Diversity and polymorphism in AHL-lactonase gene (*aiiA*) of *Bacillus*. J Microbiol Biotechnol 21:1001–1011. doi:10.4014/jmb.1105.05056

Jensen MA, Faruque SH, Mekalanos JJ, Levin BR (2006) Modeling the role of bacteriophage in the control of cholera outbreaks. Proc Natl Acad Sci U S A 12:4652–4657. doi:10.1073/pnas.0600166103

Kalia VC (2013) Quorum sensing inhibitors: an overview. Biotechnol Adv 31:224–245. doi:10.1016/j.biotechadv.2012.10.004

Kalia VC (2014a) Microbes, antimicrobials and resistance: the battle goes on. Indian J Microbiol 54:1–2

Kalia VC (2014b) In search of versatile organisms for quorum-sensing inhibitors: acyl homoserine lactones (AHL)-acylase and AHL-lactonase. FEMS Microbiol Lett 359:143. doi:10.1111/1574-6968.12585

Kalia VC (2015) Microbes: the most friendly beings? In: VC Kalia (ed) Quorum sensing vs quorum quenching: a battle with no end in sight. Springer, New Delhi, pp 1–5. doi: 10.1007/978-81-322-1982-8_1

Kalia VC, Kumar P (2015) The battle: quorum-sensing inhibitors versus evolution of bacterial resistance. In: Kalia VC (ed) Quorum sensing vs quorum quenching: a battle with no end in sight. Springer, New Delhi, pp 385–391. doi:10.1007/978-81-322-1982-8_31

Kalia VC, Raju SC, Purohit HJ (2011) Genomic analysis reveals versatile organisms for quorum quenching enzymes: acyl-homoserine lactone-acylase and—lactonase. Open Microbiol J 5:1–13. doi:10.2174/1874285801105010001

Kalia VC, Kumar P, Pandian SK, Sharma P (2014a) Biofouling control by quorum quenching. In: Kim SK (ed) Hb_25 Springer handbook of marine biotechnology, Chap. 15. Springer, Berlin, pp 431–440. doi: 10.1007/978-3-642-53971-8_15

Kalia VC, Wood TK, Kumar P (2014b) Evolution of resistance to quorum-sensing inhibitors. Microb Ecol 68:13–23. doi:10.1007/s00248-013-0316-y

Kęsik-Szeloch A, Drulis-Kawa Z, Weber-Dąbrowska B, Kassner J, Majkowska-Skrobek G, Augustyniak D, Łusiak-Szelachowska M, Żaczek M, Górski A, Kropinski AM (2013) Characterising the biology of novel lytic bacteriophages infecting multidrug resistant *Klebsiella pneumoniae*. Virol J 10:100. doi:10.1186/1743-422X-10-100

Kijjoa A, Sawangwong P (2004) Drugs and cosmetics from the sea. Mar Drugs 2:73–82

Krasowska A, Biegalska A, Augustyniak D, Los M, Richert M, Lukaszewicz M (2015) Isolation and characterization of phages infecting *Bacillus subtilis*. Biomed Res. doi:10.1155/2015/179597

Kumar P, Patel SKS, Lee JK, Kalia VC (2013) Extending the limits of *Bacillus* for novel biotechnological applications. Biotechnol Adv 31:1543–1561. doi:10.1016/j.biotechadv.2013.08.007

Kumar P, Koul S, Patel SKS, Lee JK, Kalia VC (2015) Heterologous expression of quorum sensing inhibitory genes in diverse organisms. In: Kalia VC (ed) Quorum sensing vs quorum quenching: a battle with no end in sight. Springer, New Delhi, pp 343–356. doi:10.1007/978-81-322-1982-8_28

La Scola B, Desnues C, Pagnier I, Robert C, Barrassi L, Fournous G, Merchat M, Suzan-Monti M, Forterre P, Koonin E, Raoul D (2008) The virophage as a unique parasite of the giant mimivirus. Nature 455:100–104. doi:10.1038/nature07218

Lalko J, Gunnel A (1967) A new Vi-phage-type of *Salmonella typhi* and preparation of the typing phage. Bull World Health Organ 36:227–230

Lemke PA, Nash CH (1974) Fungal viruses. Bacteriol Rev 38:29–56

Lettini AA, Saccardin C, Ramon E, Longo A, Cortini E, Dalla Pozza MC, Barco L, Guerra B, Luzzi I, Ricci A (2014) Characterization of an unusual *Salmonella* phage type DT7a and report of a foodborne outbreak of salmonellosis. Int J Food Microbiol 189:11–17. doi:10.1016/j.ijfoodmicro.2014.07.021

Mašlaňová I, Doškař J, Varga M, Kuntová L, Mužík J, Malúšková D, Růžičková V, Pantůček R (2013) Bacteriophages of *Staphylococcus aureus* efficiently package various bacterial genes and mobile genetic elements including SCC *mec* with different frequencies. Environ Microbiol Rep 5:66–73. doi:10.1111/j.1758-2229.2012.00378.x

Matsushita K, Uchiyama J, Kato S, Ujihara T, Hoshiba H, Sugihara S, Muraoka A, Wakiguchi H, Matsuzaki S (2009) Morphological and genetic analysis of three bacteriophages of *Serratia marcescens* isolated from environmental water. FEMS Microbiol Lett 291:201–208. doi:10.1111/j.1574-6968.2008.01455.x

NIID (National Institute of Infectious Diseases, Japan) (1998) *Salmonella typhi* & *Salmonella paratyphi A* phage types from human sources, Japan. http://www.nih.go.jp/niid/en/. Assessed on 21 Jul 2015 at 00:44:25.

Oliver JD (2010) Recent findings on the viable but nonculturable state in pathogenic bacteria. FEMS Microbiol Rev 34:415–425. doi:10.1111/j.1574-6976.2009.00200.x

Oren A (2010) Industrial and environmental applications of halo-philic microorganisms. Environ Technol 31:825–834. doi:10.1080/09593330903370026

Petty NK, Foulds IJ, Pradel E, Ewbank JJ, Salmond GPC (2006) A generalized transducing phage (wIF3) for the genomically sequenced *Serratia marcescens* strain Db11: a tool for functional genomics of an opportunistic human pathogen. Microbiol 152:1701–1708. doi:10.1099/mic.0.28712-0

Pickard D, Toribio AL, Petty NK, van Tonder A, Yu L, Goulding D, Barrell B, Rance R, Harris D, Wetter M, Wain J, Choudhary J, Thomson N, Dougan G (2010) A conserved acetyl esterase domain targets diverse bacteriophages to the Vi capsular receptor of *Salmonella* enterica serovar *typhi*. J Bacteriol 192:5746–5754. doi:10.1128/JB.00659-10

Plouguerné E, Hellio C, Deslandes E, Véron B, Stiger-Pouvreau V (2008) Antimicrofouling activities in extracts of two invasive algae: *Grateloupia turuturu* and *Sargassum muticuma*. Bot Mar 51:202–208. doi:10.1515/BOT.2008.026

Plouguerné E, Hellio C, Cesconetto C, Thabard M, Mason VB, Pereira RC, da Gama BAP (2010) Antifouling activity as a function of population variation in *Sargassum vulgare* from the littoral of Rio de Janeiro (Brazil). J Appl Phycol 22:717–724. doi:10.1007/s10811-010-9511-0

Plüddemann A, Van Zyl WH (2003) Evaluation of *Aspergillus niger* as host for virus-like particle production, using the hepatitis B surface antigen as a model. Curr Genet 43:439–446. doi:10.1007/s00294-003-0409-0

Refos JM, Vonk AG, Eadie K, Lo-Ten-Foe JR, Verbrugh HA, van Diepeningen AD et al (2013) Double-stranded RNA mycovirus infection of *Aspergillus fumigatus* is not dependent on the genetic make-up of the host. PLoS One 8:e77381. doi:10.1371/journal.pone.0077381

Rezanka T, Prell A, Spizek J, Sigler K (2010) Pilot-plant cultivation of *Streptomyces griseus* producing homologues of nonactin by precursor-directed biosynthesis and their identification by LC/MS-ESI. J Antibiot 63:524–529. doi:10.1038/ja.2010.93

Rodrigues E, Supriya T, Naik CG (2004) Antimicrobial activity of marine organisms collected off the coast of South East India. J Exp Mar Biol Ecol 309:121–127. doi:10.1016/j.jembe.2004.03.010

Rohwer F, Breitbart M (2005) Here a virus, there a virus, everywhere the same virus? Trends Microbiol 13:278–284. doi:10.1016/j.tim.2005.04.003

Schmidt WC, Jeffries CD (1974) Bacteriophage typing of *Proteus mirabilis, Proteus vulgaris* and *Proteus morganii*. Appl Microbiol 27:47–53

Schwartz E (2009) Microbial megaplasmids. In: Steinbüchel A (ed) Microbiology monographs, vol 11. Springer, Berlin. doi:10.1007/978-3-540-85467-8

Selvin J, Priya SS, Kiran GS, Thangavelu T, Bai NS (2009) Sponge-associated marine bacteria as indicators of heavy metal pollution. Microbiol Res 164:352–363. doi:10.1016/j.micres.2007.05.005

Souza KA, Ginoza HS, Haight RD (1972) Isolation of a polyvalent bacteriophage for *Escherichia coli, Klebsiella pneumoniae,* and *Aerobacter aerogenes*. J Virol 9(5):851–856

Srirangara D, Kali A, Charles MVP (2015) Distribution of *Salmonella typhi* phage types from Pondicherry, India. Indian J Pathomicrobiol 58:129–130. doi:10.4103/0377-4929.151220

Thabard M, Id Daoud G, Véron B, Fletcher RL, Hellio C (2009) Screening of biological activities of extracts of *Ralfsia verrucosa, Petalonia fascia* and *Scytosiphon lomentaria* (Phaeophyceae, Scytosiphonales) for potential antifouling application. Electron J Nat Subst 4:1–10

Todkar S, Todkar R, Kowale L, Karmarkar K, Kulkarni A (2012) Isolation and screening of antibiotic producing halophiles from Ratnagri coastal area, state of Maharahstra. Int J Sci Res Publ 2:1–3

Tokunaga H, Arakawa T, Tokunaga M (2008) Engineering of halophilic enzymes: two acidic amino acid residues at the carboxy terminal region confer halophilic characteristics to *Halomonas* and *Pseudomonas* nucleoside diphosphate kinases. Protein Sci 17:1603–1610

Tokunaga H, Saito S, Sakai K, Yamaguchi R, Katsuyama I, Arakawa T, Tokunaga M (2010) Halophilic β-lactamase as a new solubility-and folding-enhancing tag protein: production of native human interleukin 1α and human neutrophil α-defensin. Appl Microbiol Biotechnol 86:649–658. doi:10.1007/s00253-009-2325-9

Trung ND, Suthisarnsuntorn U, Kalambaheti T, Wonglumsom W, Tunyong W (2007) Antimicrobial susceptibility patterns and phage types of *Salmonella typhi* from Vietnam. Southeast Asian J Trop Med Public Health 38:487–492

van der Wielen PW, Bolhuis H, Borin S et al (2005) The enigma of prokaryotic life in deep hypersaline anoxic basins. Science 307:121–123. doi:10.1126/science.1103569
Velho Pereira S, Furtado I (2012) Antibacterial activity of halophilic bacterial bionts from marine invertebrates of Mandapam India. Indian J Pharm Sci 74:331–338. doi:10.4103/0250474X
Venil CK, Zakaria ZA, Ahmad WA (2013) Bacterial pigments and their applications. Process Biochem 48:1065–1079. doi:10.1016/j.procbio.2013.06.006
Wall SK, Zhang J, Rostagno MH, Ebner PD (2010) Phage therapy to reduce preprocessing *Salmonella* infections in market-weight swine. Appl Environ Microbiol 76:48–53. doi:10.1128/AEM.00785-09
Wang LC, Bu T, Zhang Y, Wang Y, Liu M, Lin X (2010) Halophilic bacteria isolated from the Weihai Solar Saltern (China) phylogenetic analysis and screening of antimicrobial and cytotoxic activities of moderately halophilic bacteria. World J Microbiol Biotechnol 26:879–888
Weynberg KD, Allen MJ, Gilg IC, Scanlan DJ, Wilson WM (2011) Genome sequence of *Ostreococcus tauri* virus OtV-2 throws light on the role of picoeukaryote niche separation in the ocean. J Virol 85:4520–4529. doi:10.1128/JVI.02131-10
Xia G, Wolz C (2014) Phages of *Staphylococcus aureus* and their impact on host evolution. Infect Genet Evol 21:593–601. doi:10.1016/j.meegid.2013.04.022
Yang Q, Su QP, Wang GY, Wen DZ, Zhang YH, Bao HZ, Wang L (2007) Production of hybrid phage displaying secreted aspartyl proteinase epitope of *Candida albicans* and its application for the diagnosis of disseminated candidiasis. Mycoses 50:165–171. doi:10.1111/j.1439-0507.2006.01349.x
Yu L, Wen Z, Yang W, Li N, Wang J, Lu J, Li J (2008) Isolation and characterization of *Serratia marcescens* phage. Wei Sheng Wu Xue Bao 48:498–502

# Part II
# Bioenergy

# Xylanases: From Paper to Fuel

**Gopalakrishnan Menon and Sumitra Datta**

**Abstract** The multifaceted hydrolytic enzyme xylanase has been found to play a pivotal role in the energy and green technology sectors. Extensive research is being carried out with the aim to produce xylanases that would fulfill industrial parameters, in the paper and pulp and fuel industries. Cellulase-free xylanase with good thermal stability finds promising use for biobleaching in paper manufacturing. In current scenario, it has been found that renewable energy is a burgeoning area, where xylanase finds a major role. Xylanase is synergistically involved with other cooperating enzymes for the deconstruction of lignocellulosic raw material, which ultimately paves way for the production of fuel ethanol. In other words, efficient utilization of lignocellulosic raw materials will positively boost the economics of alternate fuel industry.

**Keywords** Lignocellulose • Biobleaching • Bioethanol • Pretreatment • Enzymatic hydrolysis

## 1 Introduction

The world demand for energy and the generation of energy comes at a premium in today's global economic scenario. Economies are controlled by oil barons all over the world, needless to say the rapidly dwindling fossil reserves of the planet. This being a matter of global concern has drawn major attention to research focused on renewable energy or alternate energy sources. It is here that "white" biotechnology or industrial and environmental biotechnology comes into limelight which is a broad and expanding field that includes making enzymes with a variety of industrial applications ranging from the manufacture of bioplastics and biofuels and the treatment of wastes to abatement of pollution, by a process known as bioremediation employing microorganisms and plants (Sasson 2005).

G. Menon
Laboratory of Biochemistry and Molecular Biology, Tomsk State University, Prospect Lenina 36, 634050 Tomsk, Russian Federation

S. Datta (✉)
Amity Institute of Biotechnology, Amity University, Kolkata 700156, India
e-mail: sumitra.datta@gmail.com

 
V.C. Kalia, P. Kumar (eds.), *Microbial Applications Vol.1*,
DOI 10.1007/978-3-319-52666-9_7

Next to cellulose, xylan is the most abundant polysaccharide in nature apart from being the major component of hemicellulose. Hemicellulose is a heteropolymer of complex carbohydrates including xylan, xyloglucan (heteropolymer of D-xylose and D-glucose), glucomannan (heteropolymer of D-glucose and D-mannose), galactoglucomannan (heteropolymer of D-galactose, D-glucose, and D-mannose), and arabinogalactan (heteropolymer of D-galactose and arabinose) (Shallom and Shoham 2003; Collins et al. 2005). Xylanases (EC 3.2.1.8) catalyze the hydrolysis of xylan; a variety of microorganisms are reported to produce endoxylanases that can degrade β-1,4-xylan in a random fashion, yielding a series of linear and branched oligosaccharide fragments. Lignocellulosic agro- and forestry residues consist of 20–40% hemicellulose, having direct implications on viable production of second generation of biofuels. Xylanases are reported to be produced by a diverse group of organisms including bacteria, algae, fungi, protozoa, gastropods, and arthropods (Collins et al. 2005). Recently, there has been great deal of interest in insects as sources of unique microbial endo-β-1,4 xylanases with possible applications in bio-industrial applications. Similarly, invertebrates like earthworms also contain a wide variety of fibrolytic microbes in the gut; hence, it is likely that some of these organisms produce novel endoxylanases (Heo et al. 2006; Oh et al. 2008; Kim et al. 2009). One of the most interesting sources of xylanases is rumen where efficient hydrolysis of plant polysaccharides occurs. It is a complex ecosystem where a diverse group of microbes including bacteria, fungi, and protozoa coexist (Wang et al. 2011). Xylanase finds use in various industries as represented in the illustration (Fig. 1). Traditionally, it has been used in the food and feed industry, for instance, in increasing the bread volume in bakery and for reducing intestinal viscosity in chicken being fed with rye so as to increase the feed conversion efficiency. One example each for food (i.e bread volume in bakery) and feed (intestinal viscosity in chicken) industry has already been given. The other two sectors, paper and pulp and the fuel ethanol industry, are where the present research on xylanase is focused. Xylanases with high thermal and pH stabilities are required in the latter two industries, which will be covered in detail in this chapter.

**Fig. 1** Xylanase in various industries

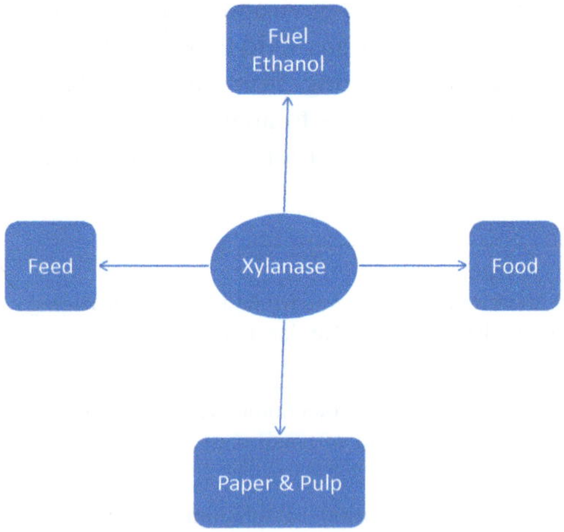

## 2 Biobleaching in Paper and Pulp Industry

India has about 515 pulp processing units and is one of the major players for the paper production like Germany, Italy, Japan, Brazil, and Sweden (http://www.forestindustries.se/; Sadhasivam et al. 2010; Passarini et al. 2014; Sharma et al. 2015a, b). Indian paper industry contributes about 1.6% of the world's total produced paper. Market demand for chlorine-free bleached pulp has increased dramatically over the last decade mainly due to its suitability toward customer satisfaction (Garcia et al. 2010; Sharma et al. 2015a, b).

The major current application of xylanases is in the pulp and paper industries where the high temperature (55–70 °C) and alkaline pH of the pulp substrate requires thermo-alkaliphilic enzymes for efficient biobleaching. Stability at high temperature and pH form the major criterion for applicability of bacterial xylanases in paper and pulp industry (Viikari et al. 1994; Beg et al. 2001; Kiddinamoorthy et al. 2008). Xylanase generally plays an important role in the pretreatment stage of current elemental chlorine-free ECF bleaching sequences. It has been proved to be economical for pulp plants of the benefits in the bleaching stage. Chlorine dioxide demand was reduced after xylanase pretreatment, suggesting decrease in the adsorbable organic halide (AOX) formation (Fillat et al. 2012; Thakur et al. 2012; Sharma et al. 2014).

Improvement in pulp bleaching of kraft pulp using xylanase from *Bacillus pumilus* has been reported (Duarte et al. 2003). Statistical optimization of cellulase-free xylanase produced by *Bacillus coagulans* BL69 from agro-residues by solid-state fermentation exhibited thermotolerant activity and hydrolyzed kraft pulp (Heck et al. 2005). Prospective biobleaching of raw banana fiber and banana pulp with cellulase-free xylanase from *Bacillus subtilis* C O1 was determined by Manimaran and Vatsala (2007). Xylanase from *Bacillus* sp. GRE7 was effective in eucalyptus kraft pulp bleaching (Kiddinamoorthy et al. 2008). Purified alkali-stable xylanase from *Aspergillus fischeri* was immobilized on polystyrene beads using diazotization method. An expanded bed bioreactor was developed with these immobilized beads to biobleach the paper pulp in continuous mode. Response surface methodology was applied to optimize the biobleaching conditions. The optimal reaction parameters for biobleaching process were reaction temperature of 60 °C, flow rate of 2 ml/min, and 5% (w/v) of pulp. There was a reduction of 87% of kappa number from 66 in the unbleached pulp to 20 (Senthilkumar et al. 2008). A thermophilic strain of *Bacillus* sp. XTR-10 producing extracellular xylanase under optimal conditions was evaluated for its potential application in biobleaching of wood kraft pulp. The xylanase-treated pulp samples showed considerable release of lignin-derived and chromophoric compounds when spectrophotometrically analyzed. Eight hours pretreatment with 40 IU of xylanase per gram of dry pulp resulted in 16.2% reduction of kappa number with 25.94% ISO increase in brightness as compared to the control. Enzymatic pretreatment of the pulp saved 15% active chlorine charges in single-step and 18.7% in multiple-step chemical bleaching with brightness level to that of the control (Saleem et al. 2009). Bacteria

*Paenibacillus campinasensis* BL11 produced xylanase with relative activities of 56.8% and 51.9 after 4 h incubation in pH 7 and pH 9 at 65 °C, respectively. Untreated and oxygen-delignified hardwood kraft pulp had increased brightness (as much as 4.4 and 3.9%) and viscosity (as much as 0.5 and 0.3 cP) with xylanase pretreatments (2.5 IU/g of oven-dry pulp) after full chlorine dioxide bleaching (Ko et al. 2010). A polyextremophilic *Bacillus halodurans* TSEV 1 produced a cellulase-free xylanase with thermophilic and alkaliphilic properties. A 7.35-fold increase in xylanase production was seen due to statistical optimization. The UV absorption spectrum of the compounds released by xylanase action on unbleached pulp showed a characteristic peak at 280 nm implying the presence of lignin-related compounds in the filtrate. The materials released after biobleaching also displayed strong absorption at 237 and 465 nm. The enzyme dose 40 U/g, pH 10.0, and 3 h of treatment were optimal for enzymatic pre-bleaching of wheat pulp that caused 14.6% reduction in kappa number and 5.6% enhancement in brightness of hand sheets without any change in viscosity (Kumar and Satyanarayana 2012). A xylanase purified from *Streptomyces rameus* L2001 and the biobleaching effect on wheat straw pulp were investigated. The purified xylanase had an optimum pH of 5.3 and was stable over pH 4.3–6.7. The stable optimal temperature of the enzyme was 70 °C. Enzymatic treatment at a charge of 20 U/g dry wheat straw pulp for 1 h prior to hypochlorite (3.8%) treatment revealed an increase in brightness index by 2.8% and increase in residual chlorine by 14.5% (Li et al. 2010). The xylanase from *Pichia pastoris* NC38 showed bleaching efficacy on bagasse pulp with a brightness of 47.4% ISO with 50 IU/g of xylanase, which was 2.1% ISO higher than the untreated pulps (Birijlall et al. 2011). Xylanase from *Bacillus subtilis* exhibited higher turnover number (K cat) and catalytic efficiency (K cat/K m) with birchwood xylan than oat spelt xylan. Bleach-boosting enzyme activity at 30 U/g dry pulp displayed the optimum bio-delignification of kraft pulp resulting in 26.5% reduction in kappa number and 18.5% ISO induction in brightness at 55 °C after 3 h treatment. The pulp properties including tensile strength and burst index were also seen to improve after the treatment (Saleem et al. 2012). Xylanase-producing *Bacillus halodurans* FNP 135 was administered for biobleaching of kraft pulp through submerged (SmF) and solid-state fermentation (SSF). SmF showed better delignification than SSF with 35 and 30% decrease in kappa number, 5.8 and 4.3% increase in brightness, 8.7 and 6.7% increase in breaking length, 13.7 and 12% increase in burst factor, 20.7 and 17.5% increase in tear factor, and 8.6 and 3.3% increase in viscosity, respectively. The chlorine consumption reduced by 20 (SmF) and 10 (SSF) % due to biobleaching (Sharma et al. 2015a, b). Bagasse is one of the main non-wood raw materials for pulping and papermaking. The hemicellulose content of the bagasse fiber is over 30% (de Freitas Branco et al. 2011). Xylanase-aided chlorine dioxide bleaching of bagasse pulp was investigated (Nie et al. 2015). The pulp was pretreated with xylanase from *Bacillus subtilis* NKC-1 and followed by a chlorine dioxide bleaching stage. The hexenuronic acid (HexA) was reduced after xylanase pretreatment, and the adsorbable organic halides (AOX) were reduced after chlorine dioxide bleaching. Compared to the control pulp, AOX was reduced by 21.4–26.6% with xylanase treatment. Chlorine dioxide demand was reduced by 12.5–22% to achieve the same brightness.

# 3 Bioethanol

Energy consumption has increased steadily over the last century as the world population has grown and more countries have become industrialized. Crude oil has been the major resource to meet the increased energy demand (Sun and Cheng 2002). At the moment, transportation sector worldwide is almost entirely dependent on petroleum-based fuels being responsible for 60% of the world oil consumption. In addition, transportation sector accounts for more than 70% of global carbon monoxide (CO) emissions and 19% of global carbon dioxide ($CO_2$) emissions. $CO_2$ emissions from a gallon of gasoline are about 8 kg. In the year 2007, there were about 806 million cars and light trucks on the road. The projected increase is 1.3 billion by 2030 and to over 2 billion vehicles by 2050 (Goldemberg 2008; Balat and Balat 2009; Balat 2011).

## *3.1 Lignocellulosic Biomass*

Corn residues and sugarcane bagasse are currently the potential sources of cellulosic biofuels (Table 1), but future prospective sources include the perennial grasses, switchgrass (McLaughlin and Kszos 2005; Parrish and Fike 2005), and *Miscanthus* due to their high yield, their minimal requirement for nutrient and water input, and them being cultivated in locations which would not compete with current food crops (Heaton et al. 2008). The degradable components, mostly cellulose and hemicellulose, resulting from these perennial grasses are expected to become significant substrates in the future for bioconversion to ethanol or other

**Table 1:** Composition of some agricultural lignocellulosic biomass (Saha 2003)

| Composition (%, dry basis) | | | |
|---|---|---|---|
| | Cellulose | Hemicellulose | Lignin |
| Corn fiber[a] | 15 | 35 | 8 |
| Corn cob | 45 | 35 | 15 |
| Corn stover | 40 | 25 | 17 |
| Rice straw | 35 | 25 | 12 |
| Wheat straw | 30 | 50 | 20 |
| Sugar cane bagasse | 40 | 24 | 25 |
| Switchgrass | 45 | 30 | 12 |
| Coastal bermuda grass | 25 | 35 | 6 |

[a]Contains 20% starch

higher-molecular-weight alcohols and hydrocarbons (Somerville 2007). The composition of dry plant biomass harvested from these perennial grasses consists primarily of the plant cell wall polymers, cellulose (31%), xylan (20%), and lignin (18%) (Dodd and Cann 2009).

### *3.2 Pretreatment Techniques for Lignocellulosic Raw Materials*

Ethanol production from lignocellulosic biomass comprises the following main steps: hydrolysis of cellulose and hemicellulose, sugar fermentation, separation of lignin residue, and, finally, recovery and purifying the ethanol to meet fuel specifications. The task of hydrolyzing lignocellulose to fermentable monosaccharides is still technically problematic because the digestibility of cellulose is hindered by many physicochemical, structural, and compositional factors. The pretreatment of lignocellulosic biomass is crucial before enzymatic hydrolysis. Various pretreatment options (Table 2) are available to fractionate, solubilize, hydrolyze, and separate cellulose, hemicellulose, and lignin components (Bungay 1992; Weil et al. 1994; Wyman 1994). These include ammonia fiber explosion (AFEX), wet oxidation, concentrated acid, dilute acid, lime, alkaline, sulfur dioxide, hydrogen peroxide, steam explosion (autohydrolysis), liquid hot water, carbon dioxide explosion, and organic solvent treatments (Dale et al. 1996; Schmidt and Thomsen 1998; Saha and Bothast 1999; Kaar and Holtzaple 2000; Fernandez-Bolanos et al. 2001; Laser et al. 2002; Jiang et al. 2015). In each option, the biomass is reduced in size and its physical structure is opened. Owing to these structural characteristics, pretreatment is an essential step for obtaining potentially fermentable sugars in the hydrolysis step (Saha and Bothast 1997; Mosier et al. 2005; Alvira et al. 2010).

### *3.3 Enzymatic Hydrolysis of Lignocellulose with Xylanase*

Endoxylanases and β-xylosidases are the main components responsible for effective conversion of xylan fraction of biomass among hemicellulases to monomeric xylose (Qing and Wyman 2011; Bhalla et al. 2015).

**Table 2:** Methods for pretreatment of lignocellulosic biomass (Saha and Bothast 1997)

| Method | Example |
|---|---|
| Thermomechanical | Grinding, milling, shearing, extruder |
| Autohydrolysis | Steam pressure, steam explosion, supercritical carbon dioxide explosion |
| Acid treatment | Dilute acid ($H_2SO_4$, HCl), concentrated acid ($H_2SO_4$, HCl) |
| Alkali treatment | Sodium hydroxide, ammonia, alkaline hydrogen peroxide |
| Organic solvent treatment | Methanol, ethanol, butanol, phenol |

**Fig. 2** Effect of enzymatic hydrolysis on wheat straw, *filled diamond* denotes untreated, *filled square* denotes alkali treated, and *filled triangle* denotes ammonia treated

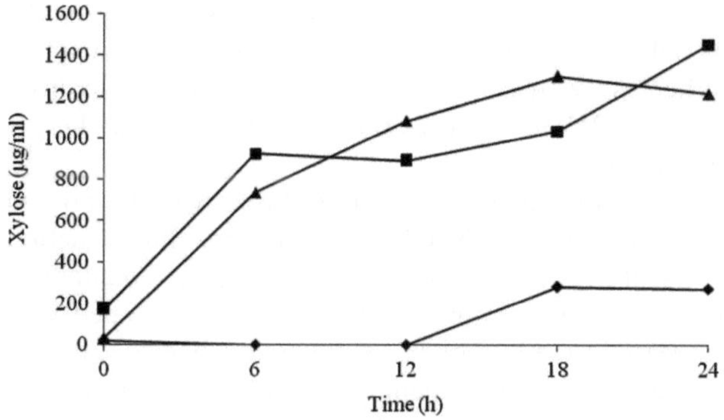


In this chapter, we present the data which one of the authors obtained while working with xylanase from *Bacillus pumilus* strain GESF-1. The enzyme was used to hydrolyze sodium hydroxide-treated and ammonia-treated wheat straw. The maximum reducing sugars released by enzymatic hydrolysis were 19.42 mg/g and 16.26 mg/g, respectively, as compared to 3.65 mg/g of reducing sugar released from untreated wheat straw. Higher yield of reducing sugars was achieved when wheat straw was pretreated with sodium hydroxide followed by ammonia as compared to untreated substrates (Fig. 2). The results demonstrated the potential of GESF-1 xylanase for the production of sugar-rich hydrolysates for various applications. It was also concluded that pretreatment of lignocellulosic substrates is necessary for better enzymatic hydrolysis (Gopalakrishnan 2010; Menon et al. 2010a).

Similarly, ammonia-treated rice husk yielded quite low, i.e., 2.32 mg/g, reducing sugar. Chapla et al. (2010) reported the enzymatic hydrolysis of different agro-residues with crude xylanase from *Aspergillus foetidus*. Maximum reducing sugars produced from dilute NaOH- and ammonia-pretreated wheat straw were 151.6 mg/g and 193.86 mg/g, respectively (Chapla et al. 2010). Rémond et al. (2010) investigated the pretreatment of wheat straw with xylanase and liquor ammonia. Both pretreatments were conducted at 15% and 20% (solid content), for SSA (soaking in aqueous ammonia) and xylanase, respectively. SSA pretreatment led to the solubilization of 38%, 12%, and 11% of acid-insoluble lignin, xylan, and glucan, respectively. In case of xylanase pretreatment, 20% of xylan was removed from native wheat straw. When pretreatments were applied consecutively (SSA and xylanase) on straw, 56% of xylans were hydrolyzed and a rapid reduction of media viscosity occurred.

A comparative analysis of enzymatic hydrolysis of hemicellulosic substrates at different temperatures with thermostable xylanase from alkalothermophilic *Thermomonospora* sp. and a commercial enzyme (Accellerase™1000) from *Trichoderma reesei* was studied by Menon et al. (2010b). At higher temperature, rapid hydrolysis of hemicellulosic substrates, oat spelt xylan (OSX) and wheat bran hemicellulose (WBH), was achieved. A hydrolysis of 62% and 50% for OSX and WBH was obtained in 36 and 48 h, respectively, using Accellerase™1000 at 50 °C. Whereas xylanase from *Thermomonospora* sp. hydrolyzed 67% (OSX) in 3 h and 58% (WBH) in 24 h at 60 °C, thus reducing the process time and quantity of enzyme used.

Pretreatment of sugarcane bagasse with ammonium hydroxide ($NH_4OH$) and sodium hydroxide (NaOH) prior to hemicellulase action was studied by Beukes and Pletschke (2011). In their investigation, they found the larger percentage of removal of lignin when sugarcane bagasse was pretreated with $NH_4OH$ and a 13.13-fold increase in the hydrolysis of sugarcane bagasse. Pedersen et al. (2011) studied the influence of pretreatment pH, temperature, time, and their interactions on the enzymatic glucose and xylose yields from mildly pretreated wheat straw in multivariate experimental designs of acid and alkaline pretreatments. The pretreatment pH was the most significant factor affecting both the enzymatic glucose and xylose yields after mild thermal pretreatments at maximum 140 °C for 10 min. The maximal enzymatic glucose and xylose yields from the solid, pretreated wheat straw fraction were obtained after pretreatments at the most extreme pH values (pH 1.0 or pH 13.0) at the maximum pretreatment temperature of 140 °C. Surface response models revealed significantly correlating interactions of the pretreatment pH and temperature on the enzymatic liberation of both glucose and xylose from pretreated, solid wheat straw. The influence of temperature was most pronounced with the acidic pretreatments, but the highest enzymatic monosaccharide yields were obtained after alkaline pretreatments. Alkaline pretreatments also solubilized most of the lignin. Pedersen et al. (2011) concluded in their study that pretreatment pH exerted significant effects and factor interactions on the enzymatic glucose and xylose released. Thus extreme pH values were necessary with mild thermal pretreatment strategies ($T \leq 140$ °C, time $\leq 10$ min). Alkaline pretreatments generally induced higher enzymatic glucose and xylose release and did so at lower pretreatment temperatures than required with acidic pretreatments. A thermophilic fungus *Scytalidium thermophilum* ATCC No. 16454 grown on corn cobs produced xylanases which hydrolyzed lignocellulosic biomass producing reducing sugars (Kocabaş et al. 2015). *Geobacillus* sp. strain WSUCF1 produced thermostable xylanase activity (crude xylanase cocktail) when grown on xylan or various cheap untreated and pretreated lignocellulosic biomasses like prairie cord grass and corn stover. The optimum pH and temperature for the crude xylanase cocktail were 6.5 and 70 °C, respectively. The WSUCF1 crude xylanase was thermostable with half-lives of 18 and 12 days at 60 and 70 °C, respectively. At 70 °C, rates of xylan hydrolysis were also found to be better with the WSUCF1 crude xylanase than the commercial enzymes, viz., Cellic HTec2 and AccelleraseXY (Bhalla et al. 2015).

## 4 Opinion

Extremozymes (viz., xylanase exhibiting unique properties in extreme temperature or pH) hold a promising future for research into fuel ethanol as well as paper and pulp industries. As the world is gearing for clean and green technologies, thermostable and alkaline xylanases will find more industrial use, be it paper and pulp or fuel ethanol. The focus is on isolating microorganisms producing xylanases with characteristics suited for the industry. Priority research work should also be done on areas pertaining to various pretreatment process involved in the use of lignocellulosic biomass for biofuel production. In second-generation biofuel production, the major bottleneck is cost-effectiveness. The biomass deconstruction steps make it a costly affair. The use of xylanases from native microorganisms or recombinant xylanases which have a wide temperature as well as pH stability for deconstruction steps, acting synergistically with other enzymes for complete degradation of xylan to monomers, would pave the way for better economics of biofuel production.

## References


Alvira P, Tomás-Pejó E, Ballesteros M, Negro MJ (2010) Pretreatment technologies for an efficient bioethanol production process based on enzymatic hydrolysis: a review. Bioresour Technol 101:4851–4861. doi:10.1016/j.biortech.2009.11.093

Balat M (2011) Production of bioethanol from lignocellulosic materials via the biochemical pathway: a review. Energy Convers Manage 52:858–875. doi:10.1016/j.enconman.2010.08.013

Balat M, Balat H (2009) Recent trends in global production and utilization of bioethanol fuel. Appl Energy 86:2273–2282. doi:10.1016/j.apenergy.2009.03.015

Beg QK, Kapoor M, Mahajan L, Hoondal GS (2001) Microbial xylanases and their industrial applications: a review. Appl Microbiol Biotechnol 56:326–338. doi:10.1007/s002530100704

Beukes N, Pletschke BI (2011) Effect of alkaline pre-treatment on enzyme synergy for efficient hemicelluloses hydrolysis in sugarcane bagasse. Bioresour Technol 102:5207–5213. doi:10.1016/j.biortech.2011.01.090

Bhalla A, Bischoff KM, Sani RK (2015) Highly thermostable xylanase production from a thermophilic *Geobacillus* sp. strain WSUCF1 utilizing lignocellulosic biomass. Front Bioeng Biotechnol 3:1–8. doi:10.3389/fbioe.2015.00084

Birijlall N, Manimaran A, Kumar KS, Permaul K, Singh S (2011) High level expression of a recombinant xylanase by *Pichia pastoris* NC38 in a 5 L fermenter and its efficiency in biobleaching of bagasse pulp. Bioresour Technol 102:9723–9729. doi:10.1016/j.biortech.2011.07.059

Bungay H (1992) Product opportunities for biomass refining. Enzyme Microb Technol 14:501–507. doi:10.1016/0141-0229(92)90145-E

Chapla D, Divecha J, Madamwar D, Shah A (2010) Utilization of agro-industrial waste for xylanase production by Aspergillus foetidus MTCC 4898 under solid state fermentation and its application in saccharification. Biochem Eng J 49:361–369. doi:10.1016/j.bej.2010.01.012

Collins T, Gerday C, Feller G (2005) Xylanases, xylanase families and extremophilic xylanases. FEMS Microbiol Rev 29:3–23. doi:10.1016/j.femsre.2004.06.005

Dale BE, Leong CK, Pham TK, Esquivel VM, Rios L, Latimer VM (1996) Hydrolysis at low enzyme levels: application of the AFEX process. Bioresour Technol 56:111–116. doi:10.1016/0960-8524(95)00183-2

de Freitas Branco R, dos Santos JC, da Silva SS (2011) A novel use for sugarcane bagasse hemicellulosic fraction: xylitol enzymatic production. Biomass Bioenergy 35:3241–3246. doi:10.1016/j.biombioe.2011.02.014

Dodd D, Cann I (2009) Enzymatic deconstruction of xylan for biofuel production. Glob Change Biol Bioenergy 18:2–17. doi:10.1111/j.1757-1707.2009.01004.x

Duarte MCT, Silva EC, Gomes IMB, Ponezi AN, Portugal EP, Vicente JR, Davanzo E (2003) Xylan-hydrolyzing enzyme system from *Bacillus pumilus* CBMAI 0008 and its effects on *Eucalyptus grandis* kraft pulp for pulp bleaching improvement. Bioresour Technol 88:9–15. doi:10.1016/S0960-8524(02)00270-5

Fernandez-Bolanos J, Felizon B, Heredia A, Rodriguez R, Guillen R, Jimenez A (2001) Steam-explosion of olive stones: hemicellulose solubilization and enhancement of enzymatic hydrolysis of cellulose. Bioresour Technol 79:53–61. doi:10.1016/S0960-8524(01)00015-3

Fillat U, Roncero MB, Sacón VM, Bassa A (2012) Integrating a xylanase treatment into an industrial-type sequence for eucalyptus kraft pulp bleaching. Ind Eng Chem Res 51:2830–2837. doi:10.1021/ie202863d

Garcia SG, Moreira T, Artal G, Maldonado L, Feijoo G (2010) Environmental impact assessment of non-wood based pulp production by soda-anthraquinone pulping process. J Clean Prod 18:137–145. doi:10.1016/j.jclepro.2009.10.008

Goldemberg J (2008) Environmental and ecological dimensions of biofuels. In: Proceedings of the conference on the ecological dimensions of biofuels, Washington, DC, 10 March

Gopalakrishnan (2010) Studies on marine microbes of west coast of India for their potential to produce industrially important enzymes. PhD Thesis. http://hdl.handle.net/10603/9199

Heaton EA, Flavell RB, Mascia PN, Thomas SR, Dohleman FG, Long SP (2008) Herbaceous energy crop development: recent progress and future prospects. Curr Opin Biotechnol 19:202–209. doi:10.1016/j.copbio.2008.05.001

Heck JX, Flores SH, Hertz PF, Ayub MAZ (2005) Optimization of cellulase-free xylanase activity produced by *Bacillus coagulans* BL69 in solid-state cultivation. Process Biochem 40:107–112. doi:10.1016/j.procbio.2003.11.044

Heo S, Kwak J, Oh H-W, Park D-S, Kyung SB, Dong HS, Park H-Y (2006) Characterization of an extracellular xylanase in Paenibacillus sp. HY-8 isolated from an herbivorous longicorn beetle. J Microbiol Biotechnol 16:1753–1759

Jiang L, Zheng A, Zhao Z, He F, Li H, Liu W (2015) Obtaining fermentable sugars by dilute acid hydrolysis of hemicellulose and fast pyrolysis of cellulose. Bioresour Technol 182:364–367. doi:10.1016/j.biortech.2015.01.032

Kaar WE, Holtzaple MT (2000) Using lime pretreatment to facilitate the enzymatic hydrolysis of corn stover. Biomass Bioenergy 18:189–199. doi:10.1016/S0961-9534(99)00091-4

Kiddinamoorthy J, Anceno AJ, Haki GD, Rakshit SK (2008) Production, purification and characterization of *Bacillus* sp. GRE7 xylanase and its application in eucalyptus kraft pulp biobleaching. World J Microbiol Biotechnol 24:605–612. doi:10.1007/s11274-007-9516-2

Kim DY, Han MK, Lee JS, Oh H-W, Park D-S, Shin D-H, Bae KS, Son K-H, Park H-Y (2009) Isolation and characterization of a cellulase-free endo-β-1,4-xylanase produced by an invertebrate-symbiotic bacterium, *Cellulosimicrobium* sp. HY-13. Process Biochem 44:1055–1059. doi:10.1016/j.procbio.2009.05.005

Ko C-H, Lin Z-P, Tu J, Tsai C-H, Liu C-C, Chen H-T, Wang T-P (2010) Xylanase production by *Paenibacillus campinasensis* BL11 and its pretreatment of hardwood kraft pulp bleaching. Int Biodeterior Biodegradation 64:13–19. doi:10.1016/j.ibiod.2009.10.001

Kocabaş DS, Güder S, Özben N (2015) Purification strategies and properties of a low-molecular weight xylanase and its application in agricultural waste biomass hydrolysis. J Mol Catal B Enzym 115:66–75. doi:10.1016/j.molcatb.2015.01.012

Kumar V, Satyanarayana T (2012) Thermo-alkali-stable xylanase of a novel polyextremophilic *Bacillus halodurans*TSEV1 and its application in biobleaching. Int Biodeterior Biodegradation 75:138–145. doi:10.1016/j.ibiod.2012.09.007

Laser M, Schulman D, Allen SG, Lichwa J, Antal MJ Jr, Lynd LR (2002) A comparison of liquid hot water and steam pretreatments of sugar cane bagasse for bioconversion to ethanol. Bioresour Technol 81:33–44. doi:10.1016/S0960-8524(01)00103-1

Li X, She Y, Sun B, Song H, Zhu Y, Lv Y, Song H (2010) Purification and characterization of a cellulase-free, thermostable xylanase from *Streptomyces rameus* L2001 and its biobleaching effect on wheat straw pulp. Biochem Eng J 52:71–78. doi:10.1016/j.bej.2010.07.006

Manimaran A, Vatsala TM (2007) Biobleaching of banana fibre pulp using Bacillus subtilis C O1 xylanase produced from wheat bran under solid-state cultivation. J Ind Microbiol Biotechnol 34:745–749. doi:10.1007/s10295-007-0248-y

McLaughlin SB, Kszos LA (2005) Development of switchgrass (*Panicum virgatum*) as a bioenergy feedstock in the United States. Biomass Bioenergy 28:515–535. doi:10.1016/j.biombioe.2004.05.006

Menon G, Mody K, Keshri J, Jha B (2010a) Isolation, purification, and characterization of haloalkaline xylanase from a marine *Bacillus pumilus* strain, GESF-1. Biotechnol Bioprocess Eng 15:998–1005. doi:10.1007/s12257-010-0116-x

Menon V, Prakash G, Prabhune A, Rao M (2010b) Biocatalytic approach for the utilization of hemicellulose for ethanol production from agricultural residue using thermostable xylanase and thermotolerant yeast. Bioresour Technol 101:5366–5373. doi:10.1016/j.biortech.2010.01.150

Mosier N, Wyman CE, Dale BD, Elander RT, Lee YY, Holtzapple M, Ladisch CM (2005) Features of promising technologies for pretreatment of lignocellulosic biomass. Bioresour Technol 96:673–686. doi:10.1016/j.biortech.2004.06.025

Nie S, Wang S, Qin C, Yao S, Ebonka JF, Song X, Li K (2015) Removal of hexenuronic acid by xylanase to reduce adsorbable organic halides formation in chlorine dioxide bleaching of bagasse pulp. Bioresour Technol 196:413–417. doi:10.1016/j.biortech.2015.07.115

Oh H-W, Heo S-Y, Kim D-Y, Park D-S, Bae K-S, Park S-Y (2008) Biochemical characterization and sequence analysis of a xylanase produced by an exo-symbiotic bacterium of *Gryllotalpa orientalis*, *Cellulosimicrobium* sp. HY-12. Antonie Van Leeuwenhoek J Microb 93:437–442. doi:10.1007/s10482-007-9210-2

Parrish DJ, Fike JH (2005) The biology and agronomy of Switch-grass for biofuels. Crit Rev Plant Sci 24:423–459. doi:10.1080/07352680500316433

Passarini KC, Pereira MA, Farias TMB, Calarge FA, Santana CC (2014) Assessment of the viability and sustainability of an integrated waste management system for the city of Campinas (Brazil), by means of ecological cost accounting. J Clean Prod 65:479–488. doi:10.1016/j.jclepro.2013.08.037

Pedersen M, Johansen KS, Meyer AS (2011) Low temperature lignocellulose pretreatment: effects and interactions of pretreatment pH are critical for maximizing enzymatic monosaccharide yields from wheat straw. Biotechnol Biofuels 4:1–10. doi:10.1186/1754-6834-4-11

Qing Q, Wyman CE (2011) Supplementation with xylanase and β-xylosidase to reduce xylo-oligomer and xylan inhibition of enzymatic hydrolysis of cellulose and pretreated corn stover. Biotechnol Biofuels 4:18. doi:10.1186/1754-6834-4-18

Rémond C, Aubry N, Crônier D, Noël S, Martel F, Roge B, Rakotoarivonina H, Debeire P, Chabbert B (2010) Combination of ammonia and xylanase pretreatments: impact on enzymatic xylan and cellulose recovery from wheat straw. Bioresour Technol 101:6712–6717. doi:10.1016/j.biortech.2010.03.115

Sadhasivam S, Savitha S, Swaminathan K (2010) Deployment of *Trichoderma harzianum* WL1 laccase in pulp bleaching and paper industry effluent treatment. J Clean Prod 18:799–806. doi:10.1016/j.jclepro.2009.11.014

Saha BC (2003) Hemicellulose bioconversion: a review. J Ind Microbiol Biotechnol 30:279–291. doi:10.1007/s10295-003-0049-x

Saha BC, Bothast RJ (1997) Enzymes in lignocellulosic biomass conversion. In: Saha BC, Woodward J (eds) Fuels and chemicals from biomass. American Chemical Society, Washington, DC, pp 46–56

Saha BC, Bothast RJ (1999) Pretreatment and enzymatic saccharification of corn fiber. Appl Biochem Biotechnol 76:65–77. doi:10.1385/ABAB:76:2:65

Saleem M, Tabassum MR, Yasmin R, Imran M (2009) Potential of xylanase from thermophilic Bacillus sp. XTR-10 in biobleaching of wood kraft pulp. Int Biodeterior Biodegrad 63:1119–1124. doi:10.1016/j.ibiod.2009.09.009

Saleem M, Aslam F, Akhtar MS, Tariq M, Rajoka MI (2012) Characterization of a thermostable and alkaline xylanase from *Bacillus* sp. and its bleaching impact on wheat straw pulp. World J Microbiol Biotechnol 28:513–522. doi:10.1007/s11274-011-0842-z

Sasson A (2005) Industrial and environmental biotechnology: achievements, prospects, and perceptions, UNU-IAS Report, pp 1–26

Schmidt AS, Thomsen AB (1998) Optimization of wet oxidation pretreatment of wheat straw. Bioresour Technol 64:139–151. doi:10.1016/S0960-8524(97)00164-8

Senthilkumar SR, Dempsey M, Krishnan C, Gunasekaran P (2008) Optimization of biobleaching of paper pulp in an expanded bed bioreactor with immobilized alkali stable xylanase by using response surface methodology. Bioresour Technol 99:7781–7787. doi:10.1016/j.biortech.2008.01.058

Shallom D, Shoham Y (2003) Microbial hemicellulases. Curr Opin Microbiol 6:219–228. doi:10.1016/S1369-5274(03)00056-0

Sharma A, Thakur VV, Shrivastava A, Jain RK, Mathur RM, Gupta R, Kuhad RC (2014) Xylanase and laccase based enzymatic kraft pulp bleaching reduces adsorbable organic halogen (AOX) in bleach effluents: a pilot scale study. Bioresour Technol 169:96–102. doi:10.1016/j.biortech.2014.06.066

Sharma AK, Anupam K, Swaroop V, Lal PS, Bist V (2015a) Pilot scale soda anthraquinone pulping of palm oil empty fruit bunches and elemental chlorine free bleaching of resulting pulp. J Clean Prod 106:422–429. doi:10.1016/j.jclepro.2014.03.095

Sharma P, Sood S, Singh G, Capalash N (2015b) An eco-friendly process for biobleaching of eucalyptus kraft pulp with xylanase producing *Bacillus halodurans*. J Clean Prod 87:966–970. doi:10.1016/j.jclepro.2014.09.083

Somerville C (2007) Biofuels. Curr Biol 17:R115–R119. doi:10.1016/j.cub.2007.01.010

Sun Y, Cheng J (2002) Hydrolysis of lignocellulosic materials for ethanol production: a review. Bioresour Technol 83:1–11. doi:10.1016/S0960-8524(01)00212-7

Thakur VV, Jain RK, Mathur RM (2012) Studies on xylanase and laccase enzymatic prebleaching to reduce chlorine-based chemicals during CEH and ECF bleaching. BioResources 7:2220–2235

Viikari L, Kantelinen A, Sundquist J, Linko M (1994) Xylanases in bleaching: from an idea to the industry. FEMS Microbiol Rev 13:335–350. doi:10.1111/j.1574-6976.1994.tb00053.x

Wang G, Luo H, Wang Y, Huang H, Shi P, Yang P, Meng K, Bai Y, Yao B (2011) A novel cold-active xylanase gene from the environmental DNA of goat rumen contents: direct cloning, expression and enzyme characterization. Bioresour Technol 102:3330–3336. doi:10.1016/j.biortech.2010.11.004

Weil J, Westgate P, Kohlmann K, Ladisch MR (1994) Cellulose pretreatments of lignocellulosic substrates. Enzyme Microb Technol 16:1002–1004. doi:10.1016/0141-0229(94)90012-4

Wyman CE (1994) Ethanol from lignocellulosic biomass: technology, economics, and opportunities. Bioresour Technol 50:3–16. doi:10.1016/0960-8524(94)90214-3

# Basic Principles of Microbial Fuel Cell: Technical Challenges and Economic Feasibility

**Soumya Pandit, Kuppam Chandrasekhar, Ramesh Kakarla, Abudukeremu Kadier, and Velpuri Jeevitha**

**Abstract** Water and energy securities are emerging as increasingly important and vital issues for today's world. Therefore, the field of wastewater management and alternative energy is one of the most unexplored fields of Biotechnology and Science. Microbial fuel cell (MFC) is emerging as a modern wastewater treatment technology which converts chemical energy stored in the bonds of organic matter present in wastewater directly into electricity using electrogenic bacteria as a catalyst, without causing environmental pollution. In this chapter, the technical know-how of MFC and biocatalyst has been depicted. A thorough understanding of the fundamental principles of microbial fuel cells would help to perceive new

S. Pandit
The Zuckerberg Institute for Water Research (ZIWR), Ben-Gurion University of the Negev, Beersheba, Israel

Department of Biotechnology, IIT-Kharagpur, Kharagpur, India
e-mail: sounip@gmail.com

K. Chandrasekhar (✉)
School of Applied Bioscience, Agriculture Department, Kyungpook National University, Building-1, Dong-Daegu 702701, South Korea

Bio-Engineering and Environmental Science (BEES), CSIR-IICT, Hyderabad, India
e-mail: chanduibt@gmail.com

R. Kakarla
Environmental Biotechnology Laboratory for Water and Energy (EBLWE), Department of Environmental Science and Engineering, Kyung Hee University, Seoul, Republic of Korea

Department of Biotechnology, College of Science and Technology, Andhra University, Visakhapatnam 530003, AP, India
e-mail: kakarlarameshbt@gmail.com

A. Kadier
Department of Chemical and Process Engineering, Faculty of Engineering & Built Environment, National University of Malaysia, Selangor, Malaysia
e-mail: abudoukeremu@163.com

V. Jeevitha
Department of Environmental Science and Technology, Jawaharlal Nehru Technological University Hyderabad, Hyderabad, Telangana, India
e-mail: jeevithavelpuri@gmail.com



V.C. Kalia, P. Kumar (eds.), *Microbial Applications Vol.1*,
DOI 10.1007/978-3-319-52666-9_8

aspects of bioenergy conversions and how such systems could be integrated with the present energy generation systems to maximize the energy recovery. In this respect, MFCs show promise to treat wastewater with simultaneous production of renewable energy. In this chapter, the theories underlying the electron transfer mechanisms, the biochemistry and the microbiology involved, and the material characteristics of anode, cathode, and the separator have been clearly described. This chapter highlights the major factors involved toward the improvement bio-electricity production processes. Advance in the design of MFC Technology and the economy of the process are also included.

**Keywords** Electroactive bacteria • Coulombic efficiency • Power density • Anode • Wastewater treatment • Bioenergy

# 1 Introduction

## *1.1 MFC as a Renewable Energy Sources*

Security for water and energy sources is gaining importance throughout the world. Increasing population and climate changes pose serious challenges that involve energy, water resources, land use, and waste issues. Throughout the world, there is intense interest in evaluating and implementing alternative energy sources (Schröder 2008). A lot of research is going in quest of renewable energy sources (Chandrasekhar et al. 2015a). The hydropower, biomass, wind, geothermal, and solar radiationare are among major sources for renewable energy generation. In recent years, microbial fuel cell (MFC) has been emerging as one of the popular wastewater treatment-based technology to provide clean water and green energy (Venkata Mohan and Chandrasekhar 2011a, b; Pant et al. 2012). MFCs are bioelectrochemical devices where organic waste is degraded to smaller molecules, releasing electrons and protons thereby generating electricity. MFCs can directly convert chemical energy into electrical energy through bioelectrochemical reactions utilizing microorganism or enzymatic catalysis (Deval et al. 2016). MFCs have several advantages compared to the traditional fuel cells and enzymatic fuel cells in some criteria. It is possible to utilize a wide range of organic or inorganic matter such as organic waste and soil sediments as a source of fuel generation with the aid of microorganism as biocatalyst. High conversion efficiency can be achieved with such devices due to the direct or a single-step conversion of substrate energy to electricity. Unlike a conventional fuel cell, MFCs can run at natural temperature and pressure (Du et al. 2007). MFCs outperformed other technologies like anaerobic digester and aerated lagoon (Logan 2008; Deval et al. 2016). First, the single-step conversion of organic waste as substrate to electricity ensures improved conversion ability. Second, an MFC can avoid additional gas treatment operation owing to its carbon dioxide-rich off-gas; third, air-cathode or single-chambered MFCs do not require any external energy input; further, it can be useful for widespread application in locations lacking electrical facilities (Stams et al. 2006).

## 1.2 *Application of MFC*

MFC showed promise to utilize it for sustainable wastewater treatment and simultaneous power harvesting from renewable biomass. MFC has the ability to treat various types of wastewater including industrial, agricultural, and municipal wastewaters (Kiran Kumar et al. 2012; Chandrasekhar et al. 2015b). Current generation has been increased with stackable MFCs. Research has been carried out worldwide to increase the power density level of MFCs and make its operation economical to bring it to large scale (Erable et al. 2010). Along with wastewater treatment, MFC can be used for various applications such as BOD biosensors, bacterial enumeration, etc. MFC enriched with electrochemically active bacteria has been used as biochemical oxygen demand (BOD) and toxicity detection biosensor. The presence of toxic substances such as metals like cadmium, lead, arsenic, chromium (VI), mercury and compound of cyanide, organophosphorus, and surfactant induces change in electric current signal and thereby facilitates in detecting toxicity level in water. At relatively low concentration, current generation was found proportional to concentration toxic and biodegradable waste. The presence of toxic substances in aqueous system can be clubbed with BOD measurement to monitor quality of wastewater (Mook et al. 2013). Recently, sediment MFCs has shown promise for constructed wetland management. Current produced during sediment MFCs operation can be successfully stored in capacitors, subsequently utilizing it to drive remote sensors through a power management system. SMFC was built to power the underwater monitoring devices. MFCs was also found to be useful to provide long-term, stable power to low-power biomedical devices implanted in the human body (Babauta et al. 2012). These *Saccharomyces cerevisiae* biocatalyzed miniatured MFCs are capable of utilizing glucose in the bloodstream to produce electricity. Food spoilage can be traced using MFC. Current generation was found to increase with increasing level of contamination. Henceforth, this kind of technology can be helpful for the rapid detection and enumeration of microorganisms in food products (Chandrasekhar et al. 2015b).

# 2 Basics of MFC: Know-how

MFC is a bioelectrochemical tool comprised of an anaerobic anode chamber and an aerobic cathode chamber physically separated by anion exchange membrane (AEM). In a typical MFC, microbes are utilized for oxidation of substrate in anode chamber; subsequently, the electron released from the microorganism goes to cathode via external wire. The anode chamber consists of microorganism (catalyst) and an electrode (anode), and it can be fed with growth media or wastewater named as anolyte and redox mediator (not required in case of mediator-less MFC), The necessary protons and electrons extracted during bacterial substrate catabolism combine with oxygen to form water on cathode (Fig. 1).

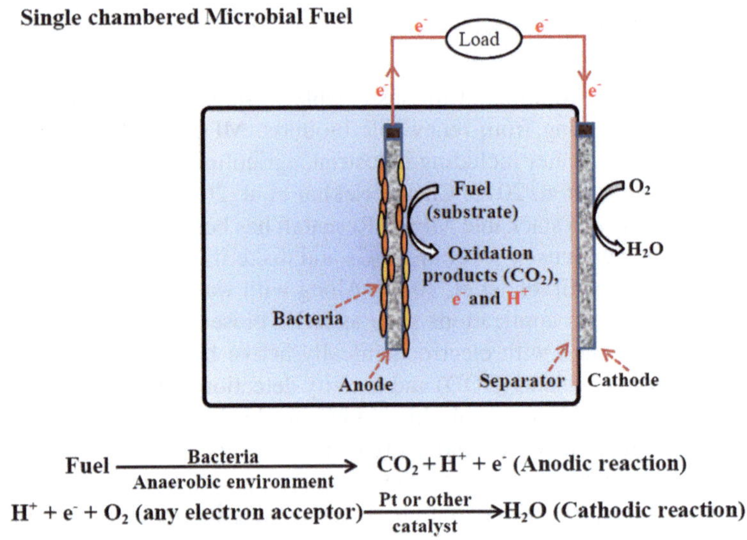


**Fig. 1** Schematics of principle of a single-chambered MFC

Usually, electrons flow to the cathode via a conductive material having an external resistance. The protons which migrate through membrane are reduced by accepting these electrons and terminal electron acceptors (e.g., $O_2$ to water) at the cathode similar to chemical fuel cell (Venkata Mohan et al. 2014) (Fig. 1). Typical electrode reactions are shown below using acetate as an example substrate:

$$\text{Anode half-cell reaction}: CH_3COO^- + 2H_2O \xrightarrow{\text{microbes}} 2CO_2 + 7H^+ + 8e^- \quad (1)$$

$$\text{Cathode half-cell reaction}: \; 2O_2 + 8H^+ + 8e^- \rightarrow 4H_2O \quad (2)$$

$$\Delta G = -847.60 \; \text{kJ/mol}; \text{EMF} = 1.10 \; \text{V} \quad (3)$$

## 2.1 *Classification of Microbial Fuel Cells (MFCs)*

MFC is usually classified into two types based on the electron transfer mechanism from bacteria to the anode. Depending on the requirement of external mediator, two types of MFCs are available, namely, mediator MFC and mediator-less MFC. In mediator MFC, supplementation of synthetic mediators mediates the electron transfer from the microorganism to the anode. Therefore, two redox couples are necessary for coupling the reduction of an $e^-$ mediator to bacterial oxidative metabolism and to recycle the $e^-$ mediator via oxidation with the reduction of the $e^-$ acceptor on the anode surface. The electron acceptor is regenerated by atmospheric oxygen on cathode surface. In this kind of MFC, microbes are unable to donate electrons directly to anode due to the nonconductive nature of the cell surface structures. Externally added suitable electron mediators can improve the electron transfer efficiencies in such MFCs. The value of oxidation reduction potential of electrochemical mediators is in between anode and NADH inside bacterial cell. Therefore, these have the ability to squeeze electrons from microbial cell and transfer it to anode. In this process, it can simultaneously reduce and oxidize itself and regenerate itself. Various organometallic compounds have been tested including phenothiazines, benzylviologen, phenazine ethosulfate, gallocyanine, resorufin, 2,6-dichlorophenolindophenol, and phenoxazine brilliant cresyl blue.

Mediator-less MFC—The mediator-less MFC can further be classified based on the ability of the microbes to transfer electron either directly or through self-synthesized electron shuttles from the microbial cell to the anode. Direct transfer of electrons in mediator-less MFCs is done by organisms called as "exoelectrogens" or electroactive bacteria (EAB), whereas any other microorganism can be used in a mediator MFC. In mediator-less MFC, the anode half-cell potential developed as microorganism can directly reduce anode through the oxidation of substrate.

## 2.2 Power Generation and Evaluation of MFC Performance

Anaerobic environment is required at the electroactive biofilm containing anode chamber to generate current through digestion of organic waste. The closest analogues to anodes for microbial metabolism in natural environments are probably Fe $^{3+}$ or $Mn^{4+}$ oxides. The resemblance lies in the fact that all are basically insoluble exocellular electron acceptors. The key difference in microbial electricity production compared to usual biogeochemical processes, such as $Mn^{4+}$ reduction, is that the electron released during microbial respiration reduces anode instead of a natural electron acceptor. In the sedimentary environments, the oxidation of organic matter is coupled to metal ($Fe^{3+}$) reduction by a consortia of fermentative microorganisms and metal-reducing organisms (Erable et al. 2010). Some nitrate- and sulfate-reducing bacteria have electron donating property in the anode of MFC. The electron donation in anode takes place in the absence of their natural electron acceptor in the MFC. Electricity is generated in an MFC since the overall reaction is thermodynamically favorable (Eq. 3). The energy output can be evaluated depending on power generation and time required for bioelectrochemical reaction:

$$V = I \times R \tag{4}$$

$$E = P \times T \tag{5}$$

$$P = V \times I \tag{6}$$

where $P$ and $T$ indicate the power (Watts) and time (s), respectively. The output operating voltage (V) and corresponding current influence power generation. The operating voltage $V$ is associated with MFC resistance, by Ohm's law in which

$R$ represents the resistance (Ohm). The operating or actual voltage due to cell resistance can be calculated by Eq. 7 (V):

$$V = E^{o} - \eta_{a} - \eta_{b} - I \times R \tag{7}$$

with $E^{o}$ the theoretical cell voltage which can be calculated using Nernst equation. The $\eta_a$ and $\eta_b$, indicated over-potential or voltage losses at the electrodes, the "*I*" is current, and "*R*" signifies ohmic loss due to both anolyte and catholyte resistances. Owing to these losses, the actual voltage output is always lower than theoretical voltage. The measured open circuit voltage is around 750–800 mV, which is lower than theoretical one of 1.1 V. Under close circuit condition, the operating voltage will reduce significantly.

## 2.3 Potential Losses in MFC

In reality, MFC always produces low operating voltage ($V_{op}$) compared to the cell electromotive force ($E_{thermo}$) that is the thermodynamically predicted potential. These potential losses are irreversible in nature (Chandrasekhar and Venkata Mohan 2012, 2014a, b). Energy loss can occur in different ways in MFCs: activation loss (maybe for initiating the reactions on both electrodes and extracellular electron transfer to the anode), bacterial metabolism loss (due to bacteria acquiring energy by oxidizing the substrate), mass transfer loss (due to limited flux of the reactants to the electrode), and ohmic losses (due to proton diffusion resistance and charge transfer resistance (Fig. 2) (Du et al. 2007).

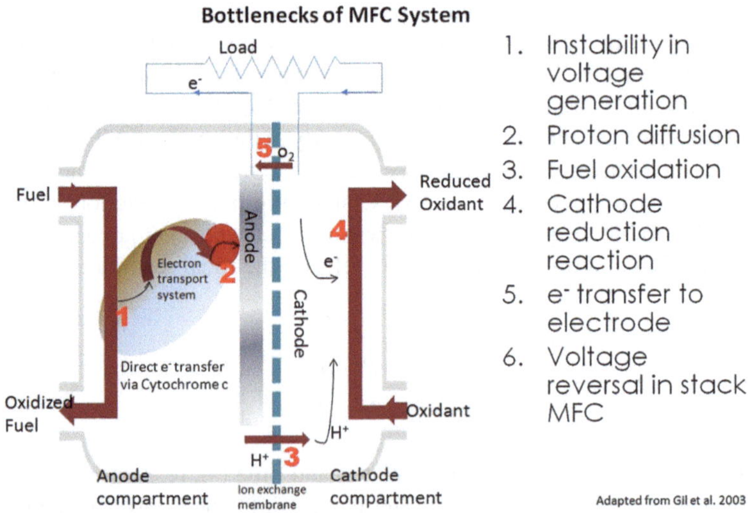


**Fig. 2** Schematics of different bottlenecks in MFC

The irreversible losses in an actual MFC are presented in the following equation (Rismani-Yazdi et al. 2008):

$$V_{op} = E_{thermo} - \left[(\eta_{act} + \eta_{ohmic} + \eta_{conc})_{cathode} + (\eta_{act} + \eta_{ohmic} + \eta_{conc})_{anode}\right] \quad (8)$$

where $\eta_{act}$, $\eta_{ohmic}$, and $\eta_{conc}$ are the activation loss, ohmic loss, and concentration loss, respectively. These voltage losses occur due to slow electrochemical reaction kinetics, development of ionic and electronic resistances, and mass transport limitations. The over-potentials of the electrodes are generally current dependent, which can be represented in a polarization curve.

**Activation Over-potential**
Activation energy is required in order to initiate the oxidation and reduction reaction at anode and cathode, respectively. The activation over-potential is the potential loss due to the activation energy required to either oxidize substrate/fuel on the anode surface or and occur during the initial phase of electron transfer from (or) to a mediator involved in chemical reaction on electrode. Various strategies were employed to reduce the activation over-potential (Rabaey et al. 2009; Chandrasekhar and Venkata Mohan 2012). It can be minimized by increasing electrode catalysis, adding mediators to facilitate efficient electron transport from microbial cell membrane to anode surface, by increasing the electrode surface area, and enriching electrogenic biofilm on anode and operational conditions inside anode and cathode compartments.

**Ohmic Over-potential**
The resistance developed owing to the migration of ions in both anolyte and catholyte and the flow of electrons between the anode and cathode causes ohmic over-potential. Therefore, by reducing the gap of anode and cathode, it is possible to minimize the ohmic loss. The current generation is proportional to ohmic over-potential. This will ensure the migration of counterion and availability of proton on cathode surface for reduction. Increasing the ionic conductivity of electrolyte is another solution to reduce ohmic over-potential (Chandrasekhar and Venkata Mohan 2012).

**Concentration Over-potential**
Limitation in transport of reactant or sluggish mass transfer rates in MFC are responsible for the concentration over-potential. The concentration loss occurs at high current densities, where rapid fall of voltage occurs with increase in current. Unavailability of substrate to the biofilm on the anode is due to mass transfer limitation which contributes toward concentration losses (Qiao et al. 2010; Chandrasekhar and Venkata Mohan 2012). Concentration over-potential can be reduced by increasing the substrate concentration of anolyte in anode chamber and terminal electron receiver like $O_2$ or proton in cathode chamber. However, substrate concentration in anode requires optimization as high concentration may inhibit the normal metabolic activity of biofilm on anode.

# 3 Performance Evaluation for MFC

MFC performance and efficiency is generally measured in terms of energetic parameters (volumetric power density, current, potential difference, cell internal resistance) and biological treatment efficiency (COD removal) (Logan 2012). Coulombic efficiency (CE) is an interesting factor that combines the two previous ones: the amount of electrons that feed the current, collected from the conversion of the organic substrates.

## 3.1 *Polarization Studies*

Polarization or voltage–current curves have become the standard method of presenting MFC performance as is the case with chemical fuel cells (Zhao et al. 2009). A polarization curve generator is a powerful experiment to analyze and characterize quality of fuel cells in terms of power generation. The polarization curve provides lot of information. Polarization curves can be obtained by varying external resistances using a resistance box or using programmed liner sweep Voltammetry. The volumetric current and power density were calculated using Eq. 9:

$$P = V \times I; \quad i_{\mathrm{d}} = V/Rv_{\mathrm{and}}; \quad P_{\mathrm{d}} = V^2/Rv_{\mathrm{and}} \tag{9}$$

where $P$, $V$, $R_{\mathrm{ext}}$, and $I$ denote volumetric power, operating cell voltage, applied external resistance, and current, respectively. Volumetric current ($i_{\mathrm{d}}$) and power density ($P_{\mathrm{d}}$) were evaluated by normalizing to anolyte volume ($v_{\mathrm{and}}$).

## 3.2 *The Current Interruption Method*

The current interrupt technique is commonly used experimental method for the evaluation of total ohmic resistance of electrochemical systems like storage cell, batteries, and fuel cells. In the current interrupt method, cell is initially kept in close circuit with a certain magnitude of external resistance. While it is observed that the MFC generated a stable current output across the terminal of applied external resistance, the circuit is opened suddenly which causes a shoot-up of the cell voltage ($V_{\mathrm{R}}$) followed by gradual further increment (Logan 2012). The immediate sharp lift of voltage is attributed to ohmic internal over-potential ($R_{\mathrm{int}}$) of the MFC. The total internal resistance of the MFC cell can be measured using following equation (Zhao et al. 2009).

$$R_{\text{int}} = \frac{V_R}{I} \tag{10}$$

## 3.3 Electrochemical Impedance Spectroscopy

Electrochemical impedance spectroscopy (EIS) is helpful to evaluate the internal resistance of the different components (solution resistance, charge transfer resistances, and diffusion resistance) in electrochemical systems. A polarization curve or current interruption method has a drawback of only measuring the total internal resistance of the MFC cell (Aelterman et al. 2006). EIS is usually performed in a three-electrode cell (with working, counter, and reference electrodes) to find out the electrochemical processes for the individual electrode or in a whole cell configuration (only working and counter electrodes). EIS helps to elucidate different important data like the electrochemical reactions that occur on electrodes and the surface and material properties of electrodes (Bard and Faulkner 2001).

## 3.4 Cyclic Voltammetry

Cyclic voltammetry (CV) is a useful method to investigate the mechanisms of oxidation or reduction reactions on the electrode surface. In CV, external specific range of applied voltage was applied and output current is monitored. From the current–voltage graph, different information can be explained (Fricke et al. 2008; Chandrasekhar and Venkata Mohan 2012). In MFC, CV experiments were carried out extensively to (1) elucidate the different type of exocellular electron transfer mechanisms of biofilm containing anode, (2) to find out the redox potentials of the mediators present in anolyte for electron transfer on anode, and (3) to evaluate the quality of the cathode catalysts for ORR (Zhao et al. 2009).

## 3.5 Coulombic Efficiency and Energy Efficiency

The Coulombic efficiency (CE) is basically the ratio of the total Coulombs transferred to the anode after bioelectro-oxidation of the substrate to maximum charge available or stored if all the substrate can be converted to current theoretically, i.e., the total electron charge stored in the substrate (Logan et al. 2006). One of the major useful indexes to evaluate the performance of an MFC is to find out coulombic as well as energy efficiency in terms of the energy recovery (Rabaey et al. 2009). The energy efficiency of an MFC can be defined as the energy recovered from the organic matter to the total energy content of the organic material. In MFCs, energy efficiencies range from 2% to more than 10% depending on the type of substrate (Logan 2008). CE can be evaluated as per Eq. 11.

$$\mathrm{CE} = \frac{M \int_0^t I \mathrm{d}t}{Fbv \Delta \mathrm{COD}} \tag{11}$$

where $v$ is the volume of the anode chamber of MFC, $M$ is 32 (MW of $O_2$), Faraday's constant ($F$) is 96485 C/mol, and b is 4, i.e., the number of electrons exchanged per mole of oxygen; ΔCOD is the subtraction of the initial to final anolyte concentration in terms of COD (g/L).

# 4 Microbes as Catalyst in MFC and Their Mode of Electron Transfer to Anode

## *4.1 Anodic Microorganisms Employed in MFCs*

The microbes in MFC draw maximum attraction in current research. They are basically of three types. In indirect MFC, microbes cannot donate electron to anode; it requires mediator from outside. *E. coli* and *Bacillus* sp. are among these categories. Pure culture of *Shewanella putrefaciens* of *Geobacter sulfurreducens* has been used widely; these kind of bacteria can donate electrons to anode directly and are known as electroactive bacteria. They are also familiar as anodophiles or exoelectrogens as they form conductive biofilm on anode. Mixed culture can be used as biocatalyst in anode chamber of MFC. Different phylogenetic group are observed to form biofilm on anode surface. Firmicutes and Acidobacteria, four of the five classes of Proteobacteria, were found effective as electron generator on anode to produce current. Further, it was reported that yeast strains *Saccharomyces cerevisiae* and *Hansenula anomala* were capable of producing electricity under anaerobic conditions. Different pure cultures capable of producing current in MFC are tabulated in Table 1. The dissimilatory metal-reducing bacteria (DMRBs) and sulfate-reducing bacteria (SRB) are major players contributing significantly on bioanode. The DMRBs can be isolated by using selective media containing solid Fe(III) or Mn(IV) oxides as insoluble electron acceptors. In natural condition, in the absence of conductive anode they will utilize oxidized metal ions [Fe(III) or Mn (IV)] as terminal electron acceptor to continue respiration process. Probably, this is the reason why the mechanism of extracellular electron transfer (EET) activity for metal reduction and anode reduction is considered to be identical. Isolation of these bacteria should ideally be done from the biofilm developed in the anode of the MFC enriched with an inoculum that is used as seed in the MFC Table 1.

**Table 1** Pure cultures used as inoculum in the anode of MFC

| Biocatalysts | Fuel used | Reactor type | Power per volume (W/m$^3$) | Power density (mW/m$^2$) | References |
|---|---|---|---|---|---|
| *Escherichia coli*, acclimated | Complex medium | Single chamber | | 600 | Zhang et al. (2006) |
| *Shewanella oneidensis* | Lactate | Miniature | 500 | 3000 | Ringeisen et al. (2006) |
| *Geobacter sulfurreducens* | Acetate | H-type | 0.356 | 13 | Bond and Lovley (2003) |
| *Rhodoferax ferrireducens* | Glucose | H-type | | 33.4 | Chaudhuri and Lovley (2003) |
| *Clostridial* isolate | Complex medium | Two-chamber | 5.62 | | Prasad et al. (2006) |

## 4.2 Exoelectrogens Isolated from MFC

A fewer number of reports are available where electroactive bacteria (EABs) are isolated directly from anode of MFC. A diverse range of EAB s and non-EABs are detected on the surface of the anode in MFC systems. Wang et al. (2010) reported isolation and culture of a Fe(III)-reducing *Bacteroides* sp. W7 from the anolyte suspension (Wang et al. 2010). An obligatory anaerobe identified *Clostridium butyricum* EG3 from an MFC containing starch processing wastewater anolyte. However, it was also observed that all the microorganisms associated with the biofilm may not interact directly with the anode but may interact indirectly through electron mediator secreted by other members of the anodophiles. For example, *Brevibacillus* sp. PTH1 was found predominantly in anode in spite of its inability to generate electricity solely. Probably, it is pyocyanin or phenoxine compound excreted from a *Pseudomonas* sp.-enabled *Brevibacillus* sp. PTH1 to donate electrons to anode when it was cocultured with *Pseudomonas* sp. Interestingly, output current was observed to be very low without being cocultured with or supernatant from an MFC run with the *Pseudomonas* sp.

## 4.3 Mixed Consortia in an MFC

Startup of MFC system for field applications such as simultaneous electricity generation and wastewater treatment involves the use of a consortium of bacterial cultures such as activated sludge or anaerobic digester sludge (Evelyn Li et al. 2014). Under suitable acclimatization conditions, the MFC system is expected to select and enrich its own electrochemically active consortia from the mixed culture in course of time. Depending on the source of the inoculum, operational conditions, substrate availability and MFC system architecture, the microbial communities

**Table 2** Mixed cultures used as inoculum in the anode of MFC

| Mixed culture type | Reactor type | substrate | Power density | Volumetric power density | References |
|---|---|---|---|---|---|
| Activated sludge | H-type | Landfill leachate | – | 2.06 | You et al. (2006a, b) |
| | H-type | Glucose | 115.6 | – | |
| | Single chamber | Landfill leachate | – | 6.82 | |
| | Bushing | Glucose | 3987 | – | |
| Biofilm from domestic wastewater | Single chamber | Glucose | 766–1540 | 51 | Cheng et al. (2006a, b) |
| | | Domestic wastewater | 464 | 15.5 | |
| Thermophilic effluent from anaerobic digestion of brewery wastewater | Two-chamber | Acetate | 1030 | – | Jong et al. (2006) |
| Activated sludge | Two-chamber | Glucose, glutamate | 560 | 102 | Moon et al. (2006) |
| Granular anaerobic sludge | Up-flow | Sucrose | | 29.2 | He et al. (2006) |

differed from each other. Table 2 enlists the various mixed microbial consortia enriched from different sources in MFCs.

## 5 Biochemistry of the Electron Transfer Mechanism in MFCs

There are three primary mechanisms (Fig. 3) by which microorganisms can transfer electrons to the electrodes. In an indirect MFC or mediator MFC, fermentative microorganism produces hydrogen, carbon dioxide, alcohols, or ammonia as end product. These are reduced products where electron generated during substrate catabolism is used to reduce intermediate product like proton or acid to form hydrogen or alcohol, respectively, in anaerobic condition. This fermentative bacterium is therefore unable to donate the electrons to anode. To utilize this bacterium in anode, external mediator is required which can shuttle between cell membrane and anode. Benzylviologen, 2,6-dichlorophenolindophenol, thionine, and 2-hydroxy-1,4-naphthoquinone are some common artificial electron mediators shuttling between anode and inside cell membrane of fermentative bacteria (Evelyn Li et al. 2014) (Fig. 3a).

Some of the microorganisms like *Pseudomonas, Lactobacillus,* and *Enterococcus* were reported to generate electron shuttles (like phenazines in case of *Pseodomonas aeruginosa*) which facilitate the electron transfer to the anode. Similar mechanism of electron shuttles was also reported in extracellular $Fe^{3+}$

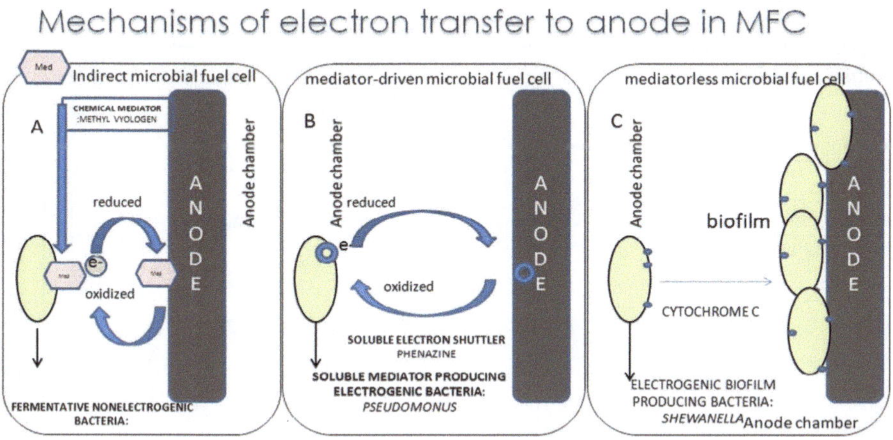


**Fig. 3** Different mechanisms of electron transfer in MFC. (**a**) Indirect MFC, (**b**) mediator-driven MFC, and (**c**) mediator-less MFC

reduction by *Shewanella oneidensis* (Rimboud et al. 2014). Producing natural mediators like pyocyanin and riboflavin is an energy-intensive process, which bacteria produce under stressed condition (Fig. 3b).

The third and the most important mechanism is the direct transport of electrons generated during respiration from the electroactive bacteria to the anode (Fig. 3c). When cultures of *Shewanella putrefaciens* produced electricity while metabolizing lactate (Pandit et al. 2015), it was first proposed that microorganisms might be able to transfer electrons to an electrode surface. The metal-reducing bacterium *Shewanella putrefaciens* MR-1 was reported to have cytochromes in its outer membrane. Under anaerobic conditions, these electron carriers (i.e., cytochromes) were able to generate anodic current in the absence of terminal electron acceptors. Further, it was also recognized that outermembrane cytochromes were important in electron shuttle reduction in *Shewanella* (Wang and Ren 2013).

## 5.1 Mediator-Less Mechanism of Electron Transfer

The mechanism of electron transport in a mediator-less MFC is also called as direct electron transfer and is based on the ability of certain microorganisms (termed “electricigens” or “exoelectrogens” or “electrogens” or “anodophiles” or “anode-respiring bacteria” or “electrochemically active bacteria, EAB”) to transfer the final electrons, coming from the oxidized organic matter, directly to the anode. Electron transfer to the electrode is the key point of an MFC technology. There are similarities between microbial reduction of insoluble metal oxides and current production in MFCs. In both the cases, the final electrons are transported to an extracellular solid substrate. This transport may occur either in direct contact between the cell

**Fig. 4** Possible mechanism of electron transfer by biofilm bacteria

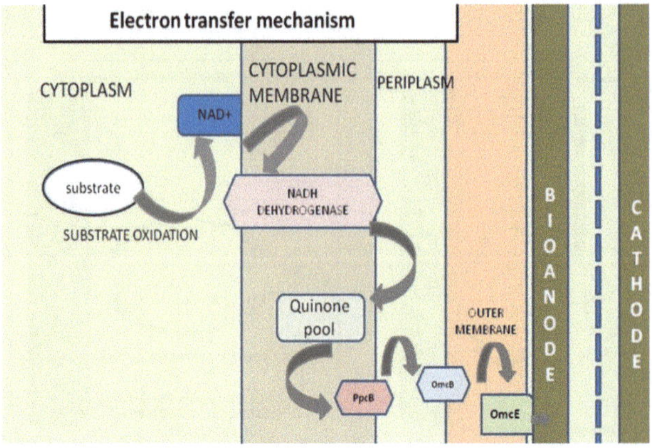


surface and the solid substrate or indirectly by endogenous mediators (Fig. 3). The direct electron transfer from microorganisms to the electrode is again of two types. The first mechanism involves endogenous mediator secretion by the exoelectrogens which mediate the electron transfer between the bacteria and the electrode, and the second mechanism involves the formation of a biofilm on the electrode by the bacteria to enable a direct electron transfer to the electrode *via* pili or "nanowires." Though the exact mechanism of electron transfer in MFCs is still not reported, it is speculated to be similar to the electron transfer mechanism observed in metal-reducing bacteria where the final electrons of substrate metabolism are accepted by the solid substrate (metal) lying outside the microorganism (Pandit et al. 2014b) (Fig. 4).

## 6 Electrode, Separator Material as Component

### 6.1 Cathode Materials

Cathode is a place where oxygen reduction reaction (ORR) occurs. In cathode, air is purged to provide oxygen which uptake electrons and proton generation during substrate metabolism in anode chamber. The performance of cathode is critical to high current generation in MFCs. The graphite, carbon felt, carbon cloth, and carbon paper have been chosen by researchers as common cathode materials. Power generation is limited by the cathode due to sluggish reduction kinetics which limits these non-catalyzed materials for the commercial use as cathode (Rismani-Yazdi et al. 2008). Large over-potential develops ($\eta_{act}$) due to noncatalytic activity of common cathode materials which reduces the current generation (Rabaey and Keller 2008). Conventional cathode like graphite plate and carbon paper has several bottlenecks as they have limited surface area. In attempts to increase effective surface area, a number of different conductive materials were applied to find out its efficacy as cathode. Carbon cloth, carbon felt, woven graphite felt, woven graphite mat, granular graphite, stainless mesh, and reticulated vitreous carbon were applied to enhance the overall reaction kinetics (Lu and Li 2012). The cathode is often modified with a highly active catalyst like Pt to reduce ORR associated over-potential and improves performance. Addition of catalyst like nanomanganese dioxide, lead oxide, polyaniline, pyrrole, phalocyanin, etc., has the ability to lower the activation energy for cathode reduction. The operating parameters associated with cathode also determine power output of MFC. The catholyte pH, oxidant concentration in cathode chamber, buffering capacity of catholyte, quantity of catalyst temperature, etc, play a significant role in developing cathode half-cell potential (Ghasemi et al. 2013).

## 6.2 Separator

Separator plays an important role in MFC. It physically divides anode chamber and cathode but ionically and electronically conjugated. In a dual-chambered MFC, separator facilitates in developing anodic and cathodic half-cell potential by splitting anolyte at the anode chamber to cathode. The electrons produced through bacterial metabolic activity are passed via an external circuit while the proton migrates via a separator, separating the anode chamber from the one in which the cathode is immersed (Leong et al. 2013). Use of separator has several advantages in MFC like transport of anolyte or substrate from anode to cathode. It contaminates cathode and generates mixed potential and consequently reduces MFC performance. Oxygen can reach anode and lower the MFC performance as oxygen will be used as electron acceptor instead of anode. There will be a possibility of short circuit in the absence of separator if both the anode and cathode are placed nearby. Although the single-chambered MFC without a separator has been recently reported to have increased current density (Chandrasekhar and Venkata Mohan 2014a, b), the absence of a separator results in increasing oxygen and substrate diffusivity and consequently reduces the CE. The use of separators however is not totally advantageous, and many problems need to be overcome. One major constraint is the retarded transfer of proton from the anodic chamber to the cathodic chamber and these result in pH splitting, phenomena where the pH of the anodic chamber decreases while that in the cathodic chamber increases (Mo et al. 2009). This lowers the system stability and bioelectrochemical performance. This also increases the overall internal resistance of the cell and overall cost of the MFC operation (Pandit et al. 2012a). Different types of separator have been explored. The types of separator materials can be differentiated into three categories on the basis of filtration ability, namely, ion exchange membrane, size selective separators, and salt bridges (Wang et al. 2010). The performances of MFCS with cation exchange membrane, anion exchange membrane, and Bipolar Membrane (BPM) were compared. The MFC with AEM showed improved performance over others in terms of voltage stability and power output. The size-selective separators includes salt bridge, glass fibers, microfiltration membrane (MFM), Ultrafiltration membrane (UFM), porous fabrics, and other coarse pore filter materials, J-Cloth, etc. In case of single-chambered MFC, separator cathode assembly (SCA) is used to reduce resistance of MFC. An SCA consists of a cathode and a membrane physically or chemically bonded together, also known as membrane cathode assembly (MCA). Rahimnejad et al. reported application of MCA in a single-chamber air-cathode MFC where membrane is hot pressed directly onto a carbon cloth cathode (Rahimnejad et al. 2014). Application of separator in MFC can reduce operational costs; further, this kind of design has potential to simplify reactor configuration by adopting passive air flow.

### *6.3 Anode*

Anode is an important component in MFC which allows EABs to form electroactive biofilm. Therefore, a quality anode material should have different attributes like superior conductivity and good biocompatibility. The other qualities like chemical stability, resistance to corrosion, mechanical strength, and toughness are also important parameter to choose anode materials. Anode should provide large specific surface area for adhesion and biofilm formation. Carbon materials are predominantly applied as anode materials which include graphite plate rod and fiber brush, carbon felt, reticulated vitreous carbon (RVC), carbon cloth, etc. (Qiao et al. 2010; Chandrasekhar and Venkata Mohan 2014a, b). Carbon cloth is much more flexible and porous than the abovementioned materials; thus, more bacteria can attach to cloth (Khilari et al. 2015). In order to promote better electroactive biofilm adhesion, the surface of anode is modified to improve current generation in MFC. Improvements were made through treatment of anode surface by physical or chemical means, by application of metal impregnated carbon electrode, or by the incorporation of highly conductive coatings. Ammonium treatment, heat treatment, and surface coating with carbon nanotube, iron oxide, etc., are among the different types of modification reported in the literature (Zhou et al. 2011). This approach is expected to help in the formation of EAB biofilm.

## 7 MFC Reactor Configuration and Architecture

The reactor configuration of the MFC affects the performance of MFC by significantly controlling the internal resistance. Many reactor configurations have been studied by many researchers. MFCs have been operated as dual-chambered MFC, single-chambered MFC, or stacked MFC (Chandrasekhar and Venkata Mohan 2012; Khilari et al. 2013).

### *7.1 Dual-Chambered MFC*

A typical dual-chambered MFC has an anode chamber and a cathode chamber separated by a membrane, which allows the proton flow from anode chamber to cathode chamber while acting as a physical barrier to the diffusion of oxygen or other oxidants from the cathode chamber to the anode chamber. The compartments of the dual-chambered MFC can be of various shapes and operated in batch or continuous mode. One of the widely used dual-chambered MFC is H-type MFC. The low power generation in an H-type reactor suffers due to very high internal resistance. This type of MFC is suitable for basic parameter research such as evaluating power generation using new electrode or separator materials, or microbial communities analysis, etc. (Pandit et al. 2011). Dual-chambered MFCs, fabricated in up-flow mode and fabricated as up-flow cylindrical MFC with inner U-shaped cathode chamber and concentric outside anode chamber with effluent recirculation, have been reported to be efficient for wastewater treatment as this type of configuration can be easily scaled up. The energy costs of pumping fluid around are much greater than their power outputs, which limit the application of such devices as net power generating devices.

### *7.2 Single-Chambered MFC*

Single-chambered MFCs can be constructed by keeping only cathode in direct contact with air. The single-chambered air-cathode MFC (sMFC) has several advantages over the dual-chambered MFC: (1) the sMFC don't require cathode chamber; it therefore reduces the size and is easy to operate, (2) no aeration is needed as direct air can be used for ORR, and (3) higher current output can be obtained because of reduced internal resistance in sMFC (Pandit et al. 2014a; Chandrasekhar and Venkata Mohan 2014a, b). In sMFC, cathode membrane assembly is applied where membrane is usually hot pressed with cathode.

## 8 Physicochemical Factors Governing MFC Performance

Operating conditions in anode chamber—Anaerobic wastewater treatment involves several groups of bacteria, having their own optimum working conditions. MFC operation in optimized anodic operating condition will facilitate proper growth of electrogenic biofilm during start-up and enhance the subsequent performance in long-term operation. Several process parameters such as pH, alkalinity, substrate type, substrate concentration, and OLR affect the activity of these anaerobic microorganisms (Gil et al. 2003). The anodic conditions which affect the performance of MFC are described below. Inoculum—Microorganisms as biocatalyst play pivotal role in current generation. MFCs have been inoculated with pure and mixed culture microorganisms. Pure culture bacteria are exoelectrogenic bacteria, which directly transport electrons outside the cell membrane. Pure culture microorganisms such as *Geobacter* sp., *Shewanella* sp., and *Rhodoferax* sp. have been studied in MFC. Domestic wastewater, soil, fresh as well as marine sediments, and activated sludge have been used as inoculums in MFC, as they are rich in mixed culture microorganisms. Pure electroactive bacterial culture systems usually grow slowly due to anaerobic nature in anode chamber. Possibilities of contamination are high with pure culture. Power output of some pure culture is relatively low compared to a mixed culture. However, mixed culture inoculum takes longer time (startup) to achieve stable current generation compared to pure culture.

Substrate—Substrate influences the performance of EAB on anode surface. It also determines dominant bacterial community when mixed culture is used as inoculum. Different types of substrates have been studied in MFCs from simple carbon source to complex carbon and nitrogen-rich biodegradable waste. Pure substrate as acetate, butyrate, glucose, sucrose, and glycerol is used widely to maintain the homogeneity. Nitrogen-rich pure substrate like cysteine and proteins is utilized by EAB in anode. Different types of wastewater are utilized to treat and simultaneously generate current in MFC such as domestic wastewater, food processing wastewater, saline sea food wastewater, real field dairy wastewater, molasses wastewater, paper wastewater, dairy wastewater, bakery wastewater, and dye wastewater (Pant et al. 2010). Influent COD concentration—Concentration of the influent chemical oxygen demand (COD) significantly affects the power generation in MFC operated both in batch feed and continuous mode. Current output is directly proportional to the concentration of substrate which can be measured in terms of COD in an MFC. Higher COD yields higher power generation (Du et al. 2007) till it causes substrate inhibition. Feed pH—The electrolyte pH plays a crucial role in MFC's power output. Power output reduces at acidic pH below 6 (Gil et al. 2003). The optimal pH range for the chosen MFC configuration was reported to be between 7 and 8 and 8 and 10 (He et al. 2006). However, there is contradiction in the available literature on the optimal setting of the operating parameters possibly due to synergistic or antagonistic interaction effect among these parameters (Jadhav and Ghangrekar 2009).

## 9 Contribution of MFC Toward Waste Treatment

MFCs cannot be on a par with conventional fuel cell as far as current generation is concerned. Therefore, efforts have been made to project MFCs as useful wastewater treatment devices. MFCs are bioelectrical devices that use microorganisms to oxidize the organic matter present in wastewaters from different sources (domestic, industrial, agricultural), generating electrical power that can be harvested for numerous purposes (Chandrasekhar and Venkata Mohan 2012, 2014a, b). Sustainability of this process enhances with utilizing the existing wastewater treatment facility which would result in reduction in energy demands and operational costs. Bacteria consume anaerobically the biodegradable components in the waste stream for metabolic energy, releasing electrons to the anode (Chandrasekhar et al. 2015b).

## *9.1 Different Bioelectrochemical System Adapted from MFC*

Recently, a number of MFC adapted technologies are drawing attraction, namely, microbial solar cells (MSC), microbial desalination cells (MDCs), microbial electrolysis cell (MEC), microbial electrochemical snorkel (MES), etc. In MDC, microbial electron-harvesting ability is used to desalinate saline water. EABs on anode created a negative electric potential gradient which allows cation and anion from a concentrated saline chamber to migrate through ion-exchange membranes (IEMs) to balance the electro-neutrality. Thus, water can be desalinating in MDC without any power consumption. Kim et al. demonstrated stackable MDCs with parallel desalination chambers and concentrated chambers which were separated by compartmental AEMs and CEMs (Kim and Logan 2013). $H_2$ production MEC is a renewable way of producing hydrogen using different types of wastewater as fuel. The principle of MEC is much alike MFC where bioorganic matter degrades in anode to generate electrons by EABs; the difference lies in cathode reduction mechanism (Kadier et al. 2015a). In the cathode of MEC, hydrogen evolution reaction (HER) occurs in strictly anaerobic condition unlike MFC where electron reduces oxygen to form water. It was observed that hydrogen gas produced in anaerobic conditions is directly proportional to the current production. Hydrogen generation in MEC is not thermodynamically favorable process unlike MFC where spontaneous reaction occurs. The theoretical reduction potential, $E_{o,}$ for $H_2$ production from proton and electron is very low at 0.414 V [vs. normal hydrogen electrode (NHE)] under standard temperature and pressure conditions (STP: 25 °C, 1 bar); therefore, additional voltage is provided to drive the HER in cathode of MEC (Kadier et al. 2015b).

In a plant microbial fuel cell (PMFC), MFCs are associated with living higher plants (Strik et al. 2011). In PMFCs, electrochemically active bacteria on anode utilize excreted rhizodeposits from plant roots to directly generate electricity. Therefore, it is possible to convert solar energy to electrical energy directly in PMFC. Rhizodeposits are rich in carbon sources like arabinose, ribose, fructose, etc. Rhizodeposits consist of a different array of substances that generate from peeled off root cells and soluble lysates and exudates. These compounds can be degraded by a wide array of EABs. Photosynthetic bacteria (e.g., *Synechocystis* pcc 6803), which are able to donate electron directly to anode, were used in PMFC.

Microbial carbon capture cells (MCC) are different types of PMFCs where microalgae can be used in cathode of a conventional dual-chambered MFC to sequester atmospheric $CO_2$ and provide oxygen to cathode via biophotolysis. Wang et al. demonstrate an MCC where off-gas $CO_2$ from the anode chamber is re-dissolved in catholyte where a photosynthetic microorganism (*Chlorella vulgaris*) utilized the $CO_2$ in presence of light (Wang et al. 2010; Pandit et al. 2012b). Microalgae biomass from the cathode chamber can be utilized for downstream processing of different value-added products like lipid extraction or ethanol production.

## 10 Conclusion

MFC is coming up as a promising technology to treat wastewater; still several challenges remain which need to overcome to commercialize this technology. The major bottleneck of MFC is low power generation. Series connection of MFC causes voltage reversal; therefore, parallel MFCs with proper power management system is required to maximize electricity generation. This also suggests that the cost of unit MFC cell should be minimized. Development of proper bacterial culture in the anode chamber of MFC requires optimization of operating condition in the anodic microenvironment. Therefore, operating conditions such as influent COD concentration, HRT, feed pH, and specific organic loading rate should be properly decided for operation in order to obtain optimum COD removal efficiency and power production.

## References

Aelterman P, Rabaey K, Pham HT, Boon N, Verstraete W (2006) Continuous electricity generation at high voltages and currents using stacked microbial fuel cells. Environ Sci Technol 40:3388–3394. doi:10.1021/es0525511

Babauta J, Renslow R, Lewandowski Z, Beyenal H (2012) Electrochemically active biofilms: facts and fiction. A review. Biofouling 28:789–812. doi:10.1080/08927014.2012.710324

Bard AJ, Faulkner LR (2001) Electrochemical methods: fundamentals and applications, 2nd edn. Wiley, New York

Bond DR, Lovley DR (2003) Electricity production by geobacter sulfurreducens attached to electrodes. Appl Environ Microbiol 69:1548–1555. doi:10.1128/AEM.69.3.1548-1555.2003

Chandrasekhar K, Venkata Mohan S (2012) Bio-electrochemical remediation of real field petroleum sludge as an electron donor with simultaneous power generation facilitates biotransformation of PAH: effect of substrate concentration. Bioresour Technol 110:517–525. doi:10.1016/j.biortech.2012.01.128

Chandrasekhar K, Venkata Mohan S (2014a) Bio-electrohydrolysis as a pretreatment strategy to catabolize complex food waste in closed circuitry: Function of electron flux to enhance acidogenic biohydrogen production. Int J Hydrogen Energy 39:11411–11422. doi:10.1016/j.ijhydene.2014.05.035

Chandrasekhar K, Venkata Mohan S (2014b) Induced catabolic bio-electrohydrolysis of complex food waste by regulating external resistance for enhancing acidogenic biohydrogen production. Bioresour Technol 165:372–382. doi:10.1016/j.biortech.2014.02.073

Chandrasekhar K, Lee YJ, Lee DW (2015a) Biohydrogen production: strategies to improve process efficiency through microbial routes. Int J Mol Sci 16:8266–8293. doi:10.3390/ijms16048266

Chandrasekhar K, Amulya K, Venkata Mohan S (2015b) Solid phase bio-electrofermentation of food waste to harvest value-added products associated with waste remediation. Waste Manage 45:57–65. doi:10.1016/j.wasman.2015.06.001

Chaudhuri SK, Lovley DR (2003) Electricity generation by direct oxidation of glucose in mediatorless microbial fuel cells. Nat Biotechnol 21:1229–1232. doi:10.1038/nbt867

Cheng S, Liu H, Logan BE (2006a) Increased performance of single-chamber microbial fuel cells using an improved cathode structure. Electrochem Commun 8:489–494. doi:10.1016/j.elecom.2006.01.010

Cheng S, Liu H, Logan BE (2006b) Increased power generation in a continuous flow MFC with advective flow through the porous anode and reduced electrode spacing. Environ Sci Technol 40:2426–2432. doi:10.1021/es051652w

Deval AS, Parikh HA, Kadier A, Chandrasekhar K, Bhagwat AM, Dikshit AK (2016) Sequential microbial activities mediated bioelectricity production from distillery wastewater using bio-electrochemical system with simultaneous waste remediation. Int J Hydrogen Energy. doi:10.1016/j.ijhydene.2016.11.114

Du Z, Li H, Gu T (2007) A state of the art review on microbial fuel cells: a promising technology for wastewater treatment and bioenergy. Biotechnol Adv 25:464–482. doi:10.1016/j.biotechadv.2007.05.004

Erable B, Duţeanu NM, Ghangrekar MM, Dumas C, Scott K (2010) Application of electro-active biofilms. Biofouling 26:57–71. doi:10.1080/08927010903161281

Evelyn Li Y, Marshall A, Gostomski PA (2014) Gaseous pollutant treatment and electricity generation in microbial fuel cells (MFCs) utilising redox mediators. Rev Environ Sci Biotechnol 13:35–51. doi:10.1007/s11157-013-9322-2

Fricke K, Harnisch F, Schröder U (2008) On the use of cyclic voltammetry for the study of anodic electron transfer in microbial fuel cells. Energy Environ Sci 1:144–147. doi:10.1039/B802363H

Ghasemi M, Daud WRW, Hassan SHA, Oh S-E, Ismail M, Rahimnejad M, Jahim JM (2013) Nano-structured carbon as electrode material in microbial fuel cells: a comprehensive review. J Alloys Compd 580:245–255. doi:10.1016/j.jallcom.2013.05.094

Gil G-C, Chang I-S, Kim BH, Kim M, Jang J-K, Park HS, Kim HJ (2003) Operational parameters affecting the performance of a mediator-less microbial fuel cell. Biosens Bioelectron 18:327–334. doi:10.1016/S0956-5663(02)00110-0

He Z, Wagner N, Minteer SD, Angenent LT (2006) An upflow microbial fuel cell with an interior cathode: assessment of the internal resistance by impedance spectroscopy. Environ Sci Technol 40:5212–5217. doi:10.1021/es060394f

Jadhav GS, Ghangrekar MM (2009) Performance of microbial fuel cell subjected to variation in pH, temperature, external load and substrate concentration. Bioresour Technol 100:717–723. doi:10.1016/j.biortech.2008.07.041

Jong BC, Kim BH, Chang IS, Liew PWY, Choo YF, Kang GS (2006) Enrichment, performance, and microbial diversity of a thermophilic mediatorless microbial fuel cell. Environ Sci Technol 40:6449–6454. doi:10.1021/es0613512

Kadier A, Simayi Y, Abdeshahian P, Azman NF, Chandrasekhar K, Kalil MS (2015a) A comprehensive review of microbial electrolysis cells (MEC) reactor designs and configurations for sustainable hydrogen gas production. Alex Eng J. doi:10.1016/j.aej.2015.10.008

Kadier A, Simayi Y, Chandrasekhar K, Ismail M, Kalil MS (2015b) Hydrogen gas production with an electroformed Ni mesh cathode catalysts in a single-chamber microbial electrolysis cell (MEC). Int J Hydrogen Energy 40:14095–14103. doi:10.1016/j.ijhydene.2015.08.095

Khilari S, Pandit S, Ghangrekar MM, Pradhan D, Das D (2013) Graphene oxide-impregnated PVA–STA composite polymer electrolyte membrane separator for power generation in a single-chambered microbial fuel cell. Ind Eng Chem Res 52:11597–11606. doi:10.1021/ie4016045

Khilari S, Pandit S, Varanasi JL, Das D, Pradhan D (2015) Bifunctional manganese ferrite/ polyaniline hybrid as electrode material for enhanced energy recovery in microbial fuel cell. ACS Appl Mater Interfaces 7:20657–20666. doi:10.1021/acsami.5b05273

Kim Y, Logan BE (2013) Microbial desalination cells for energy production and desalination. Desalination 308:122–130. doi:10.1016/j.desal.2012.07.022

Kiran Kumar A, Venkateswar Reddy M, Chandrasekhar K, Srikanth S, Venkata Mohan S (2012) Endocrine disruptive estrogens role in electron transfer: bio-electrochemical remediation with microbial mediated electrogenesis. Bioresour Technol 104:547–556. doi:10.1016/j.biortech.2011.10.037

Leong JX, Daud WRW, Ghasemi M, Liew KB, Ismail M (2013) Ion exchange membranes as separators in microbial fuel cells for bioenergy conversion: a comprehensive review. Renew Sustain Energy Rev 28:575–587. doi:10.1016/j.rser.2013.08.052

Logan BE (2008) Microbial fuel cells, 1st edn. Wiley-Interscience, Hoboken

Logan BE (2012) Essential data and techniques for conducting microbial fuel cell and other types of bioelectrochemical system experiments. ChemSusChem 5:988–994. doi:10.1002/cssc.201100604

Logan BE, Hamelers B, Rozendal R, Schröder U, Keller J, Freguia S, Aelterman P, Verstraete W, Rabaey K (2006) Microbial fuel cells: methodology and technology. Environ Sci Technol 40:5181–5192. doi:10.1021/es0605016

Lu M, Li SFY (2012) Cathode reactions and applications in microbial fuel cells: a review. Crit Rev Environ Sci Technol 42:2504–2525. doi:10.1080/10643389.2011.592744

Mo Y, Liang P, Huang X, Wang H, Cao X (2009) Enhancing the stability of power generation of single-chamber microbial fuel cells using an anion exchange membrane. J Chem Technol Biotechnol 84:1767–1772. doi:10.1002/jctb.2242

Mook WT, Aroua MKT, Chakrabarti MH, Noor IM, Irfan MF, Low CTJ (2013) A review on the effect of bio-electrodes on denitrification and organic matter removal processes in bio-electrochemical systems. J Ind Eng Chem 19:1–13. doi:10.1016/j.jiec.2012.07.004

Moon H, Chang IS, Kim BH (2006) Continuous electricity production from artificial wastewater using a mediator-less microbial fuel cell. Bioresour Technol 97:621–627. doi:10.1016/j.biortech.2005.03.027

Pandit S, Sengupta A, Kale S, Das D (2011) Performance of electron acceptors in catholyte of a two-chambered microbial fuel cell using anion exchange membrane. Bioresour Technol 102:2736–2744

Pandit S, Ghosh S, Ghangrekar MM, Das D (2012a) Performance of an anion exchange membrane in association with cathodic parameters in a dual chamber microbial fuel cell. Int J Hydrog Energy 37:9383–9392. doi:10.1016/j.ijhydene.2012.03.011

Pandit S, Nayak BK, Das D (2012b) Microbial carbon capture cell using cyanobacteria for simultaneous power generation, carbon dioxide sequestration and wastewater treatment. Bioresour Technol 107:97–102. doi:10.1016/j.biortech.2011.12.067

Pandit S, Khilari S, Bera K, Pradhan D, Das D (2014a) Application of PVA–PDDA polymer electrolyte composite anion exchange membrane separator for improved bioelectricity production in a single chambered microbial fuel cell. Chem Eng J 257:38–147. doi:10.1016/j.cej.2014.06.077

Pandit S, Khilari S, Roy S, Pradhan D, Das D (2014b) Improvement of power generation using Shewanella putrefaciens mediated bioanode in a single chambered microbial fuel cell: Effect of different anodic operating conditions. Bioresour Technol 166:451–457. doi:10.1016/j.biortech.2014.05.075

Pandit S, Khilari S, Roy S, Ghangrekar MM, Pradhan D, Das D (2015) Reduction of start-up time through bioaugmentation process in microbial fuel cells using an isolate from dark fermentative spent media fed anode. Water Sci Technol 72:106–115. doi:10.2166/wst.2015.174

Pant D, Van Bogaert G, Diels L, Vanbroekhoven K (2010) A review of the substrates used in microbial fuel cells (MFCs) for sustainable energy production. Bioresour Technol 101:1533–1543. doi:10.1016/j.biortech.2009.10.017

Pant D, Singh A, Van Bogaert G, Irving Olsen S, Singh Nigam P, Diels L, Vanbroekhoven K (2012) Bioelectrochemical systems (BES) for sustainable energy production and product recovery from organic wastes and industrial wastewaters. RSC Adv 2:1248–1263. doi:10.1039/C1RA00839K

Prasad D, Sivaram TK, Berchmans S, Yegnaraman V (2006) Microbial fuel cell constructed with a micro-organism isolated from sugar industry effluent. J Power Sources 160:991–996. doi:10.1016/j.jpowsour.2006.02.051

Qiao Y, Bao S-J, Li CM (2010) Electrocatalysis in microbial fuel cells—from electrode material to direct electrochemistry. Energy Environ Sci 3:544. doi:10.1039/b923503e

Rabaey K, Keller J (2008) Microbial fuel cell cathodes: from bottleneck to prime opportunity? Water Sci Technol 57:655. doi:10.2166/wst.2008.103

Rabaey K, Angenent L, Schroder U (2009) Bioelectrochemical systems: from extracellular electron transfer to biotechnological application. IWA Publishing, London

Rahimnejad M, Bakeri G, Najafpour G, Ghasemi M, Oh S-E (2014) A review on the effect of proton exchange membranes in microbial fuel cells. Biofuel Res J 1:7–15. doi:10.18331/BRJ2015.1.1.4

Rimboud M, Pocaznoi D, Erable B, Bergel A (2014) Electroanalysis of microbial anodes for bioelectrochemical systems: basics, progress and perspectives. Phys Chem Chem Phys 16:16349. doi:10.1039/C4CP01698J

Ringeisen BR, Henderson E, Wu PK, Pietron J, Ray R, Little B, Biffinger JC, Jones-Meehan JM (2006) High power density from a miniature microbial fuel cell using shewanella oneidensis DSP10. Environ Sci Technol 40:2629–2634. doi:10.1021/es052254w

Rismani-Yazdi H, Carver SM, Christy AD, Tuovinen OH (2008) Cathodic limitations in microbial fuel cells: an overview. J Power Sources 180:683–694. doi:10.1016/j.jpowsour.2008.02.074

Schröder U (2008) From wastewater to hydrogen: biorefineries based on microbial fuel-cell technology. ChemSusChem 1:281–282. doi:10.1002/cssc.200800041

Stams AJM, de Bok FAM, Plugge CM, van Eekert MHA, Dolfing J, Schraa G (2006) Exocellular electron transfer in anaerobic microbial communities. Environ Microbiol 8:371–382. doi:10.1111/j.1462-2920.2006.00989.x

Strik DPBTB, Timmers RA, Helder M, Steinbusch KJJ, Hamelers HVM, Buisman CJN (2011) Microbial solar cells: applying photosynthetic and electrochemically active organisms. Trends Biotechnol 29:41–49. doi:10.1016/j.tibtech.2010.10.001

Venkata Mohan S, Chandrasekhar K (2011a) Self-induced bio-potential and graphite electron accepting conditions enhances petroleum sludge degradation in bio-electrochemical system with simultaneous power generation. Bioresour Technol 102:9532–9541. doi:10.1016/j.biortech.2011.07.038

Venkata Mohan S, Chandrasekhar K (2011b) Solid phase microbial fuel cell (SMFC) for harnessing bioelectricity from composite food waste fermentation: influence of electrode assembly and buffering capacity. Bioresour Technol 102:7077–7085. doi:10.1016/j.biortech.2011.04.039

Venkata Mohan S, Velvizhi G, Annie Modestra J, Srikanth S (2014) Microbial fuel cell: Critical factors regulating bio-catalyzed electrochemical process and recent advancements. Renew Sustain Energy Rev 40:779–797. doi:10.1016/j.rser.2014.07.109

Wang H, Ren ZJ (2013) A comprehensive review of microbial electrochemical systems as a platform technology. Biotechnol Adv 31:1796–1807. doi:10.1016/j.biotechadv.2013.10.001

Wang L, Zhou X, Zhong S, Zhou N (2010) Novel materials and technologies of microbial fuel cell in environmental engineering. ECS Trans 25:311–333. doi:10.1149/1.3414027

You S, Zhao Q, Zhang J, Jiang J, Zhao S (2006a) A microbial fuel cell using permanganate as the cathodic electron acceptor. J Power Sources 162:1409–1415. doi:10.1016/j.jpowsour.2006.07.063

You SJ, Zhao QL, Jiang JQ, Zhang JN, Zhao SQ (2006b) Sustainable approach for leachate treatment: electricity generation in microbial fuel cell. J Environ Sci Health Part A Tox Hazard Subst Environ Eng 41:2721–2734. doi:10.1080/10934520600966284

Zhang T, Cui C, Chen S, Ai X, Yang H, Shen P, Peng Z (2006) A novel mediatorless microbial fuel cell based on direct biocatalysis of Escherichia coli. Chem Commun Camb Engl 2257–2259. doi:10.1039/b600876c

Zhao F, Slade RCT, Varcoe JR (2009) Techniques for the study and development of microbial fuel cells: an electrochemical perspective. Chem Soc Rev 38:1926–1939. doi:10.1039/b819866g

Zhou M, Chi M, Luo J, He H, Jin T (2011) An overview of electrode materials in microbial fuel cells. J Power Sources 196:4427–4435. doi:10.1016/j.jpowsour.2011.01.012

# Biohydrogen Production: Integrated Approaches to Improve the Process Efficiency

**Chandrasekhar Kuppam, Soumya Pandit, Abudukeremu Kadier, Chakradhar Dasagrandhi, and Jeevitha Velpuri**

**Abstract** In recent years, hydrogen ($H_2$) has emerged as a clean and attractive substitute fuel since it can be produced from renewable energy sources. Upon combustion of hydrogen, it generates only water as a major by-product. In hydrogen and fuel cell technology, hydrogen can be applied in fuel cell technology; it produces only water as a major by-product with high energy yield, hold great potential for meeting in a quite unique way by empowering the so-called hydrogen-based economy. To make hydrogen-based economy viable, it is crucial to use renewable resources in place of fossil fuels to produce hydrogen. In this direction, by considering attractive and renewable characteristics of hydrogen led us to improve a variety of biological processes for the production of hydrogen. Nonetheless, commercialization of the biological process depends on improvements in process

C. Kuppam (✉)
School of Applied Bioscience, Agriculture Department, Kyungpook National University, Building-1, Dong-Daegu, South Korea

Bio-Engineering and Environmental Science (BEES), CSIR-IICT, Hyderabad, India
e-mail: chanduibt@gmail.com

S. Pandit
Department of Desalination and Water Treatment, Ben-Gurion University of the Negev, Beersheba, Israel

Department of Biotechnology, IIT-Kharagpur, Kharagpur, India
e-mail: sounip@gmail.com

A. Kadier
Department of Chemical and Process Engineering, Faculty of Engineering & Built Environment, National University of Malaysia, Selangor, Malaysia
e-mail: abudoukeremu@163.com

C. Dasagrandhi
School of Food Science and Biotechnology, Kyungpook National University, Dong-Daegu, South Korea
e-mail: microchakri2001@yahoo.com

J. Velpuri
Department of Environmental Science and Technology, Jawaharlal Nehru Technological University, Hyderabad, Telangana, India
e-mail: jeevithavelpuri@gmail.com



V.C. Kalia, P. Kumar (eds.), *Microbial Applications Vol.1*,
DOI 10.1007/978-3-319-52666-9_9

design along with an understanding of the nature of hydrogen producing communities and process optimization. Thus, this chapter highlights the major factors involved towards the improvement of biohydrogen production processes. Environmental impact of hydrogen as carbon-neutral energy carrier is also discussed. This also includes a technical and economic analysis of the biohydrogen and its role in the proposed hydrogen economy coupled with fuel cell and in transport application. Technological advancements based on hydrogen-based fuel cell designs and process integration approaches are also discussed.

**Keywords** Biohydrogen • Biomethane • Bioenergy • Dark fermentation • Photofermentation • Microbial electrolysis cells • Biopolymers • Algae

# 1 Introduction

The rising gap between the world's energy need and inadequate supply has resulted in a sharp rise in fossil fuel usage. This led us to encounter severe limitations forced by an alarming escalation in global pollution levels as well as fossil fuel depletion. In addition, an ever-increasing level of greenhouse gas releases after the combustion of fossil-based fuels in turn aggravated the complications of global warming. At present, carbon dioxide level is exceeding 350 ppm by volume where it can potentially increase the greenhouse effect by raising the global temperature (Venkata Mohan and Pandey 2013; Chandrasekhar et al. 2015a). During the past decades, total carbon (organic) released via human accomplishments is corresponding to that gathered over millions of years. Therefore, concerns about global climate change due to GHG emissions and depletion of fossil fuel reserves have driven obvious attention to the study and improvement of carbon-neutral and renewable energy substitutes to fulfill the mounting energy requirements (Lee et al. 2013). Hence, divergence of energy and fuel selections is one of the primary necessities in the present global energy consequence (Chandrasekhar et al. 2015a). In this regard, bioenergy presents a sustainable and hopeful alternative for fossil fuels, which can defend in contrast to an energy crunch and defend the world from the verge of environmental disaster.

# 2 Biohydrogen: A Zero-Carbon Fuel

In recent years, global attention has been paid to biohydrogen as one among the carbon-neutral energy sources. Hydrogen gas is perhaps a multipurpose energy source that can change the usage of hydrocarbon-based fossil fuels since it shows the higher energy yield (122 kJ/g per unit mass), which is 2.75-fold greater than that of currently using fossil fuels (Christopher and Dimitrios 2012) and after its combustion with oxygen yields $H_2O$ (water) as the only product, which is evidently favorable for the fall of greenhouse gas emissions. In specific, in the frame of energy systems, hydrogen gas is outstanding choice of an energy transporter, more related to electrical energy than fossil fuels (Elam et al. 2003). The great electrochemical reactivity makes hydrogen an ideal for fuel cell technology in the field of suitable catalysts along with other ways to use hydrogen for electricity generation and also for energy storage (Kumar et al. 2015a). The current "merchant" and "captive" markets for hydrogen are well established in many industries (e.g., oil refinery, metal treatment, food production, and fertilizer manufacture) with the main demand in the petroleum refinery and ammonia production industries. Thus, applications such as a fuel for transportation and an energy carrier in the stationary power plant lead the hydrogen market to a rising field with huge future potential.

Currently, molecular hydrogen has been mainly generated from fossil fuel-based resources. The worldwide hydrogen production at present surpasses one billion $m^3$/day, of which 48%, 30%, and 18% is generated from natural gas, oil source, and coal source, respectively, and the outstanding 4% by water electrolysis (water-splitting) (Venkata Mohan and Pandey 2013). However, the production of hydrogen gas from fossil fuel-based resources is concomitant with GHG emission (Venkata Mohan et al. 2011; Venkateswar Reddy et al. 2011b). Alternatively, hydrogen gas production through biological routes from biomass is one of the rising technologies due to its eco-friendly and sustainable nature. A scientometric investigation using data available in the ISI Web of Knowledge (since 2001) indicated that the number of research articles published on biohydrogen accounts for 2635 records with a considerable number of citations (60,472), average citations for each item (22.95), and H-index (102). As illustrated in Fig. 1, literatures associated with biohydrogen research exhibited a steep increase after 2003, reached maximum records of 224 in 2008 (overall citations: 2883), and documented increasing trend till 2015 (records, 343) with marginal variations (Fig. 1). Average citation per year also revealed an increasing trend year by year, evidently indicating that prompt and promising research is under way to make the biohydrogen production process technologically viable.

Various organisms are known to yield hydrogen under definite conditions, including cyanobacteria that usually use carbohydrates to accumulate energy from photosynthesis mechanism to generate hydrogen from $H_2O$ and microalgae that exploit solar energy to split $H_2O$ for hydrogen production (Chandrasekhar et al. 2015a). Even with the striking advantageous features, low hydrogen production rates, less substrate conversion efficiency, and acid-rich intermediate metabolites (volatile fatty acids; VFA), production and accumulation are practical difficulties to overcome. In view of this, many researches on biological hydrogen production processes are in advancement, and various novel approaches and technologies are being investigated to overcome a few of the existing troubles and furthermore to enhance the overall process efficiency. Toward these aims, a number of sophisticated technologies for high molar hydrogen yields through metabolic engineering by providing metabolic energy to cross thermodynamic limitations, enhancing electron flux for proton ($H^+$) reduction, rerouting metabolic pathways to improve substrate consumption by expressing heterologous proteins, and so on have been well described (Hallenbeck 2012; Chandrasekhar et al. 2015a). In this chapter, we

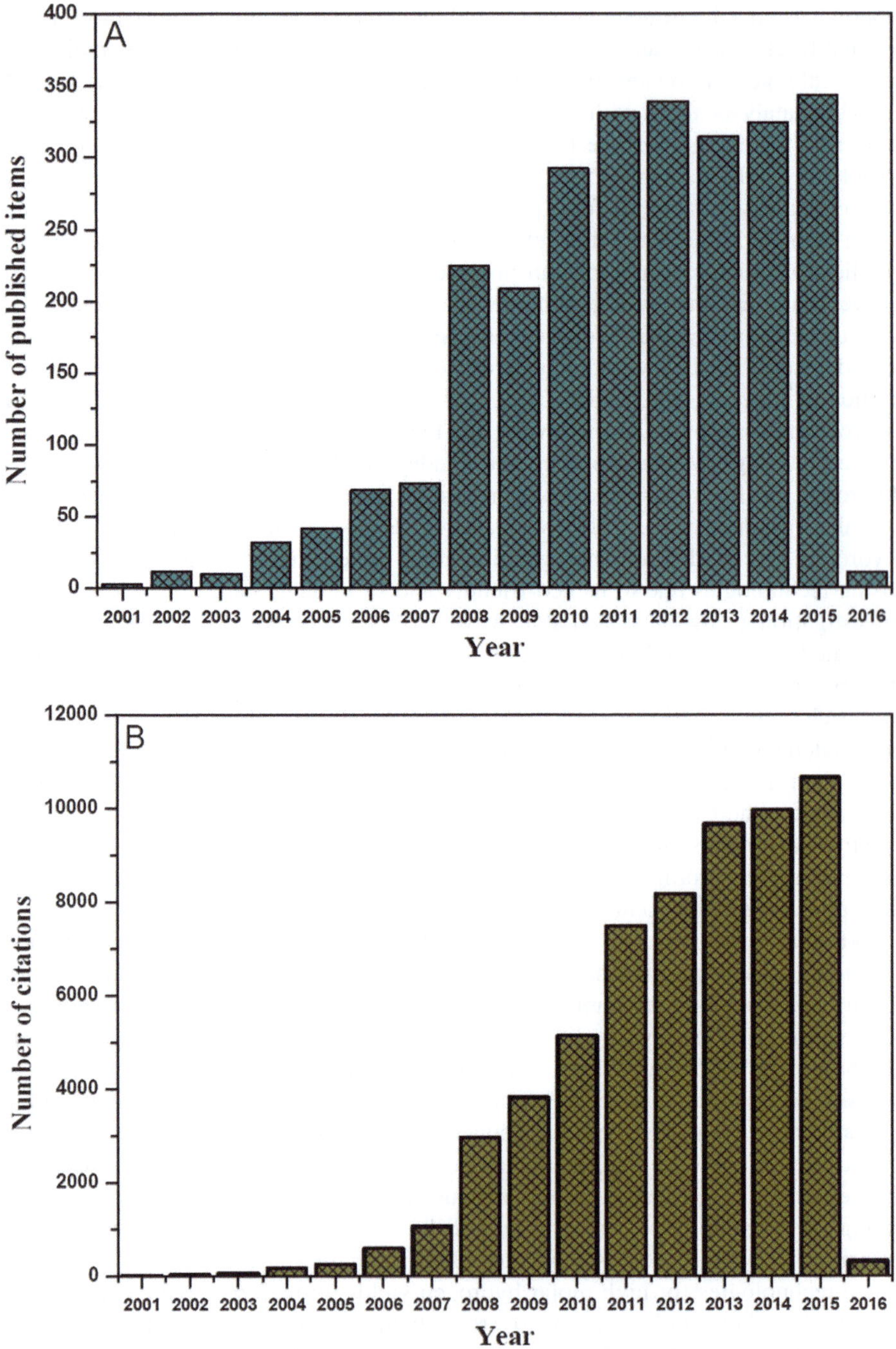


**Fig. 1** Scientometric analysis on biohydrogen. Number of published items (**a**) and number of citations (**b**) in each year (from 2001 to 2016)

focus on the biological processes/routes for hydrogen production to evaluate the efficiency and practical applicability of those processes with respect to operation factors and delineate some potentially limiting factors. In addition, alternate choices such as integration approaches and electro-fermentation to enhance process efficiency are conversed.

## 3 Diversity of Hydrogen Producers

In the environment, wide diversity of microorganisms including archaea, cyanobacteria, bacteria (facultative aerobic and anaerobic), and lower eukaryotes are reported as hydrogen producers (Boichenko et al. 2004; Kalia and Purohit 2008; Kumar et al. 2013, 2015b; Chandrasekhar et al. 2015a), which may act individually or as consortia of same types or as mixed cultures (Patel et al. 2012a, 2014; Chandrasekhar et al. 2015a) (Fig. 2). Concurrently, hydrogen consuming organisms with syntrophic association with hydrogen producing organisms contribute to balancing Earth's hydrogen. It has been noted that the molecular adaptation events may help a variety of microorganisms survive in their ecological habitats even in extreme conditions (harsh environment) such as complete darkness, extreme temperature, and presence or absence of oxygen ($O_2$). For example, typical extremophilic archaea living in extreme conditions such as salt lakes and hot springs are phototrophs, lithotrophs, or organotrophs, evolving hydrogen through their specific machineries, which is different from the bacterial hydrogen fuel cycle. Nonetheless, major hydrogen producing organisms are characteristic heterotrophs during their metabolic process. Few dark fermentative microorganisms do not necessitate sunlight as an energy source and can grow under anaerobic circumstances. These organisms are considered as obligate anaerobes, which are additionally categorized depending on their growth temperature and oxygen tolerance (Fig. 2). In practical aspects, cultivation of facultative anaerobes is more viable over obligate anaerobic microorganisms. However, by considering microbial growth temperature, they may be further subclassified into thermophiles, mesophiles, and psychrohiles. Even though the thermophiles grow at high temperatures where very intensive energy is required (Chandrasekhar et al. 2015a), their overall hydrogen yield is very closer to the theoretical yield by devastating the thermodynamic barrier. Several photo-fermentative microorganisms necessitate solar/light energy to evolve hydrogen in anoxygenic circumstances. Under anaerobic conditions, photoautotrophs including green algae and cyanobacteria can produce hydrogen through biophotolysis, by taking advantage of their specific metabolic route under specific conditions (Chandrasekhar et al. 2015a).

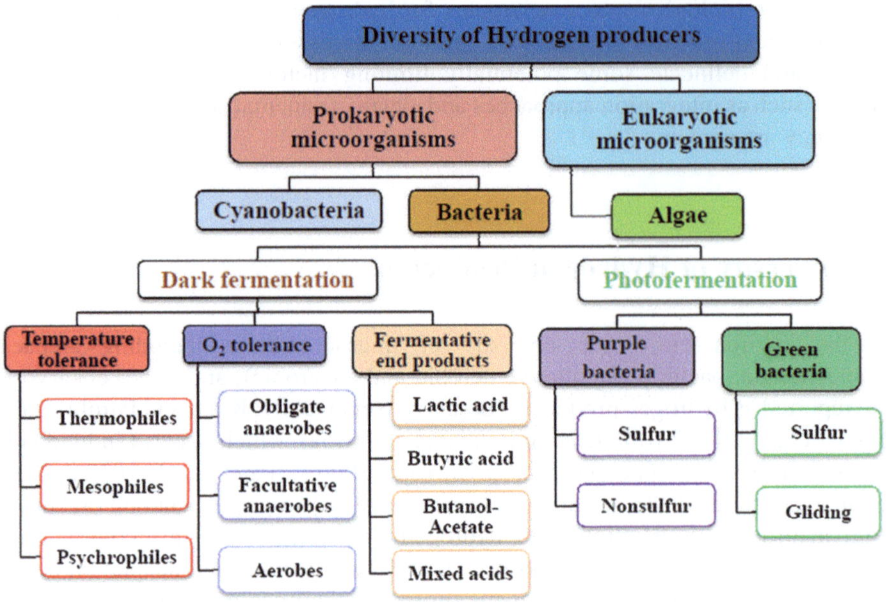


**Fig. 2** Schematic illustration of the diversity of hydrogen producers

## 4 Biohydrogen Production Processes

Biohydrogen production by either biological routes or thermochemical treatment of biomass can be described as biohydrogen. Similarly, thermochemically produced hydrogen gas is also named as biohydrogen due to the usage of biomass as a potential feedstock. However, different biological routes are available for the production of biohydrogen pertaining to anaerobic, photobiological, microbial electrolysis cell (MEC)/electro-fermentation, and enzymatic routes. The research society around the globe showed significance interest upon biological routes of hydrogen production in recent years. The passing decades above all illustrated remarkable research on both applied and basic fields.

On the basis of hydrogen evolving systems, a wide variety of diverse biological processes can be classified as follows: (1) water-splitting photosynthesis mechanism; (2) photo-fermentation mechanism; (3) dark fermentation mechanism; and (4) electro-fermentation/microbial electrolysis process (Fig. 3). Each process has its own advantages and disadvantages over other methods with respect to practicability and energy efficiency. Therefore, choice of suitable biocatalyst and/or inoculum is major subject, which is in a straight line interrelated with hydrogen production process. As expected, hydrogen can be produced by applying a single microbial type or by mixed consortia, of which few of them are hydrogen producers, whereas the rest of them utilize hydrogen for their energy supplies. Early researches on hydrogen production were typically restricted to the usage of single cultures by

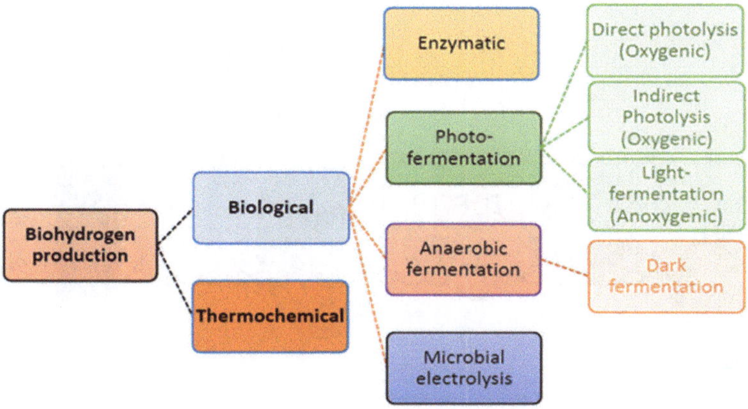


**Fig. 3** Schematic illustration concerning different routes of biohydrogen production

feeding with a defined substrate. On the other hand, when carbon-rich wastewater is used as the carbon source, mixed culture might be favorable and practically significant for the production of hydrogen at large scale (Venkateswar Reddy et al. 2011a; Mullai et al. 2013; Roy and Das 2016). In addition, mixed microbial populations are typically preferred due to operational simplicity, stability, different biochemical pathways, and also an opportunity of using a wide variety of substrates as energy source, as well as to avoid needless sterilization process (Wang and Wan 2009; Venkata Mohan et al. 2013). As a result, for the practicability of microbial hydrogen production in future, an appropriate choice of biohydrogen systems jointly with deep knowledge of their biological and physicochemical functions is a decisive factor.

## *4.1 Biophotolysis (Water-Splitting Photosynthesis)*

At present, the highly desirable hydrogen production mechanism is biophotolysis process. The oxygen producing photosynthetic biocatalysts (e.g., cyanobacteria and green microalgae) used for this process require only water and sunlight/light energy. This process has amazing theoretical advantage of an essentially unlimited supply of substrate and, potentially, the availability of incredible total energy. In green algae, a [FeFe]-hydrogenase drives the hydrogen production, whereas in the case of heterocystous cyanobacteria, nitrogenase is responsible for this process. However, this process is additionally separated as direct biophotolysis and indirect biophotolysis processes (Fig. 3). As illustrated in Fig. 4, in direct biophotolysis process, $e^-$ resulting from water-splitting mechanism are transported through PS II and PS I to ferredoxin (Fd) as an $e^-$ transporter, and consequently the reduced form of Fd($Fd_{re}$) reduces an enzyme hydrogenase which plays a key role in hydrogen production (Melis et al. 2000):

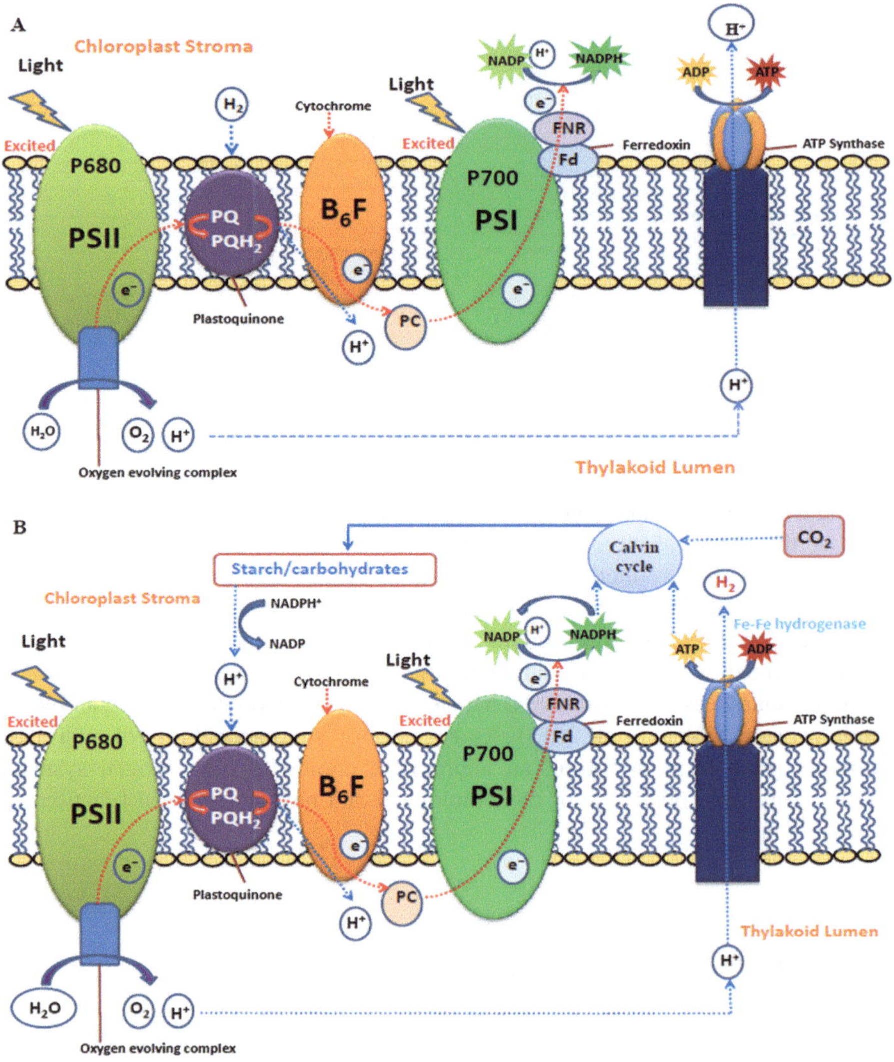


**Fig. 4** Schematic representation of biohydrogen evolution through (**a**) direct biophotolysis and (**b**) indirect biophotolysis

$$2H^+ + 2\text{ reduced ferredoxin }(Fd_{re}) \leftrightarrow H_2 + 2\text{ oxidized ferredoxin }(Fd_{ox}) \quad (1)$$

whereas in the case of indirect biophotolysis process, solar/light energy converts to chemical energy (and stored in the form of carbohydrate molecules), which are further reused to produce biohydrogen. So far, these hydrogen producing systems have been intensively explored using green algae and cyanobacteria (Hallenbeck 2012). Since the biohydrogen production by cyanobacteria takes place in the heterocyst and oxygen evolving photosynthesis as a microscopic-indirect biophotolysis process, associated with carbon dioxide fixation, extremely oxygen-sensitive nitrogenase can be protected, which leads to the formation of hydrogen.

$$\begin{aligned} &\text{Nitrogen } (N_2) + 8H^+ + 8e^- + 16\text{ATP} \\ &\quad \rightarrow \text{hydrogen } (H_2) + 2NH_3 + 16\text{ADP} + 16\text{Pi} \end{aligned} \tag{2}$$

On the other hand, hydrogen evolution by [FeFe]-hydrogenase and oxygen evolving photosynthesis mechanism cannot take place at the same time in green algae. Therefore, in order to attain sustainable hydrogen production, elemental sulfur deficiency, which caused a rigorous ($\approx$90%) decrement in photosynthesis process, was subjected to microorganisms grown on acetate as carbon source, consequential in a severe decline in oxygen evolution rate together with enhanced respiration due to the presence of leftover acetate. This situation guides the microbes to grow in anaerobic environment to produce biohydrogen via utilization of few $e^-$ from the leftover $H_2O$-splitting process (direct biophotolysis mechanism) and reserved carbon molecules (indirect biophotolysis mechanism) (Tekucheva and Tsygankov 2012).

In addition, persistent hydrogen production by single-cell, non-heterocystous cyanobacterium *Cyanothece* has been reported by growing in the glycerol-supplemented medium allowing for respiratory protection (Bandyopadhyay et al. 2010) or by replacing evolved oxygen with Argon gas during photosynthesis (Venkata Mohan and Pandey 2013). Very recently, alternative strategies to improving hydrogen production have been studied, together with a reducing the antenna size (Kosourov et al. 2011), downregulation of photosystem II proteins (Scoma et al. 2012), and alterations in operating parameters (Esquível et al. 2011).

## 4.2 *Anoxygenic Photo-fermentation*

This process similarly involves the transformation of solar/light energy in the direction of biomass with hydrogen and carbon dioxide. For this process, PNS photosynthetic bacteria as well as *Rhodobacter* species can be used to change organic acids (e.g., $CH_3COOH$, $C_3H_6O_3$, and $C_4H_8O_2$) to hydrogen and carbon dioxide in defined conditions such as anoxic and anaerobic environment. Indeed, under ammonium ions deficiency, these microorganisms capture light energy to change carbon-rich organic acids into hydrogen using nitrogenases (Azwar et al. 2014). On the other hand, oxygen-sensitive nitrogenase is not an issue for this mechanism since the PNS bacteria used to carry out this process have non-oxygenic photosynthesis mechanism (Chandrasekhar et al. 2015a). Nonetheless, nitrogenases also have numerous defects for hydrogen production due to inhibition of their expressions by $NH_4$, low catalytic action, and less photochemical efficiency (Brentner et al. 2010). In hypothesis, the photo-fermentation process is capable of completing conversion of organic compounds into hydrogen, even touching a moderately high hydrogen partial pressure; meanwhile, hydrogen production is driven by nitrogenase and leads to the formation of ATP through the capture of light energy through a photosynthesis.

## 4.3 Dark Fermentation

Dark or heterotrophic fermentation by anaerobic microorganisms as well as some microalgae (such as green algae) can produce hydrogen by utilizing carbon-rich substrate in the absence of sunlight/solar energy under anaerobic environment (Figs. 5 and 6). So far, numerous studies on biological hydrogen production through this mechanism have been achieved by employing facultative and obligate anaerobes (Chandrasekhar and Venkata Mohan 2014a, b). This process happens at a higher reaction rate than bio-photolysis and photo-fermentation process. On the other hand, low hydrogen yield on substrate due to the production and accumulation of numerous acid intermediates is considered to be a major drawback. The dark fermentation mechanism aids to produce NAD(P)H and FADH as reducing powers during metabolism, which are followed by sequential re-oxidation by terminal $e^-$ acceptor (TEA), which leads to the formation of ATP. In the case of aerobic respiration process, oxygen acts as a potential TEA that aids in generating ATP as energy-rich molecule via simultaneous regeneration of reducing powers (Chandrasekhar et al. 2015a). On the other hand, anaerobic respiration process utilizes a diversity of organic and inorganic compounds (e.g., $SO_4^{2-}$, $NO^{3-}$, etc.) as TEA (Fig. 6). Therefore, glycolysis is considered as vital metabolic process where glucose/substrate can be converted to pyruvate ($CH_3COCOO^-$), an essential intermediately metabolite (Fig. 5).

Further, $CH_3COCOO^-$ in the acidogenic metabolic pathway together with hydrogen production under anaerobic circumstances leads to the formation of acid intermediates such as $CH_3COOH$, $C_2H_5COOH$, $C_3CH_7COOH$, $C_4H_6O_5$, and so on, collectively called as volatile fatty acids (Eqs. 3–7).

$$\text{Glucose} + 2H_2O \rightarrow 2CH_3COOH + 2CO_2(\uparrow) + 4H_2(\uparrow) \tag{3}$$

$$\text{Glucose} + 2H_2 \rightarrow 2C_2H_5COOH + 2H_2O \tag{4}$$

$$\text{Glucose} \rightarrow C_3CH_7COOH + 2CO_2(\uparrow) + 2H_2(\uparrow) \tag{5}$$

$$\text{Glucose} + 2H_2 \rightarrow C_4H_6O_5 + CO_2(\uparrow) \tag{6}$$

$$\text{Glucose} \rightarrow C_2H_5OH + CO_2(\uparrow) \tag{7}$$

As affirmed on top, both facultative and obligate microorganisms can produce hydrogen with a wide range of substrates (Chandrasekhar et al. 2015a). Facultative anaerobes convert $CH_3COCOO^-$ to acetyl-CoA and further into $HCOO^-$ (formate) by pyruvate formate lyase (PFL) and then yield hydrogen by the catalysis of formate hydrogen lyase (FHL).

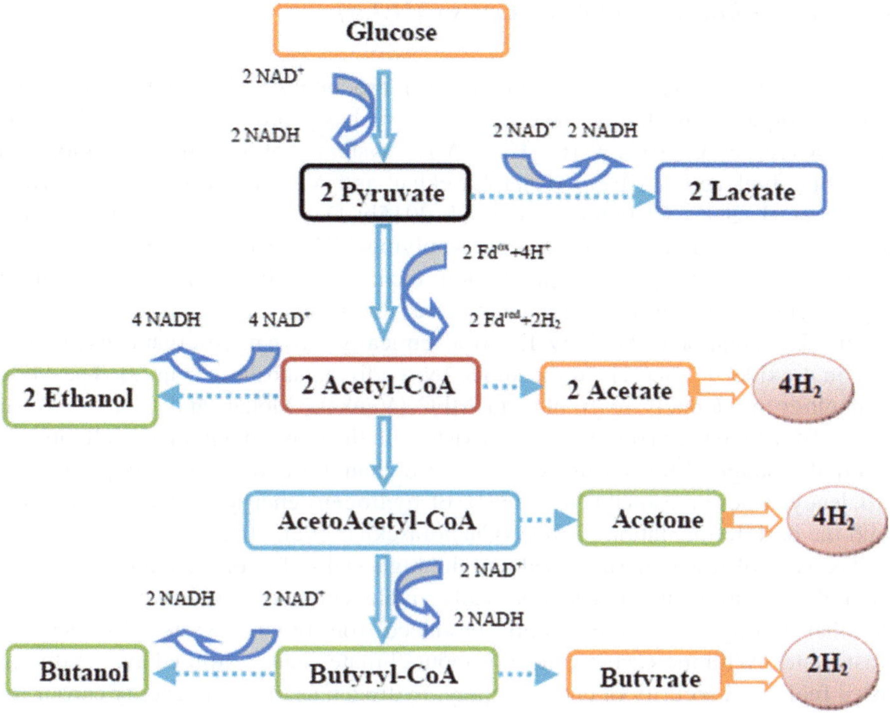


**Fig. 5** Schematic illustration of anaerobic fermentation pathways for the breakdown of glucose and production of various metabolites

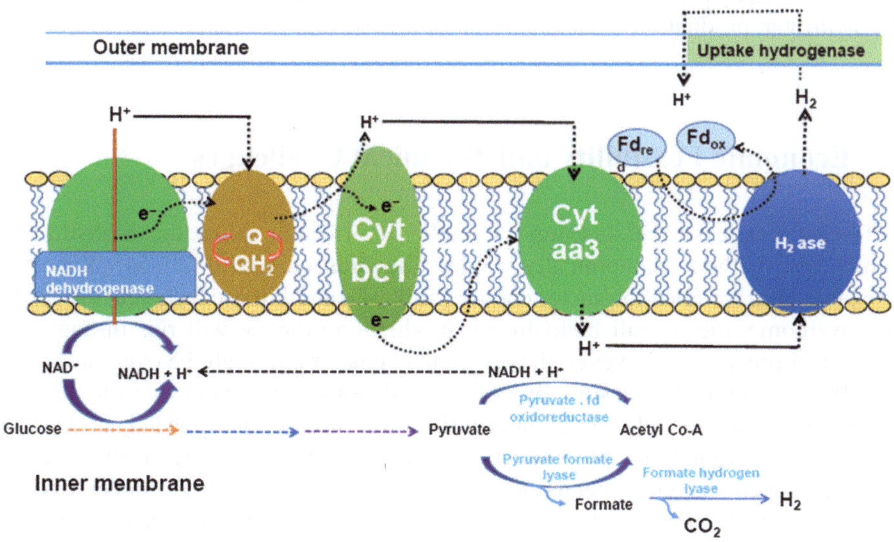


**Fig. 6** Schematic representation of hydrogen evolution through dark fermentation

### 4.4 Microbial Electrolysis Cells (MECs)

MECs, an innovative and rapid method for producing hydrogen from different types of organic substrates, have been in very hasty progress over the last few years (Kadier et al. 2015, 2016). These MEC systems are fundamentally modified microbial fuel cells (MFCs) (Chandrasekhar and Venkata Mohan 2012; Kiran Kumar et al. 2012; Chandrasekhar et al. 2015b; Deval et al. 2016), which have been under investigation for decades. Applying MEC technology as an alternate electrically driven hydrogen production technology allows the conversion of organic substrates into hydrogen. Therefore, this technology is also termed as an electro-fermentation technology. Electrochemically active microorganisms produce carbon dioxide, $e^-$, and $H^+$ at the anode by oxidizing substrate (Fig. 7). The MEC technology moderately resembles an MFC (Venkata Mohan and Chandrasekhar 2011a, b), where the major difference exists with the requirement of a small input of external voltage. MECs can achieve more than 90% of hydrogen production efficiency, where reasonably less (33% of hydrogen) energy recovery is possible with the dark fermentation process (Chandrasekhar et al. 2015a).

The type of raw materials used and the high yield of hydrogen make the MEC technology a promising and economically viable technology. Its novelty lies in the fact that hydrogen can be efficiently produced from organic waste substrates like wastewater. With increasing concerns about climate change, limited availability of fossil fuels, and security of energy supply, hydrogen as the clean energy carrier for the future has drawn intense global interests due to its unique properties. Biohydrogen from MEC is predictable to play a vital role in a non-fossil fuel based future economy. MEC is becoming popular because microorganisms can be used as catalyst and wastewater can be effectively utilized as potential substrate for hydrogen production (Chandrasekhar et al. 2015a).

## 5 Economic Feasibility and Technical Challenges

In recent years, several researchers made an effort to make the biohydrogen production methods economically viable (Chandrasekhar et al. 2015a). Nonetheless, few vital technical obstacles endure (Table 1), and if these technical problems are overcome, the overall biohydrogen production efficacy will rise through the biological processes. However, these technical challenges might be overcome with an effective bioreactor design, process amendments, assortment of suitable substrate, and with the choice of appropriate and competent biocatalyst. During biohydrogen production, by-products produced by the microorganism compete with hydrogen evolving metabolic pathways and this rerouting reduces overall hydrogen yields (Pandit et al. 2014). Henceforth, numerous investigators are endeavoring to redirect the metabolic pathways to decrease the formation of the low-end metabolic products (Roy et al. 2015, 2016). To overcome the

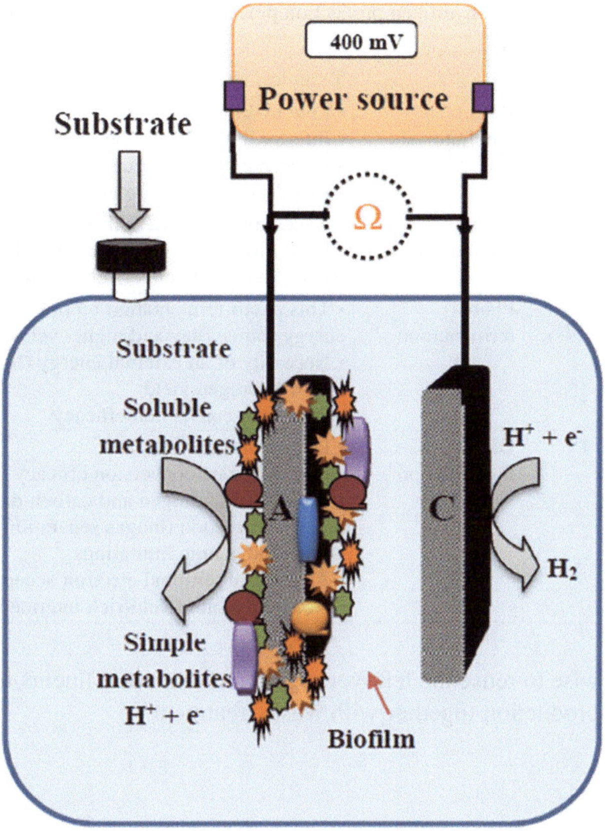


**Fig. 7** Schematic illustration of single-chambered microbial electrolysis cell (MEC)

stoichiometric restriction (to produce closely 4 mol $H_2$/mol glucose; theoretical hydrogen yield) of the dark fermentation, an efficient metabolically engineered microorganism must be examined.

## 6 Strategies to Improve the Process Efficacy

Low substrate conversion efficacy and accumulation of by-products/acid intermediates were considered as major deterrents to the conventional biohydrogen process (Chandrasekhar and Venkata Mohan 2014a). In particular, the dark fermentative hydrogen production process has a major difficulty for practical applications due to low yield (4 mol of hydrogen yield from each mole of glucose), with a conversion efficiency of 33% (Cheng et al. 2012). In addition, at the end of the dark fermentation process, significant quantities of residual organic acids are still existent in the bioreactor effluents (Kumar et al. 2016a). Consequently, further treatments are necessary prior to disposal. Considering environmental and economic factors, it is

**Table 1** Biohydrogen production processes and technical challenges

| S. No | Bioprocess type | Technical challenge |
|---|---|---|
| 1 | Direct biophotolysis | • Low hydrogen yield<br>• Inferior light conversion efficiency<br>• Oxygen production(due to the activity of PS II)<br>• Requisite for custom-made bioreactors |
| 2 | Indirect biophotolysis | • Necessity of an external light/solar energy source<br>• Lower hydrogen yield caused by hydrogenase<br>• Poor total light conversion efficiency |
| 3 | Photo-fermentation | • This photo-fermentation bioprocess is limited by sunlight as the energy source (day and night cycles)<br>• Necessity of an external energy (light energy) source<br>• Low hydrogen yield<br>• Poor light conversion efficacy |
| 4 | Dark fermentation | • Low hydrogen yield<br>• Low substrate conversion efficacy<br>• Mixture of hydrogen and carbon dioxide gases as products, which require separation (biogas separation)<br>• Thermodynamic limitations<br>• Absence of terminal electron acceptors<br>• Accumulation of acid-rich intermediate metabolites |

wise to reuse the leftover organic fraction of effluents for additional energy/biofuel production together with waste treatment.

## 6.1 *Integration Approaches*

In recent years, various process integration approaches have been proposed to address numerous process confines to increase dark fermentative hydrogen production. Utilization of the residual organic substances (such as VFAs) from the bioreactor effluents as potential feedstock for additional energy recovery is practical and an excellent idea, as in the form of integrated bioprocesses (Table 2). Several secondary bioprocesses together with dark fermentation for hydrogen production, photobiological process for hydrogen production, methanogenesis process for methane production (Laurinavichene et al. 2012; Chandrasekhar et al. 2015a), microbial electrolysis system for hydrogen production (Cheng and Logan 2007), anoxygenic nutrient-limiting process for bioplastics production, hetrotrophic algae cultivation process for lipids production, and microbial fuel cell technology for electricity generation were possible secondary bioprocess which can be integrated with dark fermentation process considering as primary hydrogen production process. These integrated approaches enable further utilization of effluents from the primary process as potential substrate in secondary process for additional energy generation; consequently, whole process becomes economically more viable and practically applicable.

**Table 2** List of few two-stage integration approaches investigated with dark fermentation

| | First stage | | Second stage | | |
|---|---|---|---|---|---|
| Substrate | Process type | Reported yield | Process type | Reported yield | References |
| Cassava wastewater | $H_2$ ($F_D$) | 54.22 ml $H_2$/g | $CH_4$ ($F_D$) | 164.87 ml $CH_4$/g | Intanoo et al. (2014) |
| Food waste | $H_2$ ($F_D$) | 85 L/kg TS | $CH_4$ ($F_D$) | 63.3 L/kg TS | Kumar et al. (2014a) |
| Vegetable waste | $H_2$ ($F_D$) | 17 L/kg TS | $CH_4$ ($F_D$) | 61.7 L/kg TS | Kumar et al. (2014a) |
| Microalgal biomass | $H_2$ ($F_D$) | 135 ± 3.11 ml $H_2$/g/VS | $CH_4$ ($F_D$) | 414 ± 2.45 ml $CH_4$/g/VS | Wieczorek et al. (2014) |
| Glucose | $H_2$ ($F_D$) | 1.20 mmol | $H_2$ ($F_P$) | 5.22 mmol | Chandra and Venkata Mohan (2011) |
| Cheese Whey wastewater | $H_2$ ($F_D$) | 2.04 mol | $H_2$ ($F_P$) | 2.69 mol | Rai et al. (2012) |
| Fruit juice industry wastewater | $H_2$ ($F_D$) | 1.4 mol $H_2$/mol hexose | Electricity ($F_D$) | 0.55 W/$m^2$ | Gonzalez del Campo et al. (2012) |
| Corn stover lignocellulose | $H_2$ ($F_D$) | 1.67 mol $H_2$/mol-glucose | $H_2$ (MEC) | 1.00 L/L-d | Lalaurette et al. (2009) |
| Spent wash | $H_2$ ($F_D$) | 39.8 L | Bioplastic | 40% dry cell weight | Amulya et al. (2014) |
| Pea shells | $H_2$ ($F_D$) | 5.2 L $H_2$ from 4 L | Bioplastic | 1685 mg PHB/L | Patel et al. (2012b) |
| Food waste | $H_2$ ($F_D$) | 69.94 mmol | Lipid | 26.4% dry cell weight of algae | Venkata Mohan and Devi (2012) |
| Food waste | Bioelectricity | 85.2 mW/$m^2$ | $H_2$ ($F_D$) | 0.91 L | Chandrasekhar and Venkata Mohan (2014b) |

$F_D$ dark fermentation, $F_P$ photo-fermentation, *TS* total solids, *MEC* microbial electrolysis cell

### 6.1.1 Photobiological Process

Residual organic acids can be readily consumed by photosynthetic bacteria (Rai et al. 2012). While metabolic intermediates from dark fermentation process can be efficiently utilized by few PNS bacteria, the two-stage integration approach of dark fermentation with anoxygenic photo-fermentation process will have double benefit of enhanced hydrogen yield together with concurrent substrate degradation. Green algae can also utilize residual metabolic intermediates from dark fermentation process, especially when acetic acid is used as a potential substrate (Bala Amutha and Murugesan 2011; Chandra and Venkata Mohan 2011). Nonetheless, photo-fermentation of residual organic substances from the hydrogen bioreactor is considered to be more difficult than dark fermentation process with respect to the process effectiveness, due to poor light diffusion, maintaining microenvironment, substrate inhibition, nutritional requirements for microbial growth, and risk of contamination (Özkan et al. 2012).

### 6.1.2 Biopolymers/Bioplastics

The organic acid-rich bioreactor reactor effluents from first-stage dark fermentation process are a promising feedstock for the production and accumulation of biopolymers [such as polyhydroxyalkanoates (PHA)] in bacterial cells at second-stage integrated bioprocess. The PHAs are a biopolyester which accumulates as cellular reserve storage materials, formed under additional nutrient and carbon-deprived conditions (Amulya et al. 2014; Kumar et al. 2014b, 2015c). These biodegradable polyesters are deposited as cytoplasmic inclusions in microbial cells, while extreme carbon source is offered and while other nutrients are growth-limiting. Production of PHA using pure strains cultures by supplementing synthetic substrates as carbon source (e.g., acetate, butyrate, etc), which is neither economically viable nor cost-effective for its production at large-scale bioreactors. VFA are simple acid-rich organic substrates with a lower number of carbons, which facilitates PHA production by the contribution of a less number of metabolic enzymes when compared to glycolysis and β-oxidation (Amulya et al. 2014). Production of PHB from different fatty acids (such as acetate, propionate, butyrate, etc) and organic effluents from a dark fermentation process were investigated under anoxic conditions using a mixed microbial population (Amulya et al. 2014; Chandrasekhar et al. 2015a; Kumar et al. 2016b). The bioplastic production coupled with $H_2$ production process and utilization of their effluents for methanogenesis enabled the whole process to be more economically viable (Patel et al. 2012b).

### 6.1.3 MEC-Driven Biohydrogen Production from Acid-Rich Effluents

Recently, considerable interest is focused on microbial electrolysis cells (MECs) due to the fusibility of integrating with other bioprocesses such as dark fermentation process (Lalaurette et al. 2009; Chandrasekhar et al. 2015a). MEC is an innovative and promising technology for hydrogen production from organic substrates, including carbon-rich wastewater and additional renewable resources. MEC as an alternate (electrically driven) biohydrogen production process can enable the biotransformation of substrate into hydrogen under applied external voltage. Certainly, this electro-fermentation process was viable for the utilization of a wide range of organic substrates to produce hydrogen together with simultaneous wastewater remediation (Wagner et al. 2009). A two-stage bioprocess was investigated to utilize organic effluents of dark fermentation bioreactor as potential feedstock for extra hydrogen production (Fig. 8).

In consideration of superior substrate conversion efficacy (90%) of MEC process, such a two-stage process integration approach, i.e., integration of MEC process with dark fermentative biohydrogen production process, could be a feasible and potential thought to attain higher hydrogen yield and to enhance substrate conversion efficacy (Chandrasekhar et al. 2015a).

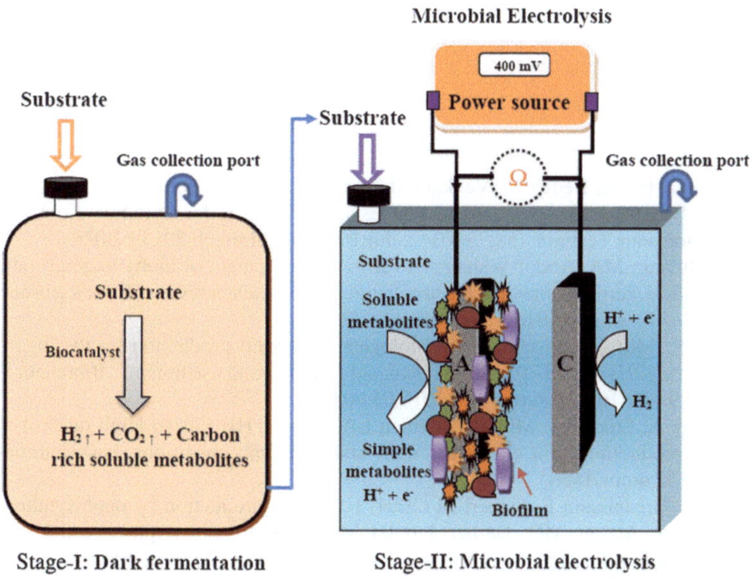


**Fig. 8** Schematic representation of a two-stage process integration approach, i.e., integration of MEC process with dark fermentative biohydrogen production process to attain higher hydrogen yield (*A* anode, *C* cathode, *Ω* resistance)

## 7 Conclusion

In recent years, hydrogen has emerged as a clean, eco-friendly, and promising substitute fuel since it can be derived from renewable energy sources. It appears as the hopeful substitute to fossil fuel-based energy sources. To make hydrogen-based economy sustainable, it is necessary to practice renewable resources as an alternative of fossil fuels to produce hydrogen. Especially, bioconversion of carbon-rich substrate or $H_2O$ into hydrogen is promising and eco-friendly technology. This chapter described the significance of biologically produced hydrogen as a clean energy carrier to build up hydrogen-based economy in the coming years. The importance of useful organisms involved in this technology was discussed toward the awareness of a stable hydrogen-based economy. Environmental impact of hydrogen as carbon-neutral energy carrier is also mentioned. Among the several bioprocesses identified to produce hydrogen, dark fermentation process has the dual advantages of gaseous energy generation and waste treatment. However, production of hydrogen from renewable resource through biological routes, using biocatalyst, is one of the potential areas to develop hydrogen-based economy. Even if numerous innovative approaches are expected in future to overcome few of the existing technical challenges, biohydrogen production process necessitates a multidisciplinary method for the biohydrogen technology to be economically viable and eco-friendly technology.

## References

Amulya K, Venkateswar Reddy M, Venkata Mohan S (2014) Acidogenic spent wash valorization through polyhydroxyalkanoate (PHA) synthesis coupled with fermentative biohydrogen production. Bioresour Technol 158:336–342. doi:10.1016/j.biortech.2014.02.026

Azwar MY, Hussain MA, Abdul-Wahab AK (2014) Development of biohydrogen production by photobiological, fermentation and electrochemical processes: a review. Renew Sustain Energy Rev 31:158–173. doi:10.1016/j.rser.2013.11.022

Bala Amutha K, Murugesan AG (2011) Biological hydrogen production by the algal biomass *Chlorella vulgaris* MSU 01 strain isolated from pond sediment. Bioresour Technol 102:194–199. doi:10.1016/j.biortech.2010.06.008

Bandyopadhyay A, Stöckel J, Min H, Sherman LA, Pakrasi HB (2010) High rates of photobiological $H_2$ production by a cyanobacterium under aerobic conditions. Nat Commun 1:139. doi:10.1038/ncomms1139

Boichenko VA, Greenbaum E, Seibert M (2004) Hydrogen production by photosynthetic microorganisms. In: Archer MD, Barber J (eds) Molecular to global photosynthesis. Imperial College Press, London, pp 397–451. doi:10.1007/978-0-585-35132-2_40

Brentner LB, Jordan PA, Zimmerman JB (2010) Challenges in developing biohydrogen as a sustainable energy source: implications for a research agenda. Environ Sci Technol 44:2243–2254. doi:10.1021/es9030613

Chandra R, Venkata Mohan S (2011) Microalgal community and their growth conditions influence biohydrogen production during integration of dark-fermentation and photo-fermentation processes. Int J Hydrogen Energy 36:12211–12219. doi:10.1016/j.ijhydene.2011.07.007

Chandrasekhar K, Venkata Mohan S (2012) Bio-electrochemical remediation of real field petroleum sludge as an electron donor with simultaneous power generation facilitates biotransformation of pah: effect of substrate concentration. Bioresour Technol 110:517–525. doi:10.1016/j.biortech.2012.01.128

Chandrasekhar K, Venkata Mohan S (2014a) Bio-electrohydrolysis as a pretreatment strategy to catabolize complex food waste in closed circuitry: function of electron flux to enhance acidogenic biohydrogen production. Int J Hydrogen Energy 39:11411–11422. doi:10.1016/j.ijhydene.2014.05.035

Chandrasekhar K, Venkata Mohan S (2014b) Induced catabolic bio-electrohydrolysis of complex food waste by regulating external resistance for enhancing acidogenic biohydrogen production. Bioresour Technol 165:372–382. doi:10.1016/j.biortech.2014.02.073

Chandrasekhar K, Lee YJ, Lee DW (2015a) Biohydrogen production: strategies to improve process efficiency through microbial routes. Int J Mol Sci 16:8266–8293. doi:10.3390/ijms16048266

Chandrasekhar K, Amulya K, Venkata Mohan S (2015b) Solid phase bio-electrofermentation of food waste to harvest value-added products associated with waste remediation. Waste Manage 45:57–65. doi:10.1016/j.wasman.2015.06.001

Cheng S, Logan BE (2007) Sustainable and efficient biohydrogen production via electrohydrogenesis. Proc Natl Acad Sci 104:18871–18873. doi:10.1073/pnas.0706379104

Cheng H, Liang YL, Xing X, Liu X, Han R, Lv Q, Mishra S (2012) Efficient misbehaving user detection in online video chat services. In: Proceedings of the fifth ACM international conference on Web search and data mining. ACM, Seattle, Washington, pp 23–32. doi:10.1145/2124295.2124301

Christopher K, Dimitrios RA (2012) Review on exergy comparison of hydrogen production methods from renewable energy sources. Energy Environ Sci 5:6640–6651. doi:10.1039/C2EE01098D

Deval AS, Parikh HA, Kadier A, Chandrasekhar K, Bhagwat AM, Dikshit AK (2016) Sequential microbial activities mediated bioelectricity production from distillery wastewater using bio-electrochemical system with simultaneous waste remediation. Int J Hydrogen Energy. doi:10.1016/j.ijhydene.2016.11.114

Elam CC, Padró CEG, Sandrock G, Luzzi A, Lindblad P, Hagen EF (2003) Realizing the hydrogen future: the international energy agency's efforts to advance hydrogen energy technologies. Int J Hydrogen Energy 28:601–607. doi:10.1016/S0360-3199(02)00147-7

Esquível MG, Amaro HM, Pinto TS, Fevereiro PS, Malcata FX (2011) Efficient h2 production via *Chlamydomonas reinhardtii*. Trends Biotechnol 29:595–600. doi:10.1016/j.tibtech.2011.06.008

Gonzalez del Campo A, Cañizares P, Lobato J, Rodrigo MA, Fernandez FJ (2012) Electricity production by integration of acidogenic fermentation of fruit juice wastewater and fuel cells. Int J Hydrogen Energy 37:9028–9037. doi:10.1016/j.ijhydene.2012.03.007

Hallenbeck P (2012) Hydrogen production by cyanobacteria. In: Hallenbeck PC (ed) Microbial technologies in advanced biofuels production. Springer, New york, pp 15–28. doi:10.1007/978-1-4614-1208-3_2

Intanoo P, Rangsanvigit P, Malakul P, Chavadej S (2014) Optimization of separate hydrogen and methane production from cassava wastewater using two-stage upflow anaerobic sludge blanket reactor (UASB) system under thermophilic operation. Bioresour Technol 173:256–265. doi:10.1016/j.biortech.2014.09.039

Kadier A, Simayi Y, Chandrasekhar K, Ismail M, Kalil MS (2015) Hydrogen gas production with an electroformed Ni mesh cathode catalysts in a single-chamber microbial electrolysis cell (MEC). Int J Hydrogen Energy 40:14095–14103. doi:10.1016/j.ijhydene.2015.08.095

Kadier A, Simayi Y, Abdeshahian P, Azman NF, Chandrasekhar K, Kalil MS (2016) A comprehensive review of microbial electrolysis cells (MEC) reactor designs and configurations for sustainable hydrogen gas production. Alex Eng J 55:427–443. doi:10.1016/j.aej.2015.10.008

Kalia VC, Purohit HJ (2008) Microbial diversity and genomics in aid of bioenergy. J Ind Microbiol Biotechnol 35:403–419. doi:10.1007/s10295-007-0300-y

Kiran Kumar A, Venkateswar Reddy M, Chandrasekhar K, Srikanth S, Venkata Mohan S (2012) Endocrine disruptive estrogens role in electron transfer: bio-electrochemical remediation with microbial mediated electrogenesis. Bioresour Technol 104:547–556. doi:10.1016/j.biortech.2011.10.037

Kosourov SN, Ghirardi ML, Seibert M (2011) A truncated antenna mutant of *Chlamydomonas reinhardtii* can produce more hydrogen than the parental strain. Int J Hydrogen Energy 36:2044–2048. doi:10.1016/j.ijhydene.2010.10.041

Kumar P, Patel SKS, Lee JK, Kalia VC (2013) Extending the limits of *Bacillus* for novel biotechnological applications. Biotechnol Adv 31:1543–1561. doi:10.1016/j.biotechadv.2013.08.007

Kumar P, Pant DC, Mehariya S, Sharma R, Kansal A, Kalia VC (2014a) Ecobiotechnological strategy to enhance efficiency of bioconversion of wastes into hydrogen and methane. Indian J Microbiol 54:262–267. doi:10.1007/s12088-014-0467-7

Kumar P, Singh M, Mehariya S, Patel SKS, Lee JK, Kalia VC (2014b) Ecobiotechnological approach for exploiting the abilities of *Bacillus* to produce co-polymer of polyhydroxyalkanoate. Indian J Microbiol 54:151–157. doi:10.1007/s12088-014-0457-9

Kumar G, Bakonyi P, Sivagurunathan P, Kim SH, Nemestóthy N, Bélafi-Bakó K (2015a) Lignocellulose biohydrogen: practical challenges and recent progress. Renew Sustain Energy Rev 44:728–737. doi:10.1016/j.rser.2015.01.042

Kumar P, Ray S, Patel SK, Lee JK, Kalia VC (2015b) Bioconversion of crude glycerol to polyhydroxyalkanoate by *Bacillus thuringiensis* under non-limiting nitrogen conditions. Int J Biol Macromol 78:9–16. doi:10.1016/j.ijbiomac.2015.03.046

Kumar P, Sharma R, Ray S, Mehariya S, Patel SKS, Lee JK, Kalia VC (2015c) Dark fermentative bioconversion of glycerol to hydrogen by *Bacillus thuringiensis*. Bioresour Technol 182:383–388. doi:10.1016/j.biortech.2015.01.138

Kumar G, Bakonyi P, Kobayashi T, Xub K-Q, Sivagurunathan P, Kim S-H, Buitrón G, Nemestóthy N, Bélafi-Bakó K (2016a) Enhancement of biofuel production via microbial augmentation: the case of dark fermentative hydrogen. Renew Sustain Energy Rev 57:879–891. doi:10.1016/j.rser.2015.12.107

Kumar P, Ray S, Kalia VC (2016b) Production of co-polymers of polyhydroxyalkanoates by regulating the hydrolysis of biowastes. Bioresour Technol 200:413–419. doi:10.1016/j.biortech.2015.10.045

Lalaurette E, Thammannagowda S, Mohagheghi A, Maness PC, Logan BE (2009) Hydrogen production from cellulose in a two-stage process combining fermentation and electrohydrogenesis. Int J Hydrogen Energy 34:6201–6210. doi:10.1016/j.ijhydene.2009.05.112

Laurinavichene TV, Belokopytov BF, Laurinavichius KS, Khusnutdinova AN, Seibert M, Tsygankov AA (2012) Towards the integration of dark- and photo-fermentative waste treatment. 4. Repeated batch sequential dark- and photofermentation using starch as substrate. Int J Hydrogen Energy 37:8800–8810. doi:10.1016/j.ijhydene.2012.01.132

Lee SJ, Lee SJ, Lee DW (2013) Design and development of synthetic microbial platform cells for bioenergy. Front Microbiol 4:92. doi:10.3389/fmicb.2013.00092

Melis A, Zhang L, Forestier M, Ghirardi ML, Seibert M (2000) Sustained photobiological hydrogen gas production upon reversible inactivation of oxygen evolution in the green alga *Chlamydomonas reinhardtii*. J Plant Physiol 122:127–136. doi:10.1104/pp.122.1.127

Mullai P, Rene ER, Sridevi K (2013) Biohydrogen production and kinetic modeling using sediment microorganisms of pichavaram mangroves, India. Biomed Res Int. doi:10.1155/2013/265618

Özkan E, Uyar B, Özgür E, Yücel M, Eroglu I, Gündüz U (2012) Photofermentative hydrogen production using dark fermentation effluent of sugar beet thick juice in outdoor conditions. Int J Hydrogen Energy 37:2044–2049. doi:10.1016/j.ijhydene.2011.06.035

Pandit S, Balachandar G, Das D (2014) Improved energy recovery from dark fermented cane molasses using microbial fuel cells. Front Chem Sci Eng 8:43–54. doi:10.1007/s11705-014-1403-4

Patel SKS, Kumar P, Kalia VC (2012a) Enhancing biological hydrogen production through complementary microbial metabolisms. Int J Hydrogen Energy 37:10590–10603. doi:10.1016/j.ijhydene.2012.04.045

Patel SKS, Singh M, Kumar P, Purohit HJ, Kalia VC (2012b) Exploitation of defined bacterial cultures for production of hydrogen and polyhydroxybutyrate from pea-shells. Biomass Bioenergy 36:218–225. doi:10.1016/j.biombioe.2011.10.027

Patel SKS, Kumar P, Mehariya S, Lee JK, Kalia VC (2014) Enhancement in hydrogen production by co-cultures of *Bacillus* and *Enterobacter*. Int J Hydrogen Energy 39:14663–14668. doi:10.1016/j.ijhydene.2014.07.084

Rai P, Singh SP, Asthana RK (2012) Biohydrogen production from cheese whey wastewater in a two-step anaerobic process. App Biochem Biotech 167:1540–1549. doi:10.1007/s12010-011-9488-4

Roy S, Das D (2016) Biohythane production from organic wastes: present state of art. Environ Sci Pollut Res Int 23:9391–9410

Roy S, Banerjee D, Dutta M, Das D (2015) Metabolically redirected biohydrogen pathway integrated with biomethanation for improved gaseous energy recovery. Fuel 158:471–478. doi:10.1016/j.fuel.2015.05.060

Roy S, Sinha P, Das D (2016) Genomic and proteomic approaches for dark fermentative biohydrogen production. Renew Sustain Energy Rev 56:1308–1321. doi:10.1016/j.rser.2015.12.035

Scoma A, Krawietz D, Faraloni C, Giannelli L, Happe T, Torzillo G (2012) Sustained $H_2$ production in a *Chlamydomonas reinhardtii* D1 protein mutant. J Biotechnol 157:613–619. doi:10.1016/j.jbiotec.2011.06.019

Tekucheva DN, Tsygankov AA (2012) Coupled biological hydrogen-producing systems: a review. Prikl Biokhim Mikrobiol 48:357–375. doi:10.1016/j.biortech.2011.03.026

Venkata Mohan S, Chandrasekhar K (2011a) Self-induced bio-potential and graphite electron accepting conditions enhances petroleum sludge degradation in bio-electrochemical system with simultaneous power generation. Bioresour Technol 102:9532–9541. doi:10.1016/j.biortech.2011.07.038

Venkata Mohan S, Chandrasekhar K (2011b) Solid phase microbial fuel cell (SMFC) for harnessing bioelectricity from composite food waste fermentation: influence of electrode assembly and buffering capacity. Bioresour Technol 102:7077–7085. doi:10.1016/j.biortech.2011.04.039

Venkata Mohan S, Devi MP (2012) Fatty acid rich effluent from acidogenic biohydrogen reactor as substrate for lipid accumulation in heterotrophic microalgae with simultaneous treatment. Bioresour Technol 123:627–635. doi:10.1016/j.biortech.2012.07.004

Venkata Mohan S, Pandey A (2013) Biohydrogen production: an introduction. In: Larroche AP-SCCH (ed) Biohydrogen. Elsevier, Amsterdam, pp 1–24. doi:10.1016/B978-0-444-59555-3.00001-5

Venkata Mohan S, Devi MP, Venkateswar Reddy M, Chandrasekhar K, Juwarkar A, Sarma PN (2011) Bioremediation of petroleum sludge under anaerobic microenvironment: influence of biostimulation and bioaugmentation. Environ Eng Manage J 10:1609–1616

Venkata Mohan S, Chandrasekhar K, Chiranjeevi P, Babu PS (2013) Biohydrogen production from wastewater. In: Larroche AP-SCCH (ed) Biohydrogen. Elsevier, Amsterdam, pp 223–257. doi:10.1016/B978-0-444-59555-3.00010-6

Venkateswar Reddy M, Chandrasekhar K, Venkata Mohan SV (2011a) Influence of carbohydrates and proteins concentration on fermentative hydrogen production using canteen based waste under acidophilic microenvironment. J Biotechnol 155:387–395. doi:10.1016/j.jbiotec.2011.07.030

Venkateswar Reddy M, Devi MP, Chandrasekhar K, Goud RK, Venkata Mohan SV (2011b) Aerobic remediation of petroleum sludge through soil supplementation: microbial community analysis. J Hazard Mater 197:80–87. doi:10.1016/j.jhazmat.2011.09.061

Wagner RC, Regan JM, Oh S-E, Zuo Y, Logan BE (2009) Hydrogen and methane production from swine wastewater using microbial electrolysis cells. Water Res 43:1480–1488. doi:10.1016/j.watres.2008.12.037

Wang JL, Wan W (2009) Factors influencing fermentative hydrogen production: a review. Int J Hydrogen Energy 34:799–811. doi:10.1016/j.ijhydene.2008.11.015

Wieczorek N, Kucuker MA, Kuchta K (2014) Fermentative hydrogen and methane production from microalgal biomass (*Chlorella vulgaris*) in a two-stage combined process. Appl Energy 132:108–117. doi:10.1016/j.apenergy.2014.07.003

# Integrated Biorefinery Approach for the Valorization of Olive Mill Waste Streams Towards Sustainable Biofuels and Bio-Based Products

**Constantina Kourmentza, Eleni Koutra, Nikolaos Venetsaneas, and Michael Kornaros**

**Abstract** Olive oil extraction and refining process generate large amounts of by-products that represent a huge environmental concern, especially for countries located within the Mediterranean region, because of their phytotoxicity against soil and aquatic environments. Their valorization is considered challenging due to their high organic content, complexity, and the presence of phenolic compounds that inhibit their biodegradation. In order to minimize their environmental impact, many research groups within the last decades have been focusing on exploring and suggesting strategies regarding their physicochemical and microbiological treatment. According to various reports, the potential of olive mill wastewater to be converted to sustainable resources of biofuels and bio-based products has been demonstrated. In the present chapter, the most significant advances concerning a variety of promising valorization scenarios have been reviewed.

**Keyword** Olive mill wastewater • Valorization • Biofuels • Bio-based compounds • Biomaterials

---

C. Kourmentza (✉)
Department of Chemistry, UCIBIO-REQUIMTE, Biochemical Engineering Group, Faculdade de Ciências e Tecnologia, Universidade Nova de Lisboa (FCT/UNL), Quinta da Torre, Campus Universitário, 2829-516 Caparica, Portugal
e-mail: ckourmentza@gmail.com; c.kourmentza@fct.unl.pt

E. Koutra • M. Kornaros
Department of Chemical Engineering, Laboratory of Biochemical Engineering and Environmental Technology (LBEET), University of Patras, 1 Caratheodory Street, 26500 Patras, Greece

N. Venetsaneas
Faculty of Engineering and the Environment, University of Southampton, Highfield, Southampton SO17 1BJ, UK



V.C. Kalia, P. Kumar (eds.), *Microbial Applications Vol.1*,
DOI 10.1007/978-3-319-52666-9_10

# 1 Introduction

The popularity of the health benefits of olive oil, due to its high nutritional and antioxidative value, turned nontraditional consuming countries to important consumers and importers increasing global trade (Mateo and Maicas 2015). Almost 75% of the global olive oil production is taking place in Europe. Mediterranean countries have dominated the world olive oil production and consumption that increased significantly within the last decades. Around 99.5% of olive oil production occurs in Spain, Italy, Greece, and Portugal reaching up to $2.4 \times 10^6$ tons in 2012. In return, large amounts of olive mill wastewater (OMW) are produced, reaching up to $300 \times 10^5$ $m^3$ in the short period of harvesting and olive oil extraction, which usually lasts 2 months (ElMekawy et al. 2014).

From an environmental point of view, disposal of OMW causes severe effects mostly because of its high organic load, acidic pH, and the high content of phytotoxic compounds such as phenols. OMW treatment and valorization can be approached as a strategy to biotransform it into valuable materials, while at the same time reduction of its organic load and toxicity will make its disposal to natural receiving bodies easier.

In the following sections, the different methods to extract oil from olives and the by-products occurring by each method are summarized. Emphasis has been given to the potential conversion of OMW to biofuels, extraction of antioxidant compounds, as well as the production of bio-based products by utilizing this complex type of wastewater.

# 2 Olive Oil Extraction Processes

Extraction of olive oil may be achieved through discontinuous or continuous processes. Discontinuous process is based on using hydraulic press in order to squeeze out most of olive's oil. The certain process is used in traditional mills and represents the oldest and most widespread method to produce olive oil (Dermeche et al. 2013). On the other hand, continuous process is based on phase separation by centrifugation. In particular, decanting systems are used in separating components (olive oil, water, and solid content) according to their density, which can be operated as three-phase or two-phase systems (Fig. 1). Three-phase systems are the most popular ones, especially in Greece, Italy, and Portugal, because they are smaller installations compared to two-phase systems and they are fully automated and result in higher quality oil, although energy requirements, water input, and OMW generated are higher. Typical physicochemical characteristics of OMW occurring during the three-phase olive oil extraction process are summarized in Table 1. Two-phase extraction systems represent a more environmentally friendly alternative because of the reduced volumes of OMW produced, and they are extremely used in Spain and Croatia. The utilization of such systems results in

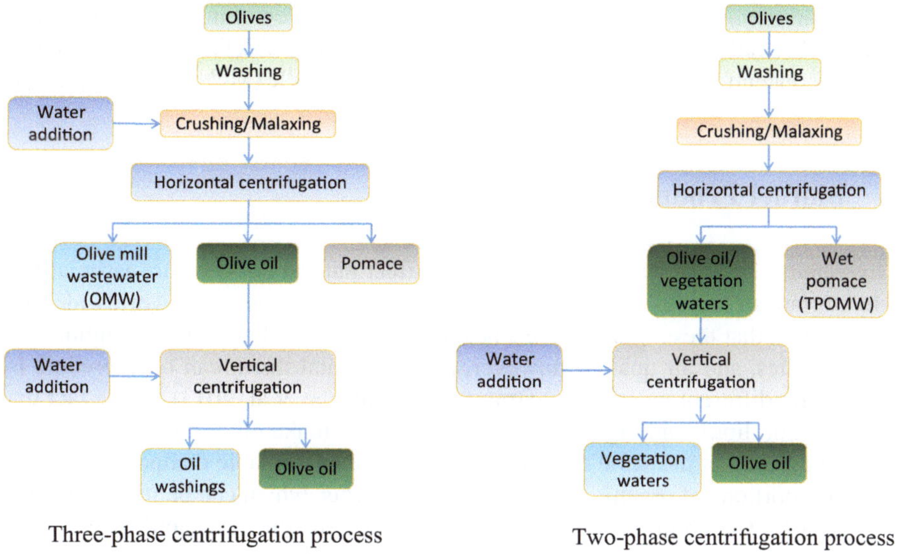


**Fig. 1** Schematic diagram of main centrifugation processes for olive oil extraction

**Table 1** Typical chemical composition of OMW (adapted by Dareioti et al. 2014)

| Parameters[a] | Units | OMW |
|---|---|---|
| pH | – | 5.02 ± 0.15 |
| TSS | g/L | 52.20 ± 0.78 |
| VSS | g/L | 49.93 ± 1.31 |
| TS | g/L | 82.53 ± 1.12 |
| VS | g/L | 65.84 ± 0.42 |
| Total COD | g/L | 140.00 ± 0.00 |
| Soluble COD | g/L | 66.37 ± 0.18 |
| TOC | g/L | 56.01 ± 1.24 |
| $BOD_5$ | g/L | 12.50 ± 0.71 |
| Total carbohydrates[b] | g/L | 30.33 ± 0.95 |
| Soluble carbohydrates[b] | g/L | 24.88 ± 0.67 |
| Total phenols[c] | g/L | 6.60 ± 0.00 |
| Total nitrogen, TKN | g/L | 0.81 ± 0.04 |
| Ammonium nitrogen | g/L | 0.12 ± 0.00 |
| Total phosphorus | mg/L | 480.40 ± 2.05 |
| Soluble phosphorus | mg/L | 309.70 ± 3.04 |
| Oil and grease | g/L | 12.76 ± 1.31 |
| Alkalinity | g $CaCO_3$/L | 0.80 ± 0.07 |
| Total VFAs | g/L | 0.23 ± 0.00 |

[a]Mean values (± standard deviation)
[b]In equivalent glucose
[c]In equivalent syringic acid

the generation of olive oil and wet pomace, which is a combination of olive husk and OMW, called two-phase olive mill waste, TPOMW. TPOMW is considered as a more concentrated OMW, thus is more difficult to handle (McNamara et al. 2008).

# 3 Biofuel Production

Untreated olive mill wastes can adversely affect natural ecosystems, especially in Mediterranean regions, which hold the lead in olive oil production worldwide. It is well known that these wastes have a negative effect on soil microbial communities, water bodies, and air quality; however, environmental threat can be overcome by employing different valorization strategies (Dermeche et al. 2013). It is a fact that high phenol, lipid, and organic acid concentrations make olive mill wastes phytotoxic; however, at the same time these wastes contain valuable compounds such as large proportions of organic substances and nutrients, which can be recycled (Roig et al. 2006). Concerning energy production, waste treatment technologies accompanied by energy recovery can reduce the environmental impact of olive oil production process while generating energy at the same time, to be used either on site, for process energy requirements, or for sale (Caputo et al. 2003).

Due to their chemical characteristics, olive mill wastes constitute effective substrates for biofuel production (Morillo et al. 2009), offering certain advantages compared to both fossil fuels and first-generation biofuels. Undoubtedly, depleting natural resources along with greenhouse gas (GHG) emissions make the use of fossil fuels unsustainable (Schenk et al. 2008). What is more, first-generation biofuels, mainly produced from food crops and oil seeds, are characterized by specific limitations, including competition for arable land and water used for agriculture and human consumption, as well as high production and processing cost, which prevent them from fulfilling global energy demand (Sims et al. 2010). In this context, second-generation biofuels, which are produced from lignocellulosic biomass and forest and non-food crop residues, appear to be highly promising renewable fuel sources. This also applies to third-generation biofuels derived from microbes and microalgae, despite the fact that more advanced technological development is still needed before these biofuels replace petroleum-based fuels (Antonopoulou et al. 2008; Venetsaneas et al. 2009; Nigam and Singh 2011; Kumar et al. 2016). Concerning bioenergy derived from OMW and TPOMW, much progress has been made in biohydrogen, methane, bioethanol, and biodiesel production.

## *3.1 Biohydrogen Production*

Hydrogen is a clean energy source which can be produced either chemically, mainly by electrolysis of water, methane steam reforming, and coal or biomass gasification, or biologically, with the latter being rather advantageous. Due to the absence of $CO_2$ emissions and an energy yield of 122 kJ/g, which is higher than that of fossil fuels, $H_2$ is considered to be the energy source of the future (Momirlan and Veziroglu 2005; Venetsaneas et al. 2009). However, certain limitations, with respect to high energy requirements, have to be overcome in order to make biological hydrogen production sustainable. Biohydrogen can be produced by anaerobic and photosynthetic microorganisms and especially through *direct biophotolysis*, *indirect biophotolysis*, *photo-fermentation*, as well as *dark fermentation* (Sen et al. 2008; Barca et al. 2015; Urbaniec and Bakker 2015). More specifically, *direct biophotolysis* includes the photosynthetic production of hydrogen from water, a process performed by green microalgae. Under anaerobic conditions microalgae can produce $H_2$ due to the fact that they possess the appropriate genetic, enzymatic, and metabolic machinery to do so (Levin et al. 2004; Ghimire et al. 2015). Cyanobacteria, having the appropriate enzymes, can also synthesize $H_2$, *indirectly* through photosynthesis. In addition, *photo-heterotrophic* bacteria (purple non-sulfur bacteria) produce $H_2$ under nitrogen deficiency. Concerning this mode, hydrogen production is higher when cells are immobilized on a solid matrix. As far as dark fermentation is concerned, biohydrogen can be formed by anaerobic bacteria which consume carbohydrates under dark conditions.

One of the major constraints in the fermentative biohydrogen production is the cost of raw materials used as substrates. Therefore, utilization of waste materials according to their availability, cost, carbohydrate content, and biodegradability makes biohydrogen production a highly promising alternative to conventional fuels (Kapdan and Kargi 2006; Arimi et al. 2015; Patel et al. 2015). OMW constitutes a substantial pollutant; thus, many studies focus both on its remediation and utilization for biohydrogen production (Table 2). Different concentrations of water-diluted OMW have been tested as the only substrate for $H_2$ production by *Rhodobacter sphaeroides* (Eroglu et al. 2004). As it was demonstrated, $H_2$ could be produced at an OMW content below 4%, with the highest production potential of 13.9 $L_{H2}/L_{OMW}$ found for 2%, while at higher concentrations both high amount of inhibitory substances and the dark color of the wastewater probably hindered the photo-heterotrophic pathway of $H_2$ production. Interestingly, nearly pure $H_2$ was produced in all experiments, which makes its use with the existing electricity-producing systems feasible. Later on, a study on coupled biological systems, including a clay treatment step prior to photo-fermentation by *Rhodobacter sphaeroides* revealed that the efficiency of photobiological $H_2$ production was substantially enhanced. This process resulted in a high hydrogen production of 35 $L_{H2}/L_{OMW}$, a light conversion efficiency of 0.42%, and a COD (chemical oxygen demand) conversion efficiency of 52% (Eroglu et al. 2006).

OMW can be used not only as a sole substrate for biohydrogen production but also as a co-substrate with other agro-wastes. That was the case in a study where a two-stage anaerobic digestion system was used to test the effect of hydraulic retention time (HRT) on biofuel production under mesophilic conditions (Dareioti and Kornaros 2014). The acidogenic reactor was started up with a waste mixture of 55% OMW, 40% cheese whey, and 5% (w/w) liquid cow manure. Afterward,

**Table 2** Potential of olive mill by-products for biohydrogen production

| Substrate | Type of reactor | pH | T (°C) | HRT | $H_2$ yield | References |
|---|---|---|---|---|---|---|
| OMW Co-substrates: cheese whey, liquid cow manure | CSTR | 6 | 37 | 0.75 d | 0.54 mol $H_2$/mol carbohydrates consumed | Dareioti and Kornaros (2014) |
| OMW Co-substrates: cheese whey, liquid cow manure | Batch | 6 | 37 | – | 0.64 mol $H_2$/mol equivalent glucose consumed | Dareioti et al. (2014) |
| OMW | Batch | 6.8–7.2 | 30 | – | 1030 ml $H_2$/L broth | Pintucci et al. (2013) |
| OMW | Batch | 7.2 | 28 | – | 150 ml $H_2$/L culture | Faraloni et al. (2011) |
| OMW | CSTR | 5.5 | 35 | 14.5 h | 196.2 mL/g carbohydrates consumed | Ntaikou et al. (2009) |
| TPOMW | CSTR | 5 | 35 | 30 h | 0.19 mmole $H_2$/g TS | Koutrouli et al. (2009) |
| TPOMW | CSTR | 5 | 35 | 30 h | 4.5 mmole $H_2$/g carbohydrates consumed | Koutrouli et al. (2006) |
| OMW | Batch | 6.7 | 32 | – | 35 $L_{H2}/L_{OMW}$ | Eroglu et al. (2006) |
| TPOMW | CSTR | 4.8 | 55 | 29 h | 0.32 mmole $H_2$/g TS | Gavala et al. (2005) |
| OMW | Batch | 6.8–8.5 | 32 | – | 13.9 $L_{H2}/L_{OMW}$ | Eroglu et al. (2004) |

different HRTs were tested in order to maximize biohydrogen production, with 0.75 d being the most effective, while the highest $H_2$ production rate of 1.72 $L/L_R$ d and $H_2$ yield of 0.54 mol $H_2$/mol carbohydrates consumed were achieved at this HRT value. The same co-substrates were used in order to evaluate the effect of pH on biohydrogen production, and the highest hydrogen production yield (0.642 mol $H_2$/mol equivalent glucose consumed) was achieved at pH 6 (Dareioti et al. 2014). Interestingly, it was observed that hydrogen productivity seemed to be primarily related to butyric acid production and lactic acid degradation.

Fermentative bio-$H_2$ production is also feasible from olive pulp (TPOMW), used as substrate. The efficiency of hydrogen production which was found in a CSTR-type anaerobic digester at 35 °C was between 2.8 and 4.5 mmole $H_2$ per g of carbohydrates consumed, depending on HRT, with the highest value observed at HRT of 30 h and the lowest at 7.5 h, respectively (Koutrouli et al. 2006). One factor which greatly affects $H_2$ production is process temperature, with thermophilic mode (55 °C) being more effective than mesophilic one. In fact, it was found that when a hydrogenic digester was fed with diluted olive pulp, hydrogen yield was

0.32 mmole $H_2$/g TS, compared to 0.19 mmole $H_2$/g TS occurring under mesophilic conditions (Gavala et al. 2005; Koutrouli et al. 2009). Diluted OMW (1:4 v/v) has also been used in a two-stage system in order to produce $H_2$ through anaerobic fermentation, and subsequently the derived effluent was used for biopolymers production (Ntaikou et al. 2009; Kourmentza et al. 2015). It was shown that not only hydrogen but also butyrate and acetate production were favored at HRT of 14.5 h.

Photobiological hydrogen production by photosynthetic microorganisms is currently of great interest as a highly promising renewable energy source, despite the fact that commercial exploitation is not yet feasible, as higher yields are still needed (Eroglu and Melis 2011). Recently, pretreated OMW at 50% dilution with a synthetic medium (TAP) was used as a substrate for $H_2$ production by *Chlamydomonas reinhardtii* (Faraloni et al. 2011). It was shown that $H_2$ production was 37% higher (150 ml $H_2$ $L^{-1}$ culture) in the TAP-OMW cultures, instead of 100 ml $H_2$ $L^{-1}$ culture, produced on TAP medium alone. A more concentrated OMW-containing medium has also been tested for biohydrogen production, after a pretreatment process (dephenolization). An OMW-based medium, including 30% of the liquid fraction of the pretreated OMW and 70% distilled water, was evaluated as an inexpensive feedstock for $H_2$ production by *Rhodopseudomonas palustris* 42OL, a purple non-sulfur photosynthetic bacterium (Pintucci et al. 2013). Both the highest amount of hydrogen production and the average hydrogen evolution rate were achieved at an irradiance of 74 W/$m^2$, while it was also found that by increasing the irradiance shorter, culture age was required.

## 3.2 *Methane Production*

Compared to methane and bioethanol, biohydrogen has a higher heating value which makes its use more promising, however, still not practical. Therefore, a higher demand for methane and bioethanol exists as they can be used directly through contemporary technology (Morillo et al. 2009). Methane, as a clean fuel, constitutes a highly advantageous renewable energy source which produces fewer atmospheric pollutants and less carbon dioxide per energy unit than other fossil fuels. Therefore, its use tends to increase in power generation, industrial applications, as well as in transportation sector (Chynoweth et al. 2001). Methane is produced through anaerobic digestion, a biological process that occurs when organic material decomposes by a microbial consortium in the absence of oxygen, or by thermal gasification of biomass, a process that is economic only at large scale (Kumar et al. 2014; Zhang et al. 2014; Pham et al. 2015). In turn, anaerobic digestion is an established waste treatment technology, the final products of which are *digestates*, residual mixtures rich in nutrients, and *biogas* which is mainly composed of methane (55–75%) and $CO_2$ (25–45%), while $H_2S$ (0–1.5%) and $NH_3$ (0–0.05%) might be present too.

Methane formation is a complex biochemical process that can be subdivided into four stages, each one characterized by the presence of different microbial consortia (Weiland 2010; Ali Shah et al. 2014). The first step is *hydrolysis*, which includes the conversion of complex biopolymers (proteins, carbohydrates, and fats) to soluble organic compounds, followed by *acidogenesis*, during which volatile fatty acids, alcohols, aldehydes, and gasses are formed by the conversion of soluble organic compounds. Subsequently, volatile fatty acids are converted to acetate, $CO_2$, and $H_2$, a step called *acetogenesis*, after which *methanogenesis* is finally taking place. This step includes the conversion of acetate, $CO_2$, and $H_2$ to methane. Biomass, irrespective of its origin, can be used for biogas production, as long as the appropriate components are present (Batstone and Virdis 2014).

Both OMW and TPOMW have been widely used as feedstock for methane production (Table 3), and much scientific research has focused on overcoming challenges derived from the chemical composition of olive mill wastes which

**Table 3** Potential of olive mill by-products for methane or biogas production

| Substrate | Type of reactor | pH | T (°C) | HRT | $CH_4$ yield or biogas productivity | References |
|---|---|---|---|---|---|---|
| OMW Co-substrates: cheese whey, liquid cow manure | CSTR | 7.8 | 37 | 25 d | 316.08 ml $CH_4$/g COD | Dareioti and Kornaros (2014) |
| OMW Co-substrate: liquid cow manure | CSTR | 7.7–8 | 35 | 19 d | 250.9 L $CH_4$ (STP)/kg COD | Dareioti et al. (2010) |
| OMW Co-substrate: olive mill solid waste | Semi-continuous feeding tubular digester | 7–8 | 35 | 12 d | 0.2 L $CH_4$/g COD | Fezzani and Cheikh Ridha (2010) |
| OMW Co-substrates: cheese whey, liquid cow manure | CSTR | 7.9 | 35 | 19 d | 243 L $CH_4$/kg $COD_{added}$ 1.35 ± 0.11 L $CH_4/L_R$ d | Dareioti et al. (2009) |
| OMW Co-substrate: diluted poultry manure | Continuously-operating reactors | 6.5–7.5 | 35 | 30 d | 0.52 L biogas/$L_R$ d | Gelegenis et al. (2007) |
| OMW Co-substrate: cheese whey | Fixed bed digester | 7–7.8 | 37 | N/R[a] | 1.25 L biogas/$L_R$ d | Martinez-Garcia et al. (2007) |
| OMW Co-substrate: manure | CSTR | 7–7.8 | 55 | 13 d | 3100 ml $CH_4$/d | Angelidaki and Ahring (1997) |
| TPOMW | CSTR | 7.62 | 35 | 20 d | 0.16 L $CH_4$/kg COD | Koutrouli et al. (2009) |

[a]N/R: not reported

hinder biogas production. Such problems stem mainly from the nutrient imbalance of these wastes due to their high C/N ratio, the low values of pH and alkalinity, as well as the presence of inhibitory substances, especially organic and phenolic compounds (Boubaker and CheikhRidha 2007). In order to avoid these constraints, several pretreatment methods can be applied to OMW before anaerobic digestion including aerobic biological pretreatment, chemical pretreatment, water dilution, and nitrogen addition. Concerning biological pretreatment, the use of fungi and yeasts has been proven effective in increasing biogas production. For instance, an aerobic detoxification step of OMW, carried out by *Aspergillus niger*, decreased wastewater toxicity through degradation of phenolic compounds and resulted in 60% COD removal and enhanced methane production (Hamdi 1991; Hamdi et al. 1992). Two different pretreatment methods, thermal pretreatment and pretreatment with the white rot fungus *Pleurotus ostreatus* P69, were employed prior to anaerobic digestion of OMW in a stirred tank reactor under mesophilic conditions (Blika et al. 2009). The use of fungus was proved to be more effective than sterilization, achieving process stability at HRT of 30 d. Apart from fungi, effective aerobic pretreatment can also be achieved with yeasts. Pretreatment of an OMW mixture with cheese whey by *Candida tropicalis* resulted in effective COD and phenol reduction. In addition, a high organic loading rate and a satisfactory biogas production rate of 1.25 $L_{biogas}\ {L_{reactor}}^{-1}\ day^{-1}$ were finally recorded (Martinez-Garcia et al. 2007). However, these practices currently applied for OMW pretreatment require inputs that increase cost-benefit ratio and decrease the organic load, thus the overall methanogenic potential of this feedstock (Sampaio et al. 2011).

As it concerns digesters configuration, an effective way to improve methane yield is through two-phase anaerobic digestion, compared to conventional one-phase systems. In two-phase systems physical separation of different microorganisms gives the opportunity to maximize their performance by separately achieving optimum conditions in each tank. As a matter of fact, through two-phase systems, the imbalance of acidogenesis and methanogenesis is successfully averted, resulting in excellent robustness, effective control, and optimization of AD process. Applying to anaerobic digestion of OMW, two-phase systems have been reported to enhance biogas production (Koutrouli et al. 2009). Usually, another technology of improving OMW bioconversion is concurrently applied: co-digestion with other substrates including diluted olive mill solid wastes (OMSW), poultry manure, cheese whey, and liquid cow manure (Gelegenis et al. 2007; Azbar et al. 2008; Dareioti et al. 2009; Fezzani and CheikhRidha 2010).

Co-digestion is an innovative waste treatment technology, where different organic substrates are combined and digested together in one anaerobic reactor. This practice offers a number of significant advantages, including improvement of plant profitability, increased methane yield, efficient use of plant facilities, and stable operation throughout a year which is often characterized by discontinuous production of specific waste streams such as OMW. Certain challenges encountered during OMW treatment derive mainly from inhibitory effect of polyphenols, lack of nitrogen, and low alkalinity of this waste stream. It has been demonstrated that such problems can be overcome via co-digestion of olive mill effluents with manure, resulting in approximately 40 L biogas/kg OMW, when 1:5 diluted OMW is used (Angelidaki and Ahring 1997). Also, a methane production rate of 0.91 L $CH_4$ $L^{-1}$ reactor $d^{-1}$ was achieved when a mixture containing 20% OMW and 80% liquid cow manure was used. Digestion took place in two stages under mesophilic conditions with HRT of 19 days, proving this method sustainable and environmentally attractive for the valorization of such wastes (Dareioti et al. 2010). Lastly, stable methanogenesis with a high methane production rate of 0.33 L $CH_4/L_R$ d was achieved when a two-stage anaerobic digestion system was fed with a co-mixture of OMW, cheese whey, and cow manure, operated at HRT of 25 days at 37 °C (Dareioti and Kornaros 2014).

## *3.3 Bioethanol Production*

Depleting natural resources, industrialization, global warming, and climate change have shifted international interest into renewable energy sources. Among them, bioethanol is receiving increasing attention due to its potential as a valuable substitute of gasoline in the market of transport fuels (Sarkar et al. 2012). Bioethanol can be used as a modern biofuel, applied directly as a gasoline improver or subsistent, or in order to reduce exhaust gasses emissions. Ethanol production from traditional feedstocks, including sucrose- and starch-containing materials such as sugar substances, corn, wheat, and rice, is not desirable due to their high feed value, and alternative sources must be employed (Sarris and Papanikolaou 2016). Currently, the use of substrates such as crop residues and other biodegradable waste materials for low-cost bioethanol makes biofuel production sustainable (Li et al. 2007). However, in terms of economics, bioethanol production needs to become more cost-effective in order to outperform fossil fuels.

Ethanol can be produced from sugar-containing materials, through fermentation processes. The available raw materials can be categorized into three groups, with each one treated appropriately in order to produce ethanol (Lin and Tanaka 2006). The first group includes sugars mainly derived from sugarcane, molasses, or fruits, which can be directly converted to ethanol. Starches need first to be enzymatically hydrolyzed to fermentable sugars before producing ethanol. Likewise, lignocellulosic materials need to be converted into sugars, before microbial enzymes ferment them. Sugars from cellulose and hemicellulose can be converted to ethanol by either a simultaneous saccharification or fermentation process or by a separated enzymatic hydrolysis followed by fermentation process (Romero-García et al. 2014). However, bioconversion of this type of biomass to bioethanol is rather challenging. The reason is the resistance to breakdown, while a great content of sugars occurring from cellulose and hemicellulose polymers subsequently need suitable microorganisms to convert them and also due to the cost of both collection and storage of low-density lignocellulosic materials (Balat 2011). Concerning the recalcitrance of lignocellulosic biomass, physical, chemical, or biological pretreatment is always needed in order to make cellulose accessible to enzymes prior to hydrolysis (Zheng et al. 2009).

Among waste materials that can be converted to ethanol, olive mill wastes represent interesting substrates for bioethanol production, due to their high content of organic matter (Morillo et al. 2009). However, lowering the phenolic content prior to fermentation might be necessary in order to enhance process performance resulting in higher bioethanol yield (Zanichelli et al. 2007). The effect of such a pretreatment of OMW with the white rot fungus *Pleurotus sajor-caju* has been studied, subsequently evaluating ethanol production after anaerobic fermentation with the yeast *Saccharomyces cerevisiae* L-6 (Massadeh and Modallal 2008). An increase in ethanol production, which reached the maximum value of 14.2 g/L, was demonstrated after 48 h fermentation using 50% diluted and pretreated OMW. Enhanced glucose and xylose bioavailability was also observed after wet oxidation and enzymatic hydrolysis pretreatment of olive pulp, prior to fermentation by *S. cerevisiae* and *Thermoanaerobacter mathranii* (Haagensen et al. 2009). However, enzymatic pretreatment was proved to be more effective. OMW have been used also as co-substrates with molasses for bioethanol production by *S. cerevisiae*, as in large-scale processes OMW could replace water used for molasses dilution reducing in that way the cost of the process (Sarris et al. 2014). Effective decolorization and 28% of phenolic compound removal were observed, along with satisfactory ethanol production, despite the fact that yeast growth was performed under aerated conditions.

Apart from OMW, olive oil mill solid residue and olive pulp have been tested as potential feedstocks for ethanol production. In the first case, bioethanol production was investigated using the yeast *Pachysolen tannophilus*, and it was shown that carbohydrate biotransformation and higher ethanol yields were facilitated by thermochemical pretreatment of the substrate (Senkevich et al. 2012). Also, when enzymatic hydrolysis was employed for OP pretreatment, followed by fermentation with *S. cerevisiae*, the traditional baker's yeast, a maximum ethanol production of 11.2 g/L was observed, without any nutrient addition or indication of yeast toxicity (Georgieva and Ahring 2007). However, it was suggested that incomplete conversion of olive pulp to ethanol constitutes this process not viable, and economic feasibility would improve only by simultaneous production of other added-value products, such as methane. Bioethanol production from olive mill solid wastes (OMSW) was also considered to be ineffective, even when endogenous yeasts grown on OMSW were used for fermentation, as it was found that xylitol was produced from xylose, instead of ethanol (Tayeh et al. 2014).

### *3.4 Biodiesel Production*

Biodiesel is a highly attractive alternative diesel fuel that is considered to be one of the most important near-market biofuels, due to the fact that all industrial vehicles are diesel based (Schenk et al. 2008; Rico and Sauer 2015). Biodiesel derives from vegetable oils or animal fats, through a catalyzed chemical reaction between triglycerides present in the oil and fats with a monohydric alcohol resulting in monoalkylesters (Gerpen 2005). Biodiesel is a biodegradable, nontoxic, and clean biofuel that is traditionally obtained from different fuel crops, including soybean, rapeseed, canola, and palm, while terpenoid products from *Copaifera* species can also constitute an alternative biodiesel source (Yuan et al. 2008). Although the processing of biodiesel is rather simple, the produced fuel can greatly vary in quality.

The major challenge faced by first-generation biofuels is the competition for arable land, which is also the case for biodiesel production from vegetable oils. A great alternative to current biodiesel production is based on lipids produced by microalgae (Schenk et al. 2008). Microalgae have the ability not only to grow rapidly, as they double their biomass within 24 h, but also have high oil content, usually 20–50%, resulting in high oil productivities that are desired for biodiesel production (Chisti 2007). Especially, when microalgae cultivated for biodiesel production consume carbon dioxide as carbon source, either atmospheric or from power plants, the whole cycle generates zero carbon dioxide emissions to the atmosphere (Cheng and Timilsina 2011). Also, microalgae have the potential to grow in conditions where no freshwater input is required, for instance, in saline or brackish water and in wastewaters due to their ability to utilize abundant organic carbon and inorganic nitrogen and phosphorus present in effluents (Pittman et al. 2011).

Phenol-resistant microalgae have been used in order to evaluate phenol removal from OMW (Pinto et al. 2003). A limited reduction of 12% was demonstrated, suggesting that phenolic degradation could be enhanced by OMW co-treatment with suitable algae and fungi, or through a two-stage process where ligninolytic fungi would be employed prior to phenol-resistant microalgae. Additionally, OMW was used as a co-substrate with urban wastewater from secondary treatment (UWST) in order to evaluate *Scenedesmus obliquus* growth and its subsequent use for biofuels. Composition analysis of the fatty acids accumulated, when *S. obliquus* was cultivated in a mixture of UWST and OMW, showed that the obtained lipid fraction could result in a good quality biodiesel, as specified by the European Standard (Hodaifa et al. 2013).

Microbial lipid production generally offers great advantages in biodiesel technology, when oleaginous microorganisms that produce more than 20% of their weight lipids are used. Combining this technology with waste materials has been proved as an effective way to reduce cost and make such processes sustainable. In case of OMW, the presence of phenolics hinders microbial growth; however, several microorganisms have been able to grow on olive mill wastes (Lanciotti et al. 2005). The oleaginous yeast *Lipomyces starkeyi* has shown the ability to grow effectively on OMW reducing total organic carbon and phenolics present in olive mill wastewater (Yousuf et al. 2010). Most significantly, an increased lipid concentration of 28.6% was observed when *L. starkeyi* was cultivated in 50% diluted OMW. Among identified fatty acids, oleic acid was the most abundant (49.1%), which is rather satisfactory since oleic acid is considered ideal for biodiesel due to its better cold flow properties.

# 4 Bio-Based Products

## 4.1 Antioxidants

Phenolic compounds are molecules that have been reported to show antioxidant, anti-inflammatory, anti-allergic, antimutagenic, antiaging, as well as antibacterial activity (Lule and Xia 2005; Taguri et al. 2006; Larif et al. 2015). Due to their remarkable characteristics, they may find potential applications in the development of functional foods (Wildman and Kelley 2007); in pharmaceuticals, i.e., as active agents for cancer prevention and treatment (Huang et al. 2010); and in cosmetic formulations (Padilla et al. 2005). OMW is a waste by-product characterized by its high polyphenol content and therefore raises issues of environmental concern since, after disposal, high polyphenol concentrations result in toxic matter and environmental degradation (Kavvadias et al. 2015). Recovery of phenolic compounds from OMW, prior to its disposal, is considered advantageous since it reduces its phytotoxicity, while at the same time high added-value products are obtained (Barbera et al. 2014). The high content of phenolic compounds, with a wide range between low and high molecular weights (MW), is responsible for its black-brownish color and depends on the ratio between low and high MW polyphenols (Borja et al. 2006). For the recovery of polyphenols from OMW, such as hydroxytyrosol and tyrosol, different methodologies have been proposed and studied such as membrane separation, liquid-liquid extraction, solid-liquid extraction, cloud point extraction, and polymer incompatibility (Rahmanian et al. 2014), while some of them have been patented (López et al. 2008; Villanova et al. 2010; De Magalhães et al. 2011).

Membrane processes have been successfully used in the food and beverage field. Integrated membrane systems for phenolic compound recovery, including microfiltration (MF), ultrafiltration (UF), nanofiltration (NF), reverse osmosis (RF), as well as membrane bioreactors, have been tested by many researchers during the last three decades (Paraskeva et al. 2007; El-Abbassi et al. 2009, 2012; Garcia-Castello et al. 2010; Petrotos et al. 2014; Ochando-Pulido and Martinez-Ferez 2015). Several combinations of membrane processes for the fractionation of OMW to different by-products using UF, NF, and/or RO processes have been studied (Paraskeva et al. 2007). According to results, a fraction rich in phenols may be obtained after NF, whereas the residual effluent could be disposed in aquatic environments according to National and European regulations, or to be used for irrigation (around 75–80% of the initial volume). The application of MF to OMW, subsequent NF, of the MF, and concentration of the NF permeate by osmotic distillation (OD) was evaluated in a recent study (Garcia-Castello et al. 2010). MF allowed suspended solids and total organic carbon (TOC) removal up to 91 and

26%, respectively, and 78% polyphenol recovery from the raw material. After NF, of the MF permeate, almost all polyphenols were recovered, while TOC content was further reduced by 37%. OD resulted in a concentrated solution rich in high molecular weight polyphenols with hydroxytyrosol representing 56% of it. In another study, the application of direct contact membrane distillation (DCMD) using commercial polytetrafluoroethylene (PTFE) membranes of different pore sizes at different temperatures was examined, in order to treat and concentrate OMW (El-Abbassi et al. 2012). With their methodology, after 8 h, they were able to concentrate a permeate characterized by a concentration factor of 1.8. They also concluded that treatment at high temperatures (up to 80 °C) had no negative effect on the total phenolic fraction and its antioxidant activity, as a respective increase of 16 and 15% was achieved. HPLC analysis of the recovered monocyclic phenolic compounds of OMW showed that hydroxytyrosol was the dominant compound, by 70%, while gallic acid (11%), para-coumaric acid (10%), tyrosol (4%), hydroxytyrosol-4-β-glucoside (3%), and caffeic acid (1.7%) were also present.

Efforts have been made to recover phenolic compounds through liquid–liquid extraction from centrifuged OMW, and subsequent anaerobic digestion in order to minimize OMW toxicity (Khoufi et al. 2008). Ethyl acetate was used as solvent while phenol recovery reached up to 90%. Phenolic compounds present in the extract were identified to be hydroxytyrosol, tyrosol, homovanillic acid, caffeic acid, *para*-coumaric acid, and ferulic acid, by employing GC-MS analysis. Tyrosol and caffeic acid are characterized as natural antioxidants, whereas ferulic acid is considered to be nutraceutically positive. Another research team (Kalogerakis et al. 2013) studied the recovery of phenolic compounds from TPOMW. Due to the high solid content, the effluent was first filtered, through mess gauge filters, and subsequently centrifuged. After supernatant fat and solid removal, the aqueous phase was subjected again to filtration by a 0.45 μm filter. Liquid-liquid extraction was performed using a solvent to TPOMW ratio of 2:1 v/v, at ambient temperature, and continuous stirring at 120 rpm for 30 min. The solvents tested were ethyl acetate, diethyl ether, and a mixture of chloroform/isopropyl alcohol 7:3 v/v, which resulted in total phenols recovery of 57%, 47%, and 56%, respectively. In addition LCA methodology was applied in order to identify the environmental footprint of the process. It was shown that the use of ethyl acetate or diethyl ether has a similar environmental footprint concerning their impacts on ecosystem and fossil fuel resources, contribution to global warming, as well as on human health. However, ethyl acetate was considered advantageous, due to higher extraction efficiency. On the other hand, chloroform/isopropyl alcohol mixture was shown to pose detrimental environmental effects.

As it regards solid phase extraction, various adsorbent resins have been employed in order to achieve deodorizing and decolorization of OMW, polyphenol, and lactone recovery. A process in which the effluent is at first subjected to successive filtration steps to remove suspended solids, followed by adsorption using XAD16 and XAD7HP resins, subsequent thermal evaporation and recovery of the organic fraction, and finally, separation of the polyphenols through fast centrifuge partition chromatography has been proposed (Agalias et al. 2007).

According to the results obtained, an odorless wastewater is characterized by 99.99 and 98% of polyphenols and COD removal, respectively, a rich in polyphenols and lactones extract, an extract which contained the coloring substances of the OMW and pure hydroxytyrosol. In another study, four resins, viz., XAD7, XAD16, IRA96, and ISOLUTE ENV+, were tested as solid adsorbing phases in two types of TPOMW obtained from different olive mills (Bertin et al. 2011). The solvents employed in desorption experiments were water, methanol, ethanol, and acidified ethanol (0.5 w/w HCl 37%). According to their results, ENV+ showed to be promising in terms of process productivity, adsorbing 84% of total phenols, hydroxytyrosol, and tyrosol, while with IRA96 the highest phenol adsorption ratios were achieved from the water phase. They concluded that in general nonpolar resins are more efficient. In more detail, the efficiency of the resins to adsorb phenols present in TPOMW decreased according to the following sequence: ENV +<XAD16<IRA96<XAD7.

Cloud point extraction (CPE) is a process of transferring a nonionic surfactant from one liquid phase to another by heating, taking advantage of the ability of surfactant molecules to form micelles. When temperature increases above the cloud point, micelles dehydrate and aggregate, which leads to macroscopic phase separation into two distinguished phases: solvent and surfactant-rich phase. CPE requires less time, labor cost, and equipment and represents a simpler and cleaner technology, compared to liquid-liquid and solid-liquid extraction. Therefore, the concentrated antioxidants obtained are suitable for food and pharmaceutical and cosmetic applications. Various low toxicity surfactants, such as Span 20, PEG 400, Tween 80, and Tween 20, have been examined for the separation of phenols from OMW (Katsoyannos et al. 2012). Among them, Tween 80 showed the highest recovery at a concentration of 5%, incubated at 55 °C for 30 min. One step CPE extraction resulted in 86.8% of phenols recovery, whereas when a double-step CPE extraction was performed, using 5 + 5% Tween 80, phenols recovery increased to 94.4%. In another study, CPE was performed to TPOMW, which was pretreated by ultrafiltration in order to remove the suspended solids present (El-Abbassi et al. 2014). Ultrafiltration resulted in color intensity reductions of 80 and 87%, estimated at 395 and 465 nm, respectively, while the COD and the dry residue were reduced by 31.4 and 27.6%. As it regards CPE, a range of concentrations, 0–10% of the surfactant Triton X-100 in OMW, incubated at different temperatures, 70 °C, 80 °C, and 90 °C, for 30 min were studied. After they examined the efficiency of the process, they observed that the highest yield of phenols extraction, 66.5%, was achieved using 10% of Triton X-100 at 90 °C.

Lately, polymer incompatibility has been suggested as a potential tool for polyphenol recovery from OMW (Hajji et al. 2014). The certain methodology is based on thermodynamic incompatibility between polymers and concerns the use of aqueous two-phase systems (ATPS). In their study, the ATPS consisted of solutions containing protein (6.33 or 7.45 wt.% caseinate, 18 wt.% ovalbumin) and a certain concentration of polysaccharide (0.75–2 wt.% alginate and methylcellulose) adjusted to neutral pH. In particular caseinate–alginate, caseinate–methylcellulose, and ovalbumin–methylcellulose systems were tested at ambient temperature. They concluded that caseinate–alginate systems, mixed with OMW, were more efficient in terms of separation, resulting to an upper polysaccharide-rich phase and a bottom protein-polyphenol-rich phase characterized by a polyphenol recovery yield of 85.8%.

In general, dephenolization and detoxification of OMW results in the recovery of high value-added compounds, while at the same time it presents an opportunity to reduce wastewater treatment cost. Furthermore, polyphenols are compounds that usually occur by chemical reactions, which is the main reason for their high cost. OMW and other types of wastewater, with high phenolic content, can be valorized toward the recovery of such remarkable compounds. Last, but not least, their recovery is considered advantageous because their presence in OMW inhibits its microbiological treatment.

## *4.2 Biosurfactants*

Within the last years, OMW has been exploited for the production of biopolymers and fine chemicals, by employing pure as well as mixed microbial consortia, combining microbiological treatment with the production of high added-value products.

Biosurfactants are biologically derived surface-active agents, not associated with bacterial growth; therefore, they are characterized as secondary metabolites. They are amphiphilic compounds consisting of hydrophilic "heads" and hydrophobic "tails," and they have the ability to decrease the surface tension of water and interfacial tension between water and hydrophobic substances. For those reasons, hydrophobic substrates are used in order to induce biosurfactant production, as bacteria secrete them in order to increase nutrient availability and grow on hydrophobic substrates. Studies have demonstrated the potential of OMW to be utilized as carbon source for biosurfactant production in the form of rhamnolipids, which are classified as glycolipid biosurfactants, as well as in the form of surfactin, a lipopeptide biosurfactant. This is due to the fact that residual oil and the polysaccharide content in OMW constitute the precursors of biosurfactant production.

The first attempt on investigating the production of rhamnolipids using this complex wastewater was performed by employing *Pseudomonas* sp. JAMM (Mercade et al. 1993). OMW was used as the sole carbon source whereas $NaNO_3$ (2.5 g/L) was supplemented in the effluent to enhance rhamnolipids production. According to the results, rhamnolipids conversion yield reached up to 0.058 g/g of OMW, calculated on a COD basis, while at the same time a 50% reduction of OMW COD and a 55% reduction in the total phenol content was achieved after 3 days. A total bioconversion yield of 14 g of rhamnolipids per kg of OMW was estimated after 150 h of fermentation. Later on, OMW was used for the production of rhamnolipids from *P. aeruginosa* ATCC 10145 and its recombinant strain expressing *Vitreoscilla* hemoglobin gene *vgb* (Colak and Kahraman 2013). The maximum production of rhamnolipids reached up to 0.4 g/L, for both wild type and recombinant strain, when grown at 37 °C and 100 rpm after 3 days.

TPOMW was recently used as carbon source for the production of surfactin, from *Bacillus subtilis* DSM 3256 (Maass et al. 2015). After 36 h the maximum surfactin concentration was achieved, reaching up to 0.25 g/L, characterized by a productivity of 0.17 g/L/d, while the surface tension of the culture's medium decreased to around 30 mN/m. Finally, in another study performed recently, the production of rhamnolipids and surfactin, by *P. aeruginosa* and *B. subtilis*, respectively, was investigated utilizing OMW (Ramírez et al. 2015). It was shown that rhamnolipids production ranged between 8.78 and 191.46 mg/L, with the highest concentration obtained by using 10% w/v OMW. On the other hand, surfactin production reached 3.12 mg/L using 2% w/v OMW, and it dropped to 0.57 mg/L when a more concentrated OMW solution of 10% w/v was used. Although OMW may be considered inhibitory for the production of biosurfactants, pretreatment regarding its dephenolization and detoxification may be proven beneficial and establish the appropriate conditions within the fermentation medium in order to enhance biosurfactant production.

### *4.3 Polyhydroxyalkanoates*

Polyhydroxyalkanoates (PHAs) are polymers of hydroxyalkanoates produced by a wide variety of bacteria. They are produced as intracellular inclusions, called granules, which bear a diameter of 0.2–0.9 μm. Their monomeric structure may vary according to the carbon atoms present. The most common forms are polyhydroxybutyrate (PHB) and polyhydroxybutyrate-polyhydroxyvalerate (PHBV) copolymers. PHAs have attracted interest since they bear similar properties to polypropylene (PP) and low-density polyethylene (PE/LDPE) and represent an attractive alternative to replace those petrochemical plastics (Gao et al. 2011). It is worth mentioning that PP and PE/LDPE have gained a major position in the global plastic market that accounts for almost 50% of the total plastic demand in Europe (Plastics-The Facts 2015). Therefore, there is a wide application field for PHA biopolymers, and due to their biodegradability, their use can minimize the detrimental impact of persistent plastics on the environment (Kourmentza et al. 2015).

Within the last decade, researchers have been studying the valorization of OMW toward the production of PHAs. In particular, the feasibility of anaerobic fermented effluents, rich in volatile fatty acids (VFAs) serving as precursors for PHA production by enriched mixed cultures, has been investigated (Dionisi et al. 2005; Beccari et al. 2009; Kourmentza et al. 2009a; Ntaikou et al. 2009, 2014). Other studies concerning OMW utilization, as the sole carbon source, by pure cultures have also demonstrated the potential valorization of OMW for PHA production (Kourmentza et al. 2009b, 2015; Martinez et al. 2015).

VFAs are the most common carbon sources used for PHA production. Investigation on the anaerobic fermentation of OMW, with and without pretreatment steps, at different concentrations, and subsequent PHA production using the fermented effluent has been performed (Dionisi et al. 2005). It was shown that VFA concentration and yield were significantly increased after centrifugation, although centrifugation led to differences in acid distribution, which resulted in different hydroxyvalerate (HV) content in the PHBV copolymer. For PHA production, a mixed culture, enriched in an aerobic SBR under feast and famine conditions, was employed. After testing centrifuged fermented and not fermented OMW, they concluded that PHA production is feasible in both cases, although fermented OMW showed much higher potential, characterized by an initial specific rate of around 0.42 g COD/g COD/h, while the final PHA concentration reached up to 0.54 g PHA/g volatile suspended solids (VSS). The polymer obtained was a PHBV copolymer consisting 11% HV, on a molar basis.

The performance of a process consisting of three stages for PHA production from OMW has been also studied (Beccari et al. 2009). In the first stage, the OMW was anaerobically fermented, in a packed bed biofilm reactor, in order to produce an effluent rich in VFAs. In the second stage, the VFA-rich effluent was used as a feed for an aerobic SBR where mixed cultures were enriched to PHA-producing bacteria. Finally, in the third stage of the proposed process, PHA production was tested, by employing the enriched cultures, under aerobic batch conditions. They observed that during anaerobic fermentation of the wastewater, VFA content increased from 18 to 32%, in a COD basis, while during the second stage, an enriched culture with high PHA-storing capacity was formed, characterized by a maximum production rate and yield of 0.15 g COD/g COD/h and 0.36 COD/COD, respectively. In the final stage, they observed that PHA concentration increased almost linearly as a function of the organic loading rate (OLR), revealing the possibility to design a process operation using higher OLR. The highest PHA content in the biomass achieved was around 20% g COD/g COD, while storage yield was almost the same, 0.35 COD/COD. This scenario of microbiological treatment resulted in approximately 85% of COD removal; thus, it was considered effective for both OMW treatment and OMW valorization toward PHA production.

Combination of OMW anaerobic fermentation, employed for the biological production of hydrogen, and subsequent feeding of an SBR reactor, used for PHAs production, with the obtained effluent has been recently proposed (Ntaikou et al. 2009). In that way the VFAs present in the acidogenic reactor's effluent serve as the precursors for PHA production while at the same time COD reduction occurs. For the anaerobic fermentation, a continuous stirred tank reactor (CSTR) was used at different HRT, ranging from 7.5 to 60 h. HRTs were examined in terms of biohydrogen and VFA production. It was shown that at an HRT of 27–33 h propionate production was favored while at lower HRT acetate and butyrate were dominant. Biohydrogen productivity rates and yields were severely affected in lower HRT with the most effective ones achieved at 24 h calculated at 165 mL/d and 330 mL/L of OMW (diluted 1:4), respectively. The effluent of the anaerobic fermenter was forwarded to an SBR used for the enrichment of activated sludge to

PHA-forming bacteria and also the production of PHAs. During the growth phase acetate, propionate and butyrate were fully consumed, whereas during the PHA accumulation phase butyrate was preferably consumed, followed by propionate and acetate. Substrate preference of the specific culture was identical when fed with a mixture of synthetic VFAs (Kourmentza et al. 2009a, b). This conclusion is of great importance since during anaerobic fermentation, manipulation of operating conditions may lead to a desirable VFA profile that will eventually favor PHA production and can also result to certain HV content in the PHBV copolymer, due to the presence of propionate. In this study PHA production capacity reached up to 9% g PHA/g cell dry weight (CDW) and a PHBV copolymer was obtained.

Recently, scaling up of the abovementioned proposed process in a 20 L reactor has been investigated (Ntaikou et al. 2014). The highest PHAs capacity achieved was increased from 9%, in lab scale, to 24.6% g PHAs/g VSS. Previous studies performed indicated that the enriched PHA-forming bacteria culture mainly consisted of strains that belong to *Pseudomonas* sp., in particular *P. putida* (Kourmentza et al. 2009a, b, 2015). As a result a copolymer consisting of both PHB and polyhydroxyoctanoate (PHO) was produced characterized by a weight average molecular weight of 490 kDa and a polydispersity index (PI) of 5.15. The PI of the polymer is considered high for a biological polymer (usually ~1) indicating its heterogenicity probably due to the diversity of the carbon sources consumed and/or the microbial consortium eventually formed.

As it regards the production of PHAs by pure cultures lately, the potential of fermented OMW for PHA production from an enriched culture and strains that consisted this mixed consortium was studied (Kourmentza et al. 2015). The isolated strains were identified to belong to *Pseudomonas* genus. Batch experiments were conducted, under nitrogen-limiting conditions, which revealed that the specific strains had the ability to accumulate 0.9–6.2% g PHAs/g VSS. Under conditions of dual nitrogen-oxygen limitation, PHA production ranged from 0.6 to 11.5% g PHAs/g VSS.

Finally, different concentrations of OMW solutions (25, 50, 75 and 100% v/v) have been exploited for the production of PHAs by *Cupriavidus necator* DSM 545 (Martinez et al. 2015). It was shown that the maximum PHA capacity, 60% g PHAs/g CDW, was obtained by using 75% diluted OMW, with a conversion yield of 0.26 g PHAs/g VFAs resulting in the formation of a PHBV copolymer consisting of 20% mol HV. Although, they concluded that the presence of phenols significantly contributes to inhibitory effects.

### 4.4 *Polysaccharides*

Exopolysaccharides (EPS) are extracellular polymeric substances produced by numerous bacterial species. EPS are a structurally diverse class of biological macromolecules with a wide broad of applications in pharmaceuticals, cosmetics, and bioremediation (Liang and Wang 2015).

Xanthan gum is an exopolysaccharide that is mainly secreted by *Xanthomonas campestris* and is considered one of the most commercially important microbial polysaccharides due to its remarkable physical properties. Xanthan gum is composed by repeated pentasaccharide units, consisting of glucose, mannose, and glucuronic acid at a molar ratio of 2:2:1. It has been extensively used as a food supplement and rheology modifier, as an emulsion stabilizer in cosmetic formulations, and as a thickening agent in salad dressings (Petri 2015). Production of xanthan gum, using OMW as the sole source of nutrients, was firstly described by *X. campestris* NRRL B1459-S4L41 (López and Ramos-Cormenzana 1996). Results showed that biomass and xanthan production were inhibited at OMW concentrations above 60%. Although, maximum xanthan production reached up to 4 g/L when *X. campestris* was fed with 30% v/v OMW solution. Xanthan production was increased when phosphate buffer was added in the medium, resulting to a final xanthan concentration of 6.3 g/L. In a further study, four strains of *X. campestris* using different % v/v OMW solutions were tested, and according to the results obtained, the highest xanthan production was achieved using 30% v/v OMW (López et al. 2001a). Xanthan concentrations ranged from 3.48 to 7.01 g/L while the viscosity of the broth varied from 3890 to 4710 mPa s. In another study, xanthan production was achieved when nitrogen was supplemented in a TPOMW resulting to the production of 3.5 g/L of xanthan (López et al. 2001b).

The production of another type of EPS produced by *Paenibacillus* strains has also been reported when using OMW. *Paenibacillus jamilae* sp., isolated from corn compost treated with OMW, was able to grow on a 100% v/v OMW solution, at 30 °C and pH 7, and produce EPS (Aguilera et al. 2001). The heteropolysaccharide produced consisted of fucose, xylose, rhamnose, arabinose, mannose, galactose, and glucose. Production of EPS from the bacteria *P. jamilae* CP-7 when grown on a medium containing 80% v/v OMW has also been reported (Ruiz-Bravo et al. 2001). After 72 h of incubation at 30 °C, a water-soluble EPS in the form of white powder was obtained with a yield of 5.5 g/L. In another study TPOMW was investigated as the substrate of EPS production by *P. jamilae* CECT 5266 (Morillo et al. 2006). They observed that maximum EPS yield of 2 g/L was obtained using 20% v/v TPOMW, whereas nutrient supplementation, in the form of nitrate, phosphate, and other inorganic nutrients, did not favor EPS production. In another study performed by the same research group (Morillo et al. 2007), 5.1 g/L EPS were obtained by *P. jamilae* CECT 5266 using 80% v/v OMW, while inhibition on growth and EPS production was observed for increased OMW concentration of 100% v/v OMW, resulting in 2.7 g/L of EPS. Characterization of the EPS showed that it was composed of two EPS fractions of different molecular weights, above 2000 kDa and 500 kDa. Both fractions consisted mainly of the carbohydrates glucose, galactose, mannose, arabinose, rhamnose, as well as hexosamines and uronic acid. Later on, researchers isolated 60 different strains from compost treated with OMW (Aguilera et al. 2008). From those, ten strains were selected due to their ability to produce EPS-utilizing OMW as the sole carbon source. Initial experiments performed in shake flasks showed that *P. jamilae* CP-38 was characterized by the maximum yield, as it was able to produce 4.2 g/L EPS, within 48 h, using 80% v/v

OMW. Further tests in a 2 L bioreactor showed that the yield of EPS was increased to 5.2 g/L after 72 h of incubation. Although the most interesting conclusion was that the toxicity of OMW was decreased by 75% within the first 24 h, with a subsequent decrease until the end of the fermentation. High phenolic degrading activity is a characteristic of *Paenibacillus* strains (Raj et al. 2007) and so is the production of extracellular enzymes able to degrade polysaccharides (Ko et al. 2007). Taking those into account, it was assumed that the production of EPS plays an important role on OMW detoxification.

## 5 Perspectives

Olive oil processing constitutes one basic economic activity, especially in Mediterranean regions, which offers significant advantages along with serious environmental issues that have to be handled. Millions of tons of olive mill wastewaters and by-products which are produced annually may serve as renewable sources for green biomaterials and energy production, toward a more sustainable and environmentally friendly economy. Undoubtedly, challenges of our times demand immediate action in terms of depleting energy sources, energy insecurity, pollution and climate change, and wastes including those produced by olive mill industry can contribute to this direction. Effective exploitation of OMW and TPOMW can substantially reduce the environmental impact of olive oil production, while concurrently biopolymers, and other valuable natural components, as well as gas and liquid biofuel production can successfully materialize. Additionally, it is of high significance that nothing is wasted and all residues can be recycled. Despite the fact that further research and technological development are needed in order to establish such an economical industrial scale production, great effort and progress has been done so far, and it is only a matter of time for this venture to be widely established.

**Acknowledgements** Dr. Constantina Kourmentza would like to thank the European Commission for providing support through the FP7-PEOPLE-2013-IEF-Marie-Curie Action: Intra-European Fellowships for Career Development (grant no. 625774).

## References

Agalias A, Magiatis P, Skaltsounis AL, Mikros E, Tsarbopoulos A, Gikas E, Spanos I, Manios T (2007) A new process for the management of olive oil mill wastewater and recovery of natural antioxidants. J Agric Food Chem 55:2671–2676. doi:10.1021/jf063091d

Aguilera M, Monteoliva-Sanchez M, Suarez A, Guerra V, Lizama C, Bennasar A, Ramos-Cormenzana A (2001) *Paenibacillus jamilae* sp. nov., an exopolysaccharide-producing bacterium able to grow in olive-mill wastewater. Int J Syst Evol Microbiol 51:1687–1692. doi:10.1099/00207713-51-5-1687

Aguilera M, Quesada MT, del Aguila VG, Morillo JA, Rivadeneyra MA, Ramos-Cormenzana A, Monteoliva-Sanchez M (2008) Characterisation of *Paenibacillus jamilae* strains that produce exopolysaccharide during growth on and detoxification of olive mill wastewaters. Bioresour Technol 99:5640–5644. doi:10.1016/j.biortech.2007.10.032

Ali Shah F, Mahmood Q, Maroof Shah M, Pervez A, Ahmad Asad S (2014) Microbial ecology of anaerobic digesters: the key players of anaerobiosis. Sci World J 2014:1–21. doi:10.1155/2014/183752

Angelidaki I, Ahring BK (1997) Codigestion of olive oil mill wastewaters with manure, household waste or sewage sludge. Biodegradation 8:221–226. doi:10.1023/A:1008284527096

Antonopoulou G, Stamatelatou K, Venetsaneas N, Kornaros M, Lyberatos G (2008) Biohydrogen and methane production from cheese whey in a two-stage anaerobic process. Ind Eng Chem Res 47:5227–5233. doi:10.1021/ie071622x

Arimi MM, Knodel J, Kiprop A, Namango SS, Zhang Y, Geiben SU (2015) Strategies for improvement of biohydrogen production from organic-rich wastewater: a review. Biomass Bioenergy 75:101–118. doi:10.1016/j.biombioe.2015.02.011

Azbar N, Keskin T, Yuruyen A (2008) Enhancement of biogas production from olive mill effluent (OME) by co-digestion. Biomass Bioenergy 32:1195–1201. doi:10.1016/j.biombioe.2008.03.002

Balat M (2011) Production of bioethanol from lignocellulosic materials via the biochemical pathway: a review. Energy Convers Manage 52:858–875. doi:10.1016/j.enconman.2010.08.013

Barbera AC, Maucieri C, Ioppolo A, Milani M, Cavallaro V (2014) Effects of olive mill wastewater physico-chemical treatments on polyphenol abatement and Italian ryegrass (*Lolium miltiflorum* Lam.) germinability. Water Res 52:275–281. doi:10.1016/j.watres.2013.11.004

Barca C, Soric A, Ranava D, Giudici-Orticoni MT, Ferrasse JH (2015) Anaerobic biofilm reactors for dark fermentative hydrogen production from wastewater—a review. Bioresour Technol 185:386–398. doi:10.1016/j.biortech.2015.02.063

Batstone DJ, Virdis B (2014) The role of anaerobic digestion in the emerging energy economy. Curr Opin Biotechnol 27:142–149. doi:10.1016/j.copbio.2014.01.013

Beccari M, Bertin L, Dionisi D, Fava F, Lampis S, Majone M, Valentino F, Vallini G, Villano M (2009) Exploiting olive oil mill effluents as a renewable resource for production of biodegradable polymers through a combined anaerobic-aerobic process. J Chem Technol Biotechnol 84:901–908. doi:10.1002/jctb.2173

Bertin L, Ferri F, Scoma A, Marchetti L, Faba F (2011) Recovery of high added value natural polyphenols from actual olive mill wastewater through solid phase extraction. Chem Eng J 171:1287–1293. doi:10.1016/j.cej.2011.05.056

Blika PS, Stamatelatou K, Kornaros M, Lyberatos G (2009) Anaerobic digestion of olive mill wastewater. Global Nest J 11:364–372

Borja R, Rincon B, Raposo F (2006) Anaerobic biodegradation of two-phase olive mill solid wastes and liquid effluents: kinetic studies and process performance. J Chem Technol Biotechnol 81:1450–1462. doi:10.1016/j.biortech.2013.03.155

Boubaker F, Cheikh Ridha B (2007) Anaerobic co-digestion of olive mill wastewater with olive mill solid waste in a tubular digester at mesophilic temperature. Bioresour Technol 98:769–774. doi:10.1016/j.biortech.2006.04.020

Caputo AC, Scacchia F, Pelagagge PM (2003) Disposal of by-products in olive oil industry: waste-to-energy solutions. Appl Therm Eng 23:197–214. doi:10.1016/S1359-4311(02)00173-4

Cheng JJ, Timilsina GR (2011) Status and barriers of advanced biofuel technologies: a review. Renew Energy 36:3541–3549. doi:10.1016/j.renene.2011.04.031

Chisti Y (2007) Biodiesel from microalgae. Biotechnol Adv 25:294–306. doi:10.1016/j.biotechadv.2007.02.001

Chynoweth DP, Owens JM, Legrand R (2001) Renewable methane from anaerobic digestion of biomass. Renew Energ 22:1–8. doi:10.1016/S0960-1481(00)00019-7

Colak AK, Kahraman H (2013) The use of raw cheese whey and olive oil mill wastewater for rhamnolipid production by recombinant *Pseudomonas aeruginosa*. Environ Exp Biol 11:125–130

Dareioti MA, Kornaros M (2014) Effect of hydraulic retention time (HRT) on the anaerobic co-digestion of agro-industrial wastes in a two-stage CSTR system. Bioresour Technol 167:407–415. doi:10.1016/j.biortech.2014.06.045

Dareioti MA, Dokianakis SN, Stamatelatou K, Zafiri C, Kornaros M (2009) Biogas production from anaerobic co-digestion of agroindustrial wastewaters under mesophilic conditions in a two-stage process. Desalination 248:891–906. doi:10.1016/j.desal.2008.10.010

Dareioti MA, Dokianakis SN, Stamatelatou K, Zafiri C, Kornaros M (2010) Exploitation of olive mill wastewater and liquid cow manure for biogas production. Waste Manage 30:1841–1848. doi:10.1016/j.wasman.2010.02.035

Dareioti MA, Vavouraki AI, Kornaros M (2014) Effect of pH on the anaerobic acidogenesis of agroindustrial wastewaters for maximization of bio-hydrogen production: a lab-scale evaluation using batch tests. Bioresour Technol 162:218–227. doi:10.1016/j.biortech.2014.03.149

De Magalhães ML, Cardador SJL, Figueiredo AA, Marques NAV, Duarte CM, Goulão CJP (2011) Method of obtaining a natural hydroxytyrosol-rich concentrate from olive tree residues and subproducts using clean technologies. Patent: US8066881

Dermeche S, Nadour M, Larroche C, Moulti-Mati F, Michaud P (2013) Olive mill wastes: biochemical characterizations and valorization strategies. Process Biochem 48:1532–1552. doi:10.1016/j.procbio.2013.07.010

Dionisi D, Carucci G, Papini MP, Riccardi C, Majone M, Carrasco F (2005) Olive oil mill effluents as a feedstock for production of biodegradable polymers. Water Res 39:2076–2084. doi:10.1016/j.watres.2005.03.011

El-Abbassi A, Hafidi A, García-Payo MC, Khayet M (2009) Concentration of olive mill wastewater by membrane distillation for polyphenols recovery. Desalination 245:670–674. doi:10.1016/j.desal.2009.02.035

El-Abbassi A, Kiai H, Hafidi A, Garcia-payo MC, Khayet M (2012) Treatment of olive mill wastewater by membrane distillation using polytetrafluoroethylene membranes. Sep Purif Technol 98:55–61. doi:10.1016/j.seppur.2012.06.026

El-Abbassi A, Kiai H, Raiti J, Hafidi A (2014) Cloud point extraction of phenolic compounds from pretreated olive mill wastewater. J Environ Chem Eng 2:1480–1486. doi:10.1016/j.jece.2014.06.024

ElMekawy A, Diels L, Bertin L, De Wever H, Pant D (2014) Potential biovalorization techniques for olive mill biorefinery wastewater. Biofuel Bioprod Bioref 8:283–293. doi:10.1002/bbb.1450

Eroglu E, Melis A (2011) Photobiological hydrogen production: recent advances and state of the art. Bioresour Technol 102:8403–8413. doi:10.1016/j.biortech.2011.03.026

Eroglu E, Gündüz U, Yücel M, Turker L, Eroglu I (2004) Photobiological hydrogen production by using olive mill wastewater as a sole substrate source. Int J Hydrogen Energy 29:163–171. doi:10.1016/S0360-3199(03)00110-1

Eroglu E, Eroglu I, Gündüz U, Turker L, Yücel M (2006) Biological hydrogen production from olive mill wastewater with two-stage processes. Int J Hydrogen Energy 31:1527–1535. doi:10.1016/j.ijhydene.2006.06.020

Faraloni C, Ena A, Pintucci C, Torzillo G (2011) Enhanced hydrogen production by means of sulfur-deprived *Chlamydomonas reinhardtii* cultures grown in pretreated olive mill wastewater. Int J Hydrogen Energy 36:5920–5931. doi:10.1016/j.ijhydene.2011.02.007

Fezzani B, Cheikh Ridha B (2010) Two-phase anaerobic co-digestion of olive mill wastes in semi-continuous digesters at mesophilic temperature. Bioresour Technol 101:1628–1634. doi:10.1016/j.biortech.2009.09.067

Gao X, Chen JC, Wu Q, Chen GQ (2011) Polyhydroxyalkanoates as a source of chemicals, polymers, and biofuels. Curr Opin Biotechnol 22:768–774. doi:10.1016/j.copbio.2011.06.005

Garcia-Castello E, Cassano A, Criscuoli A, Conidi C, Drioli E (2010) Recovery and concentration of polyphenols from olive mill wastewaters by integrated membrane system. Water Res 44:3883–3892. doi:10.1016/j.watres.2010.05.005

Gavala HN, Skiadas IV, Ahring BK, Lyberatos G (2005) Potential for biohydrogen and methane production from olive pulp. Water Sci Technol 52:209–215

Gelegenis J, Georgakakis D, Angelidaki I, Christopoulou N, Goumenaki M (2007) Optimization of biogas production from olive-oil mill wastewater, by codigesting with diluted poultry-manure. Appl Energy 84:646–663. doi:10.1016/j.apenergy.2006.12.001

Georgieva TI, Ahring BK (2007) Potential of agroindustrial waste from olive oil industry for fuel ethanol production. Biotechnol J 2:1547–1555. doi:10.1002/biot.200700128

Gerpen JV (2005) Biodiesel processing and production. Fuel Process Technol 86:1097–1107. doi:10.1016/j.fuproc.2004.11.005

Ghimire A, Frunzo L, Pirozzi F, Trably E, Escudie R, Lens PNL, Esposito G (2015) A review on dark fermentative biohydrogen production from organic biomass: process parameters and use of by-products. Appl Energy 144:73–95. doi:10.1016/j.apenergy.2015.01.045

Haagensen F, Skiadas IV, Gavala HN, Ahring BK (2009) Pre-treatment and ethanol fermentation potential of olive pulp at different dry matter concentrations. Biomass Bioenergy 33:1643–1651. doi:10.1016/j.biombioe.2009.08.006

Hajji F, Kunz B, Weissbrodt J (2014) Polymer incompatibility as a potential tool for polyphenol recovery from olive mill wastewater. Food Chem 156:23–28. doi:10.1016/j.foodchem.2014.01.068

Hamdi M (1991) Effects of agitation and pretreatment on the batch anaerobic digestion of olive mill wastewater. Bioresour Technol 36:173–178. doi:10.1016/0960-8524(91)90176-K

Hamdi M, Garcia JL, Ellouz R (1992) Integrated biological process for olive mill wastewater treatment. Bioprocess Eng 8:79–84. doi:10.1007/BF00369268

Hodaifa G, Sanchez S, Martinez ME, Orpez R (2013) Biomass production of *Scenedesmus obliquus* from mixtures of urban and olive-oil mill wastewaters used as culture medium. Appl Energy 104:345–352. doi:10.1016/j.apenergy.2012.11.005

Huang WY, Cai YZ, Zhang Y (2010) Natural phenolic compounds from medicinal herbs and dietary plants: potential use for cancer prevention. Nutr Cancer 62:1–20. doi:10.1080/01635580903191585

Kalogerakis N, Politi M, Foteinis S, Chatzisymeon E, Mantzavinos D (2013) Recovery of antioxidants from olive mill wastewaters: a viable solution that promotes their overall sustainable management. J Environ Manage 128:749–758. doi:10.1016/j.jenvman.2013.06.027

Kapdan IK, Kargi F (2006) Bio-hydrogen production from waste materials. Enzyme Microb Technol 38:569–582. doi:10.1016/j.enzmictec.2005.09.015

Katsoyannos E, Gortzi O, Chatzilazarou A, Athanasiadis V, Tsaknis J, Lalas S (2012) Evaluation of the sustainability of low hazard surfactants for the separation of phenols and carotenoids from red-flesh orange juice and olive mill wastewater using cloud point extraction. J Sep Sci 35:2665–2670. doi:10.1002/jssc.201200356

Kavvadias V, Doula M, Papadopoulou M, Theocharopoulos S (2015) Long-term application of olive-mill wastewater affects soil chemical and microbial properties. Soil Res 53:461–473. doi:10.1071/SR13325

Khoufi S, Aloui F, Sayadi S (2008) Extraction of antioxidants from olive mill wastewater and electro-coagulation of exhausted fraction to reduce its toxicity on anaerobic digestion. J Hazard Mater 151:531–539. doi:10.1016/j.jhazmat.2007.06.017

Ko CH, Chen WL, Tsai CH, Jane WN, Liu CC, Tu J (2007) *Paenibacillus campinasensis* BL11: a wood material-utilizing bacterial strain isolated from black liquor. Bioresour Technol 98:2727–2733. doi:10.1016/j.biortech.2006.09.034

Kourmentza C, Ntaikou I, Kornaros M, Lyberatos G (2009a) Production of PHAs from mixed and pure cultures of *Pseudomonas* sp. using short-chain fatty acids as carbon source under nitrogen limitation. Desalination 248:723–732. doi:10.1016/j.desal.2009.01.010

Kourmentza C, Mitova E, Stoyanova N, Ntaikou I, Kornaros M (2009b) Investigation of PHAs production from acidified olive oil mill wastewater (OOMW) by pure cultures of *Pseudomonas* spp. strains. New Biotechnol 25:S269. doi:10.1016/j.nbt.2009.06.604

Kourmentza C, Ntaikou I, Lyberatos G, Kornaros M (2015) Polyhydroxyalkanoates from *Pseudomonas* sp. using synthetic and olive mill wastewater under limiting conditions. Int J Biol Macromol 74:202–210. doi:10.1016/j.ijbiomac.2014.12.032

Koutrouli EC, Gavala HN, Skiadas IV, Lyberatos G (2006) Mesophilic biohydrogen production from olive pulp. Process Saf Environ Prot 84:285–289. doi:10.1205/psep.05165

Koutrouli EC, Kalfas H, Gavala HN, Skiadas IV, Stamatelatou K, Lyberatos G (2009) Hydrogen and methane production through two-stage mesophilic anaerobic digestion of olive pulp. Bioresour Technol 100:3718–3723. doi:10.1016/j.biortech.2009.01.037

Kumar P, Pant DC, Mehariya S, Sharma R, Kansal A, Kalia VC (2014) Ecobiotechnological strategy to enhance efficiency of bioconversion of wastes into hydrogen and methane. Indian J Microbiol 54:262–267. doi:10.1007/s12088-014-0467-7

Kumar P, Mehariya S, Ray S, Mishra A, Kalia VC (2016) Biotechnology in aid of biodiesel industry effluent (glycerol): biofuels and bioplastics. In: Kalia VC (ed) Microbial cell factories. Springer, New Delhi, pp 105–119. isbn: 978-81-322-2597-3. doi: 10.1007/978-81-322-2598-0_7

Lanciotti R, Gianotti A, Baldi D, Angrisani R, Suzzi G, Mastrocola D, Guerzoni ME (2005) Use of *Yarrowia lipolitica* strains for the treatment of olive mill wastewaters. Bioresour Technol 96:317–322. doi:10.1016/j.biortech.2004.04.009

Larif M, Ouhssine M, Soulaymani A, Elmidaoui A (2015) Potential effluent oil mills and antibacterial activity polyphenols against some pathogenic strains. Res Chem Intermed 41:1213–1225. doi:10.1007/s11164-013-1267-0

Levin DB, Pitt L, Love M (2004) Biohydrogen production: prospects and limitations to practical application. Int J Hydrogen Energy 29:173–185. doi:10.1016/S0360-3199(03)00094-6

Li A, Antizar-Ladislao B, Khraisheh MAM (2007) Bioconversion of municipal solid waste to glucose for bio-ethanol production. Bioproc Biosyst Eng 30:189–196. doi:10.1007/s00449-007-0114-3

Liang TW, Wang SL (2015) Recent advances in exopolysaccharides from *Paenibacillus* spp.: production, isolation structure, and bioactivities. Mar Drugs 13:1847–1863. doi:10.3390/md13041847

Lin Y, Tanaka S (2006) Ethanol fermentation from biomass resources: current state and prospects. Appl Microbiol Biotechnol 69:627–642. doi:10.1007/s00253-005-0229-x

López MJ, Ramos-Cormenzana A (1996) Xanthan production from olive-mill wastewaters. Int Biodeterior Biodegrad 38:263–270. doi:10.1016/S0964-8305(96)00059-5

López MJ, Moreno J, Ramos-Cormenzana A (2001a) *Xanthomonas campestris* strain selection for xanthan production from olive mill wastewaters. Water Res 35:1828–1830. doi:10.1016/S0043-1354(00)00430-9

López MJ, Moreno J, Ramos-Cormenzana A (2001b) The effect of olive mill wastewaters variability on xanthan production. J Appl Microbiol 90:829–835. doi:10.1046/j.1365-2672.2001.01326.x

López MJA, Streitenberger SA, Mellado MP, Ortiz PM (2008) Process and apparatus for the production of hydroxytyrosol containing extract from olives and solids containing residues of olive oil extraction. Patent: WO20008090460 A1

Lule SU, Xia W (2005) Food phenolics, pros and cons: a review. Food Rev Int 21:367–388. doi:10.1080/87559120500222862

Maass D, Ramirez IM, Roman MG, Alameda EJ, de Souza AAU, Valle JAB, Vaz DA (2015) Two-phase olive mill waste (alpeorujo) as carbon source for biosurfactant production. J Chem Technol Biotechnol 91(7):1990–1997. doi:10.1002/jctb.4790

Martinez GA, Bertin L, Scoma A, Rebecchi S, Braunegg G, Fava F (2015) Production of polyhydroxyalkanoates from dephenolised and fermented olive mill wastewaters by employing a pure culture of *Cupriavidus necator*. Biochem Eng J 97:92–100. doi:10.1016/j.bej.2015.02.015

Martinez-Garcia G, Johnson AC, Bachmann RT, Williams CJ, Burgoyne A, Edyvean RGJ (2007) Two-stage biological treatment of olive mill wastewater with whey as co-substrate. Int Biodeterior Biodegrad 59:273–282. doi:10.1016/j.ibiod.2007.03.008

Massadeh MI, Modallal N (2008) Ethanol production from olive mill wastewater (OMW) pretreated with *Pleurotus sajor-caju*. Energy Fuel 22:150–154. doi:10.1021/ef7004145

Mateo JJ, Maicas S (2015) Valorization of winery and oil mill wastes by microbial technologies. Food Res Int 73:13–25. doi:10.1016/j.foodres.2015.03.007

McNamara CJ, Anastasiou CC, O'Flaherty V, Mitchell R (2008) Bioremediation of olive mill wastewater. Int Biodeterior Biodegrad 61:127–134. doi:10.1016/j.ibiod.2007.11.003

Mercade ME, Manresa MA, Robert M, Espuny MJ, de Andres C, Guinea J (1993) Olive oil mill effluent (OOME). New substrate for biosurfactant production. Bioresour Technol 43:1–6. doi:10.1016/0960-8524(93)90074-L

Momirlan M, Veziroglu TN (2005) The properties of hydrogen as fuel tomorrow in sustainable energy system for a cleaner planet. Int J Hydrogen Energy 30:795–802. doi:10.1016/j.ijhydene.2004.10.011

Morillo JA, Aguilera M, Ramos-Cormenzana A, Monteoliva-Sanchez M (2006) Production of a metal-binding exopolysaccharide by *Paenibacillus jamilae* using two-phase olive-mill waste as fermentation substrate. Curr Microbiol 53:189–193. doi:10.1007/s00284-005-0438-7

Morillo JA, del Aguila VG, Aguilera M, Ramos-Cormenzana A, Monteoliva-Sanchez M (2007) Production and characterization of the exopolysaccharide produced by *Paenibacillus jamilae* grown on olive mill-waste waters. World J Microbiol Biotechnol 23:1705–1710. doi:10.1007/s11274-007-9418-3

Morillo JA, Antizar-Ladislao B, Monteoliva-Sánchez M, Ramos-Cormenzana A, Russell NJ (2009) Bioremediation and biovalorisation of olive-mill wastes. Appl Microbiol Biotechnol 82:25–39. doi:10.1007/s00253-008-1801-y

Nigam P, Singh A (2011) Production of liquid biofuels from renewable resources. Prog Energy Combust Sci 37:52–68. doi:10.1016/j.pecs.2010.01.003

Ntaikou I, Kourmentza C, Koutrouli EC, Stamatelatou K, Zampraka A, Kornaros M, Lyberatos G (2009) Exploitation of olive oil mill wastewater for combined biohydrogen and biopolymers production. Bioresour Technol 100:3724–3730. doi:10.1016/j.biortech.2008.12.001

Ntaikou I, Valencia Peroni C, Kourmentza C, Ilieva VI, Morelli A, Chiellini E, Lyberatos G (2014) Microbial bio-based plastics from olive-mill wastewater: generation and properties of polyhydroxyalkanoates from mixed cultures in a two-stage pilot scale system. J Biotechnol 188:138–147. doi:10.1016/j.jbiotec.2014.08.015

Ochando-Pulido JM, Martinez-Ferez A (2015) On the recent use of membrane technology for olive mill wastewater purification. Membranes 5:513–531. doi:10.3390/membranes5040513

Padilla M, Palma M, Barroso CG (2005) Determination of phenolics in cosmetic creams and similar emulsions. J Chromatogr A 1091:83–88. doi:10.1016/j.chroma.2005.07.041

Paraskeva CA, Papadakis VG, Tsarouchi E, Kanellopoulou DG, Koutsoukos PG (2007) Membrane processing for olive mill wastewater fractionation. Desalination 213:218–229. doi:10.1016/j.desal.2006.04.087

Patel SKS, Kumar P, Singh M, Lee JK, Kalia VC (2015) Integrative approach to produce hydrogen and polyhydroxybutyrate from biowaste using defined bacterial cultures. Bioresour Technol 176:136–141. doi:10.1016/j.biortech.2014.11.029

Petri DFS (2015) Xanthan gum: a versatile biopolymer for biomedical and technological applications. J Appl Polym Sci 132:42035. doi:10.1002/app.42035

Petrotos KB, Lellis T, Kokkora MI, Gkoutsidis PE (2014) Purification of olive mill wastewater using microfiltration membrane technology. J Membr Sep Technol 3:50–55

Pham TPT, Kaushik R, Parshetti GK, Mahmood R, Balasubramanian R (2015) Food waste-to-energy conversion technologies: current status and future directions. Waste Manage 38:399–408. doi:10.1016/j.wasman.2014.12.004

Pinto G, Pollio A, Previtera L, Stanzione M, Temussi F (2003) Removal of low molecular weight phenols from olive oil mill wastewater using microalgae. Biotechnol Lett 25:1657–1659. doi:10.1023/A:1025667429222

Pintucci C, Giovannelli A, Traversi ML, Ena A, Padovani G, Carlozzi P (2013) Fresh olive mill waste deprived of polyphenols as feedstock for hydrogen photo-production by means of *Rhodopseudomonas palustris* 42OL. Renew Energy 51:358–363. doi:10.1016/j.renene.2012.09.037

Pittman JK, Dean AP, Osundeko O (2011) The potential of sustainable algal biofuel production using wastewater resources. Bioresour Technol 102:17–25. doi:10.1016/j.biortech.2010.06.035

Plastics-The Facts 2015 An analysis of European latest plastics production, demand and waste data. Plastics Europe: Association of Plastics Manufacturers. http://www.plasticseurope.org/Document/plastics—the-facts-2015.aspx?FolID=2

Rahmanian N, Jafari SM, Galanakis CM (2014) Recovery and removal of phenolic compounds from olive mill wastewater. J Am Oil Chem Soc 91:1–18. doi:10.1007/s11746-013-2350-9

Raj A, Reddy MMK, Chandra R (2007) Decolourisation and treatment of pulp and paper mill effluent by lignin-degrading *Bacillus* sp. J Chem Technol Biotechnol 82:399–406. doi:10.1002/jctb.1683

Ramírez IM, Tsaousi K, Rudden M, Marchant R, Alameda EJ, Roman MG, Banat IM (2015) Rhamnolipid and surfactin production from olive oil mill waste as sole carbon source. Bioresour Technol 198:231–236. doi:10.1016/j.biortech.2015.09.012

Rico JAP, Sauer IL (2015) A review of Brazilian biodiesel experiences. Renew Sust Energy Rev 45:513–529. doi:10.1016/j.rser.2015.01.028

Roig A, Cayuela ML, Sanchez-Monedero MA (2006) An overview on olive mill wastes and their valorisation methods. Waste Manage 26:960–969. doi:10.1016/j.wasman.2005.07.024

Romero-García JM, Niño L, Martínez-Patiño C, Álvarez C, Castro E, Negro MJ (2014) Biorefinery based on olive biomass. State of the art and future trends. Bioresour Technol 159:421–432. doi:10.1016/j.biortech.2014.03.062

Ruiz-Bravo A, Jimenez-Valera M, Moreno E, Guerra V, Ramos-Cormenzana A (2001) Biological response modifier activity of an exopolysaccharide from *Paenibacillus jamilae* CP-7. Clin Diagn Lab Immunol 8:706–710. doi:10.1128/CDLI.8.4.2001

Sampaio MA, Goncalves MR, Marques IP (2011) Anaerobic digestion challenge of raw olive mill wastewater. Bioresour Technol 102:10810–10818. doi:10.1016/j.biortech.2011.09.001

Sarkar N, Kumar Ghosh S, Bannerjee S, Aikat A (2012) Bioethanol production from agricultural wastes: an overview. Renew Energy 37:19–27. doi:10.1016/j.renene.2011.06.045

Sarris D, Papanikolaou S (2016) Biotechnological production of ethanol: biochemistry, processes and technologies. Eng Life Sci 16:307–329. doi:10.1002/elsc.201400199

Sarris D, Matsakas L, Aggelis G, Koutinas AA, Papanikolaou S (2014) Aerated vs non-aerated conversions of molasses and olive mill wastewaters blends into bioethanol by *Saccharomyces cerevisiae* under non-aseptic conditions. Ind Crops Prod 56:83–93. doi:10.1016/j.indcrop.2014.02.040

Schenk PM, Thomas-Hall SR, Stephens E, Marx UC, Mussgnug JH, Posten C, Kruse O, Hankamer B (2008) Second generation biofuels: high-efficiency microalgae for biodiesel production. Bioenergy Res 1:20–43. doi:10.1007/s12155-008-9008-8

Sen U, Shakdwipee M, Banerjee R (2008) Status of biological hydrogen production. J Sci Ind Res 67:980–993

Senkevich S, Ntaikou I, Lyberatos G (2012) Bioethanol production from thermochemically pre-treated olive mill solid residues using the yeast *Pachysolen tannophilus*. Global Nest J 14:118–124

Sims REH, Mabee W, Saddler JN, Taylor M (2010) An overview of second generation biofuel technologies. Bioresour Technol 101:1570–1580. doi:10.1016/j.biortech.2009.11.046

Taguri T, Tanaka T, Kound I (2006) Antibacterial spectrum of plants polyphenols and extracts depending upon hydroxyphenyl structure. Biol Pharm Bull 29:2226–2235. doi:10.1248/bpb.29.2226

Tayeh HA, Najami N, Dosoretz C, Tafesh A, Azaizeh H (2014) Potential of bioethanol production from olive mill solid wastes. Bioresour Technol 152:24–30. doi:10.1016/j.biortech.2013.10.102

Urbaniec K, Bakker RR (2015) Biomass residues as raw materials for dark hydrogen fermentation—a review. Int J Hydrogen Energy 40:3648–3658. doi:10.1016/j.ijhydene.2015.01.073

Venetsaneas N, Antonopoulou G, Stamatelatou K, Kornaros M, Lyberatos G (2009) Using cheese whey for hydrogen and methane generation in a two-stage continuous process with alternative pH controlling. Bioresour Technol 100:3713–3717. doi:10.1016/j.biortech.2009.01.025

Villanova L, Villanova L, Fasiello G, Merendino A (2010) Process for the recovery of tyrosol and hydroxytyrosol from oil mill wastewaters and catalytic oxidation method in order to convert tyrosol in hydroxytyrosol. Patent: EP2147898 A1

Weiland P (2010) Biogas production: current state and perspectives. Appl Microbiol Biotechnol 85:849–860. doi:10.1007/s00253-009-2246-7

Wildman REC, Kelley M (2007) Nutraceuticals and functional foods. In: REC W (ed) Handbook of nutraceuticals and functional foods, 2nd edn. CRC Press, Taylor & Francis Group, London, New York, pp 1–22

Yousuf A, Sannino F, Addorisio V, Pirozzi D (2010) Microbial conversion of olive oil mill wastewaters into lipids suitable for biodiesel production. J Agric Food Chem 58:8630–8635. doi:10.1021/jf101282t

Yuan JS, Tiller KH, Al-Ahmad H, Stewart NR, Stewart CN Jr (2008) Plants to power: bioenergy to fuel the future. Trends Plant Sci 13:421–429. doi:10.1016/j.tplants.2008.06.001

Zanichelli D, Carloni F, Hasanaj E, D'Andrea N, Filippini A, Setti L (2007) Production of ethanol by an integrated valorization of olive oil byproducts. The role of phenolic inhibition. Environ Sci Pollut Res 14:5–6. doi:10.1065/espr2006.06.316

Zhang C, Su H, Baeyens J, Tan T (2014) Reviewing the anaerobic digestion of food waste for biogas production. Renew Sust Energy Rev 38:383–392. doi:10.1016/j.rser.2014.05.038

Zheng Y, Pan Z, Zhang R (2009) Overview of biomass pretreatment for cellulosic ethanol production. Int J Agric Biol Eng 2:51–68. doi:10.3965/j.issn.1934-6344.2009.03.051-068

# Algae—The Potential Future Fuel: Challenges and Prospects

**Ramesh Kakarla, Chandrasekhar Kuppam, Soumya Pandit, Abudukeremu Kadier, and Jeevitha Velpuri**

**Abstract** Algae are single or multicellular photosynthetic organisms that can fix the atmospheric carbon into valuable lipids, proteins, carbohydrates, and fats. These algae are also capable of growing vigorously in different habitats from freshwater to brackish water environments and wastewater streams with nutrient uptake ability. These features make the algae themselves uniquely important in biofuel generation and wastewater treatment process along with $CO_2$ sequestrations without competing with food crop land. The algae can also be used as a substrate for various biofuel generations, bioethanol, bio-butanol, hydrogen, methane, and many commercially valuable products. The applicability and renewability of algal fuel are most promising for the future biotechnological applications. Optimization of algal growth conditions

R. Kakarla
Department of Environmental Science and Engineering, Environmental Biotechnology Laboratory for Water and Energy (EBLWE), Kyung Hee University, Seoul, Republic of Korea

Department of Biotechnology, College of Science and Technology, Andhra University, Visakhapatnam 530003, Andhra Pradesh, India
e-mail: kakarlarameshbt@gmail.com

C. Kuppam (✉)
School of Applied Bioscience, Agriculture Department, Kyungpook National University, Building-1, Dong-Daegu 702-703, South Korea

Bio-Engineering and Environmental Science (BEES), CSIR-IICT, Hyderabad, India
e-mail: chanduibt@gmail.com

S. Pandit
Department of Desalination and Water Treatment, Ben-Gurion University of the Negev, Beersheba, Israel

Department of Biotechnology, IIT, Kharagpur, India
e-mail: sounip@gmail.com

A. Kadier
Department of Chemical and Process Engineering, Faculty of Engineering and Built Environment, National University of Malaysia, Bangi, Selangor, Malaysia
e-mail: abudoukeremu@163.com

J. Velpuri
Department of Environmental Science and Technology, Jawaharlal Nehru Technological University, Hyderabad, Telangana, India
e-mail: jeevithavelpuri@gmail.com



V.C. Kalia, P. Kumar (eds.), *Microbial Applications Vol.1*,
DOI 10.1007/978-3-319-52666-9_11

and harvesting technology with desired biofuel generations at low processing cost can make the algae as one of the best sources of energy for future generations. Genetically modified algae which are capable to grow rapidly with generating high cellular lipids and carbohydrate content can be crucial for future energy demand.

**Keywords** Algae • Biodiesel • Algae cultivation • Harvesting technologies • Microbial fuel cells • Nutrient removal

## 1 Introduction

Growing energy needs and transportation sectors were linked with total greenhouse gas (GHG) emissions producing up to 60% and 20%, subsequently, whereas the agriculture countries like India and China account to about 9% in total GHG emissions (Mata et al. 2010). The usage of a large amount of fossil fuels leads to increased production of GHG emission which showed adverse effects on the environment and ecosystem (Vologni et al. 2013; Chandrasekhar et al. 2015a). Moreover, the increased GHG leads to global warming, and it was estimated that the excess soluble bicarbonate ($CO_2$) in seawater could alter the pH and damage both sea and land ecosystem as well, which in turn damage the sea food chain ecosystem and food supply to human needs. The use of fossil fuels is not a continuing process as they are not renewable source of energy. In beneficial to minimize the emission of GHG and fulfill the future energy requirement, an alternative energy is needed. In this regard a clean and renewable energy is need for better environment. Initially the clean energy fuels (biofuels) were extracted from the energy crops like corn and soya bean (first generation). However, the purpose of energy crops in production of biofuels has been limited due to human consumption (Mata et al. 2010). Moreover, growing the biodiesel crops like Jatropha in land that are used as food crops are also not feasible which ultimately decrease the food availability and increase prices. Then the algae came to focus as a third-generation fuels, which are rich in carbohydrates, proteins, and lipids, which can grow rapidly in a variety of habitats without competing with agriculture land (Mata et al. 2010). The growing algae are easy and economic with use of naturally available sunlight and nutrients that are present in different kinds of wastewaters and water streams. Moreover, the increased usage of fertilizers led to contamination of groundwater with nutrients ammonium, nitrate, nitrite, and phosphorus that can be uptaken with microalgae growth. Finally, the fourth-generation fuel (bioelectricity) came into consideration which produces direct electricity with microorganisms (exoelectrogens) that are capable of transferring externally (Venkata Mohan and Chandrasekhar 2011a). These systems have great advantage with production of direct electricity rather than producing fuels (bio-hydrogen, methane) that can be converted to electricity (Venkata Mohan and Chandrasekhar 2011b).

In this chapter a detailed description about beneficial aspects of algae in various biotechnological fields includes about microalgae and its potentials, current research, algae growing and harvesting methods, biofuel generation, application in microbial fuel cells as feedstock (anode), and oxygen supplier (cathode) for generation of electricity.

## 2 Microalgae and Its Potentials

Algae are single (prokaryotic) or multicellular (eukaryotic) organisms that can be found ubiquitous in nature from freshwater to brackish water habitats. Most of algaeare chlorophyll containing photosynthetic (photoautotrophic) organisms; however, some are heterotrophic in nature, which extract energy from organic carbon such as cellulosic materials. Microalgae can be seen in various parts of earth ecosystems, including aquatic and also in terrestrial habitats, consisting of different species found in a broad range of environmental regions. Moreover, there were more than 50,000 algae species found to exist (Mata et al. 2010). Algae are naturally able to replicate rapidly with production of large biomass generation which is rich in oils and proteins (SCP); alcohols (butanol) have drawn the consideration of several industrial applicability and researchers (Fig. 1a). Algae growth depends on availability of inorganic carbon ($CO_2$) or organic carbon compounds along with nutrients like nitrogen and phosphorus. The change in growing environment and the abundance of daylight, carbon, and nutrients could possibly result in production of varied compounds like lipids and carbohydrates. In fact, with growing of algae in nutrients limited conditions, the production of desired end product was not significantly impressive. In fact, naturally the algae are capable of producing hydrocarbon compound that can be alterative for today's petroleum fuels. Current biotechnology research particularly focused on investigating the possible ways to increase the algae biomass growth rate with enhanced production of commercially valuable oils, biofuels, alcohols, and proteins. Moreover, it was found that many algae species are amenable for genetic engineering with expression of desired recombinant proteins in great quantities.

## 3 Ongoing Research

The field of biotechnology has initiated the sequencing and illustrates the genomes of algae species for understanding the metabolic pathways. This study helps the researchers to know how the algae can synthesize desired lipids and carbohydrates from available carbon and nutrients. A better and clear understanding of metabolic pathways and growth pattern in algae could obviously help in better implementation of future research. The technology that involves the discipline of genetic engineering which is currently used in plant metabolic engineering, microbial biotechnology, and synthetic biology was used for production of recombinant lipids (Ahmad et al. 2015), proteins, and carbohydrates. In this way biotechnology

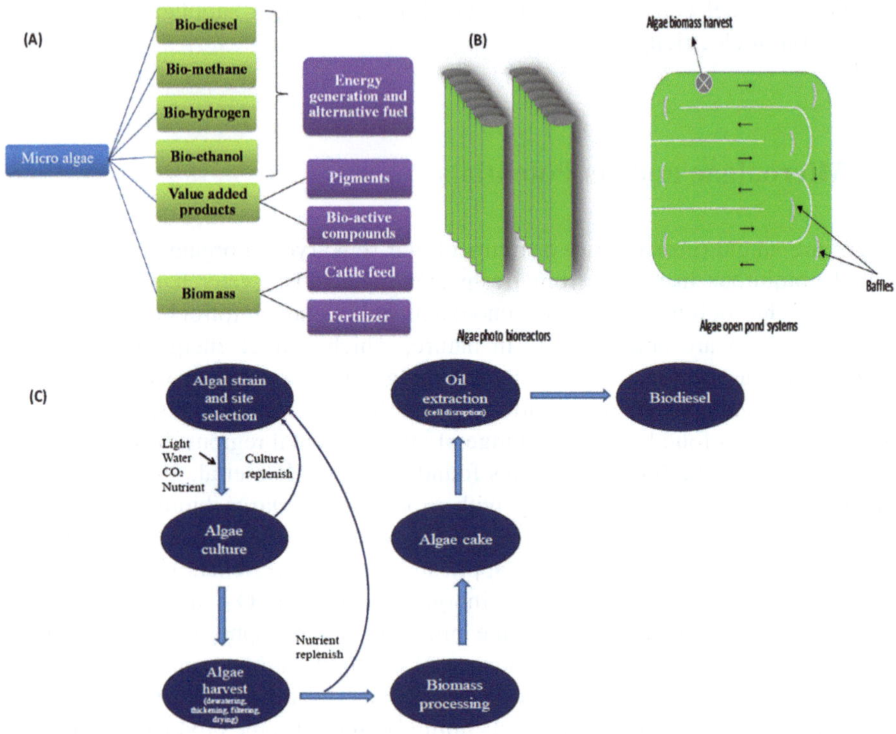


**Fig. 1** Schematic illustration of potential, cultivation, and biodiesel from algae (**a**). Flowchart of various value-added products that can be generated from microalgae biomass (**b**). Cultivation systems that are used to generated algae biomass in closed (photo-bioreactors) and open cultivation (open pond) systems (**c**). Complete steps that are involved in production of biodiesel from algae

research plays a key role in development and optimization of algae biofuel generation and algal strains for better and increased biofuel generation.

## 4 Algae Cultivation and Harvest for Biodiesel Production

### *4.1 Algae Cultivation*

In order to use the commercially developed algal strains successfully, they must fit to the industrial processes including harvest and processing technology. Biofuel companies are intended to scale commercial production of algae fuels using closed, open pond and hybrid systems as cost-effective way for algae biomass production (Fig. 1b, Table 1). In closed type systems, the algae growth conditions can be regulated well including nutrient and light availability for better algal growth. In fact, the maintenance and operation is not easy and economical; however, the

**Table 1** Algae cultivation systems and their merits and demerits

| Cultivation system | Merits | Demerits |
|---|---|---|
| Open pond | • Cost effectiveness<br>• Maintenance is easy<br>• Lower energy consumption | • Less biomass productivity<br>• High contamination<br>• Large land area is needed |
| Photo-bioreactors | • Moderately economic<br>• Good biomass productivity<br>• Less contamination<br>• Better light usage | • Difficult temperature control<br>• Low hydrodynamic stress<br>• Algae growth on walls<br>• Gradient of pH, dissolved oxygen, and $CO_2$ can be found |
| Hybrid | • Lower contamination<br>• Better algae productivity<br>• Low energy consumption<br>• High biomass productivity | • Often susceptible for contamination<br>• Effected by environmental factors |

product purity and contamination of algal species can be minimized. These closed type systems do not need to repeat inoculations as contamination is minimal. Closed type systems include photo-bioreactors for growth of photosynthetic algae and traditional bioreactors for algae that depend on carbohydrates, whereas open pond-type systems have been used in many ways, but these are sensitive to several environmental factors, which include contamination by other algae strains or change in nutrient or heat and light availability, as the microalgae can be dispersed from open pond type systems to the surrounding environment easily by environmental factors like wind and heavy rain. However, the maintenance and cost of operation are less compared to closed type systems. In some case a combined approach of these two of systems (hybrid systems) were employed for better productivity with minimized contamination. In this regard pond type systems covered with a thin plastic film in combination of both closed and open type systems are more beneficial with high product yield and less contamination. However, the development of an economical harvesting system and recycling of residual biomass after the biofuel is extracted are essential factors in commercialization. Generally the biomass cultivation and harvest can include some preliminary steps that include selection of strain and site of cultivation, algae biomass generation, harvesting, biomass processing, preparing algae cake, oil extraction, and biodiesel production as shown in Fig. 1c.

## *4.2 Algae Harvesting Technologies*

Algae biomass can be harvested in several different ways based on type of algae whether it is a macro- or microalgae and cultivation system including photobioreactors, open ponds, or industrial purpose. Generally after settling several strategies will be followed to concentrate the algae present in a large amount of water, which include membrane filtration, centrifugation, flotation, and flocculation. These technologies were chosen based on the type of algae (micro or macro), cultivation system, and volume. In the filtration method, membranes were used to separate the algae from the rest of the liquid; however the selection of membrane is based on the size of algae that need to be filtered. Generally membranes are made of polyvinyl chloride (PVC), ultrafiltration membranes, and modified cellulose membranes. The filtration is very effective when combined with vacuum and limited to small quantity. Even though the membrane filtration is very effective, it has several limitations like membranes are expensive, membrane fouling, clogging of membrane pores with algae, and membranes need backwash repeatedly (Zhang et al. 2010). Use of membranes is good to concentrate small volume of algae culture in a short time.

The efficient technologies that can segregate and separate algae from its liquid medium is centrifugation. In centrifugation processes the algae are prone to centrifugal forces which make them settle down at the bottom of a centrifuge tube. Even though the process of centrifugation and drying is not economic for personal use, however it can be applicable in commercial and industrial scale. Centrifuge can be helpful in both lipid separation and chemical extraction of biodiesel from algae. Coupled with a homogenizer, one may be able to separate bio-lipids and other useful materials from algae. Continuous-flow centrifugation with the classical Foerst rotor is more efficient, but it is not suitable for delicate algal cells which may damage while rotation. The other one is variant on zonal centrifugation that offers a series of theoretical advantages in the condensation and distillation of particles. Flotation is usually used in combination of flocculation while harvesting algae from wastewaters. It is a very easy process in which it makes algae float on the surface of medium as a froth to be removed. Dissolved air flotation (DAF) makes use of both flocculation and froth flotation features to separate algae from its culture. It uses alum (trivalent sulfate of metals including chromium or iron and aluminum) salts to flocculate algae (algae trapped in air bubble) in air mixture, through foam (small bubbles) created from an air compressor. In froth flotation technique, froth is created by adjusting pH and air bubbling through a column where algae entrapped in froth float above liquid level. The collected algae froth on the liquid surface can be removed by suction. At present froth flotation and drying are currently considered to be noneconomic in commercial application. The cost of froth flotation was estimated to be too high for commercial use. In future, the cost of operation can be minimized with further research and experimentation.

Flocculation technology also can be used to separate algae from the medium in which algae were force to form lumps. The flocculation can be done in many ways including chemical, biological, physical, and auto-flocculation. Nevertheless, the difficulty in removing the unwanted chemicals from the separated algae is the main drawback in this method of separation, making it uneconomical process.

Flocculants are chemical substances that initiate the process by initiating colloids in liquid to aggregate, ultimately developing a flock. Ferric chloride and alum are common chemical flocculants that are useful to harvest algae. Moreover, "chitosan" is a commercially available biopolymer which is commonly useful for drinking water purification technology as well as a flocculent which is a costly polymer. The shells of crustaceans (shrimps and crabs) are crushed into powder and treated to form a polysaccharide called chitin, which is found in the shells. Then this powdered chitin undergoes a deacetylation process to form chitosan, but this chitosan works good only at low pH, where microalgae grow at high pH. Physical flocculation is the only method that the contamination of biomass can be avoided using physical forces. For example, flocculation of microalgae can be done by applying a field of standing ultrasound waves. Although this method is not applicable to apply on larger scales, it works well in laboratory scale. Bio-flocculation is often successfully used for harvesting microalgae in wastewater treatment facilities (Craggs et al. 2012). Some microalgal species flocculate more quickly than others, and such naturally bio-flocculating microalgae can be mixed with other species to induce flocculation (Taylor et al. 2012). The auto-flocculation often occurs spontaneously when pH increases above 9 in microalgal cultures. This type of flocculation is said to be auto-flocculation because it occurs automatically in microalgal cultures as a result of increased pH due to depletion of $CO_2$ caused by photosynthetic activity (Spilling et al. 2011).

## 5 Algal Biofuels

The term biofuel can be referred to solid, liquid, or gaseous fuels that are originated from bio-renewable feedstocks that can be used for humankind. Microalgae have been emerged as a promising and ecofriendly third-generation feedstock for biofuels as they cultivate in fresh or marine water and without competing with the agriculture lands (Sharma et al. 2011). The biofuels include biodiesel, biomethane, bioethanol, bio-hydrogen, and value-added products (pigments, bioactive compounds). The algae are rich in lipids, carbohydrates, and proteins which make them a feasible candidate for generation of biofuels. Cellular composition of several microalgae species based on (%) dry matter basis was explained in Table 2 (Zhu et al. 2014). Moreover, the International Energy Agency (IEA) evaluated that biofuels could contribute 27% of total energy demand in the transport industry by year 2050 owing to its rising importance and demand (Fornell et al. 2013).

In biofuel extraction the algal biomass was subjected to pretreatment; afterward it will be used for biofuel generation. However, the pretreatment of algae sludge might vary including physical, chemical, biological, or enzymatic treatments in order to convert complex substrates to simple substrates. Physical treatment methods make use of heat and pressure (autoclave) and sonication, while the chemical treatments make use of acid and alkali to break down the large molecule to simple molecules. In some studies involvement of both physical and chemical

**Table 2** Micro- and macroalgal cellular lipid, carbohydrate, and protein constitutions among different algal species

| Microalgal species | Lipids (%) | Carbohydrates (%) | Proteins (%) |
|---|---|---|---|
| *Arthrospira maxima* | 6–7 | 13–16 | 60–71 |
| *Chlamydomonas reinhardtii* | 21 | 17 | 48 |
| *Chlorella* sp. | 2–46 | 12–28 | 11–58 |
| *Euglena gracilis* | 14–20 | 14–18 | 39–61 |
| *Scenedesmus* sp. | 1.9–40 | 10–52 | 8–56 |
| *Spirogyra* sp. | 11–21 | 33–64 | 6–20 |
| *Spirulina* sp. | 4–9 | 8–16 | 46–63 |
| *Dunaliella* sp. | 6–8 | 4–32 | 49–57 |
| *Synechococcus* sp. | 11 | 13 | 63 |

process is done for better outcome. The use of physical treatment with the use of heat and pressure is very easy and inexpensive, but the formed end product sometimes needs further treatments or needs a long time to degrade to simple molecules, whereas chemical treatments which make use of acid and alkali (Harun et al. 2011) are most efficient, but the use of chemical may leave some residues in treated algae sludge. On the other hand, the use of enzymes for treatment of algae sludge is quite fast and efficient, but enzymes are very costly to use in large scale (Harun and Danquah 2011). Finally, the biological treatment can be only an alternate to solve above problems, even the process of degradation which takes a bit longer.

# 6 Algae in MFC

## 6.1 Algae as an Electron Generator in Anode

Electricity generation from microbes (bioelectricity) has recently gained more attention as a renewable energy source which is considered to be a fourth-generation fuel. Bioelectricity can be defined as the electricity that was produced by microorganisms mostly by anaerobic bacteria. Microbial fuel cells (MFCs) are one of the innovative and renewable systems, which can mimic the bacteria as a biocatalyst to produce direct electricity from organics (Kakarla and Min 2014a). Algae can grow well in natural habitats producing a large amount of biomass that can be used as potential feedstock for MFC electricity generation (Fig. 2), whereas in some previous reports, some primitive algae like cyanobacteria are capable of external electron transfer (Fig. 3). Due to these reasons, several researchers had implemented algae in MFC anode for the generation of electricity. In some other studies, algae were used as a feedstock (carbon source) for the generation of electricity in treated (Kondaveeti et al. 2014) or untreated form (Walter et al. 2015) for the generation of electricity. The use of untreated algae as energy can

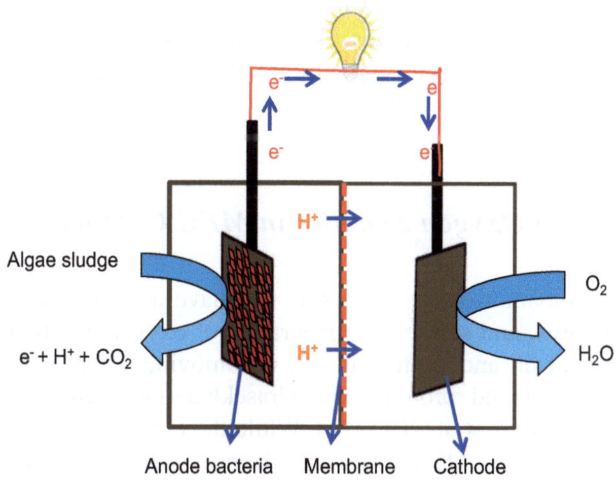


**Fig. 2** Algae sludge used as a feedstock (carbon source) in MFC anode for electricity generation

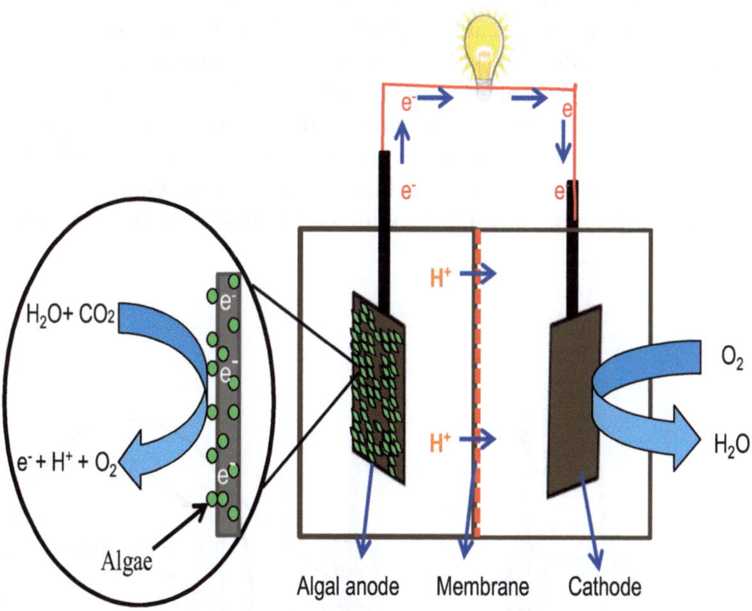


**Fig. 3** Live algae used as an electron provider to MFC anode for electricity generation

be easy and economic compared to treated algae, but the efficiency of electricity generation can be lower and needs a long time to remove total algal COD. On the other hand, in some other cases, live algae were used as an electron donor with or without mediator involvement (Xu et al. 2015). The involvement of mediators is not suggestible as mediators can pollute the environment and carcinogenic. Moreover, even without the use of mediator, electron transfer from anode to cathode can be limited significantly due to generation of oxygen from algae. Considering these merits and optimizing challenges in the future, the use of algae in MFC anode can be beneficial.

## 6.2 *Algae as an Oxygen Supplier in MFC Cathode*

The algae are photoautotrophic organisms that have a great ability in removing nitrogen, phosphorus, and sulfur with generation of oxygen and biomass. In MFC the bacteria present on anode are efficient in removing COD with generation of electrons (electricity) and protons (Chandrasekhar and Venkata Mohan 2012, 2014a, b; Chandrasekhar et al. 2015a, b). While the cathode doesn't have any sort of mechanism to remove COD or nutrients, the necessity of oxygen supplement to cathode could increase the operational cost. However, the implementation of algae in the cathode could solve these problems with generation of biomass as a by-product, which can be further used for biofuel generation (Fig. 4). The use of algae in the cathode could provide higher oxygen (dissolved oxygen) compared to normal oxygen saturated levels at STP (100% or 7.6 mg/l); this also helps in getting higher cathode potentials than with the use of normal atmospheric air (20.8% $O_2$) (Kakarla and Min 2014b; Kakarla et al. 2015). This increased cathode potentials could increase over all MFC performance. There were several studies in these recent years with the use of algae as oxygen supplier with increased MFC performance, removal of nutrients, and algal biomass generation (Gajda et al. 2015). In fact, there are several more advantages with use of algae in MFC cathode like $CO_2$

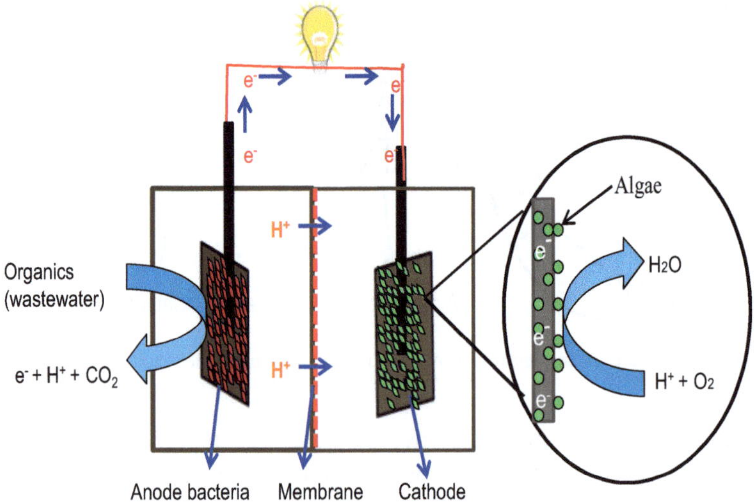


**Fig. 4** Algae used as an oxygen supplier in MFC cathode for electricity generation

sequestration and good catholyte buffering due to soluble $CO_2$ ($HCO_3^-$) (Pandit et al. 2012). In some studies the MFC cathode was linked to a photo-bioreactor with providing algae to cathode and harvested algal biomass used as a feedstock for MFC anode (Gajda et al. 2015). Wang et al. use anodic off-gas as a carbon source for the algae in cathode for oxygen production (Wang et al. 2010).

## 7 Other Beneficial Aspects of Algae: Carbon Sequestration and Nutrient Removals

Due to the human efforts and increasing population, the usage of natural resources (fossil fuels) has been increased greatly with production of a large amount of $CO_2$, waste, and wastewaters from both domestic and industrial sectors led to nutrient rich polluted ground wasters. The $CO_2$ (GHG) gas can be uptaken by sparkling wastewater which is rich in $NH_4^+$, $NO_3^-$, $NO_2^-$, and $PO_4^{3-}$ with the growth of algae, and the generated biomass can be further used for several other purposes including biodiesel generation other than biofuels. The waste residue leftover after oil has been extracted from the algae can be processed into bio-hydrogen, bioethanol, biomethane, and feedstock and can also be utilized as carbon-rich organic fertilizer (for the reason that it is high in N:P ratio) (Fig. 5). Considering the microalgae species, there are several other valuable compounds that can be extracted including simple chemicals and bulk products including oil, fats, polyunsaturated fatty acids, sugars, pigments, antioxidants, natural dyes, and finally waste as fertilizer for agriculture. Due to a high variety of valuable products, microalgae can be potential organisms in large biotechnological fields including food additives and nutrition, cosmetics, pharmaceuticals, aquaculture, and pollution control.

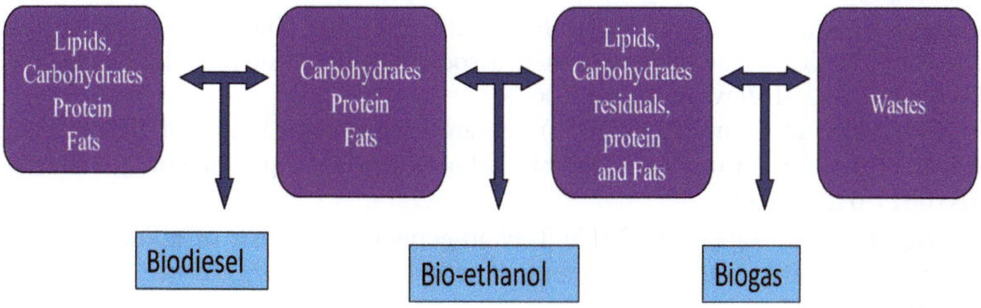


**Fig. 5** Series of steps that are involved in production of different biofuel products from microalgae cellular contents

## 8 Conclusion

The algae's ability to grow in different habitats with natural available sunlight, nutrients in wastewaters, and $CO_2$ sequestrations makes it as a potential organism for future energy needs. The applicability of algae in direct electricity-producing systems like MFC as a feedstock or electron provider in the anode while as oxygen supplier in the cathode is a renewable process. The algae play an important role in treatment of wastewater with nutrient uptake and biomass generation which can be used for synthesis of various valued products like biofuels, commercial products, protein, and feedstocks for various fuel generations. However, the harvesting and processing can be a big challenge which need to over to make algae fuels as future energy.

## References

Ahmad I, Sharma AK, Daniell H, Kumar S (2015) Altered lipid composition and enhanced lipid production in green microalga by introduction of brassica diacylglycerol acyltransferase 2. Plant Biotech J 13:540–550. doi:10.1111/pbi.12278

Chandrasekhar K, Venkata Mohan S (2012) Bio-electrochemical remediation of real field petroleum sludge as an electron donor with simultaneous power generation facilitates biotransformation of PAH: effect of substrate concentration. Bioresour Technol 110:517–525. doi:10.1016/j.biortech.2012.01.128

Chandrasekhar K, Venkata Mohan S (2014a) Bio-electrohydrolysis as a pretreatment strategy to catabolize complex food waste in closed circuitry: function of electron flux to enhance acidogenic biohydrogen production. Int J Hydrogen Energy 39:11411–11422. doi:10.1016/j.ijhydene.2014.05.035

Chandrasekhar K, Venkata Mohan S (2014b) Induced catabolic bio-electrohydrolysis of complex food waste by regulating external resistance for enhancing acidogenic biohydrogen production. Bioresour Technol 165:372–382. doi:10.1016/j.biortech.2014.02.073

Chandrasekhar K, Lee YJ, Lee DW (2015a) Biohydrogen production: strategies to improve process efficiency through microbial routes. Int J Mol Sci 16:8266–8293. doi:10.3390/ijms16048266

Chandrasekhar K, Amulya K, Venkata Mohan S (2015b) Solid phase bio-electrofermentation of food waste to harvest value-added products associated with waste remediation. Waste Manage 45:57–65. doi:10.1016/j.wasman.2015.06.001

Craggs R, Sutherland D, Campbell H (2012) Hectare-scale demonstration of high rate algal ponds for enhanced wastewater treatment and biofuel production. J Appl Phycol 24:329–337. doi:10.1007/s10811-012-9810-8

Fornell R, Berntsson T, Åsblad A (2013) Techno-economic analysis of a kraft pulp-mill-based biorefinery producing both ethanol and dimethyl ether. Energy 50:83–92. doi:10.1016/j.energy.2012.11.041

Gajda I, Greenman J, Melhuish C, Ieropoulos I (2015) Self-sustainable electricity production from algae grown in a microbial fuel cell system. Biomass Bioenerg 82:87–93. doi:10.1016/j.biombioe.2015.05.017

Harun R, Danquah MK (2011) Enzymatic hydrolysis of microalgal biomass for bioethanol production. Chem Eng J 168:1079–1084. doi:10.1016/j.cej.2011.01.088

Harun R, Jason WSY, Cherrington T, Danquah MK (2011) Exploring alkaline pre-treatment of microalgal biomass for bioethanol production. Appl Energy 88:3464–3467. doi:10.1016/j.apenergy.2010.10.048

Kakarla R, Min B (2014a) Evaluation of microbial fuel cell operation using algae as an oxygen supplier: carbon paper cathode vs. carbon brush cathode. Bioproc Biosyst Eng 37:1–9. doi:10.1007/s00449-014-1223-4

Kakarla R, Min B (2014b) Photoautotrophic microalgae *Scenedesmus obliquus* attached on a cathode as oxygen producers for microbial fuel cell (MFC) operation. Int J Hydrogen Energy 39:10275–10283. doi:10.1016/j.ijhydene.2014.04.158

Kakarla R, Kim JR, Jeon BH, Min B (2015) Enhanced performance of an air–cathode microbial fuel cell with oxygen supply from an externally connected algal bioreactor. Bioresour Technol 195:210–216. doi:10.1016/j.biortech.2015.06.062

Kondaveeti S, Choi K, Kakarla R, Min B (2014) Microalgae *Scenedesmus obliquus* as renewable biomass feedstock for electricity generation in microbial fuel cells (MFCs). Front Environ Sci Eng 8:784–791. doi:10.1007/s11783-013-0590-4

Mata TM, Martins AA, Caetano NS (2010) Microalgae for biodiesel production and other applications: a review. Renew Sustain Energy Rev 14:217–232. doi:10.1016/j.rser.2009.07.020

Pandit S, Nayak BK, Das D (2012) Microbial carbon capture cell using cyanobacteria for simultaneous power generation, carbon dioxide sequestration and wastewater treatment. Bioresour Technol 107:97–102. doi:10.1016/j.biortech.2011.12.067

Sharma YC, Singh B, Korstad J (2011) A critical review on recent methods used for economically viable and eco-friendly development of microalgae as a potential feedstock for synthesis of biodiesel. Green Chem 13:2993–3006. doi:10.1039/c1gc15535k

Spilling K, Seppälä J, Tamminen T (2011) Inducing autoflocculation in the diatom *Phaeodactylum tricornutum* through $CO_2$ regulation. J Appl Phycol 23:959–966. doi:10.1007/s10811-010-9616-5

Taylor R, Rand J, Caldwell G (2012) Treatment with algae extracts promotes flocculation, and enhances growth and neutral lipid content in nannochloropsis oculata a candidate for biofuel production. Mar Biotechnol 14:774–781. doi:10.1007/s10126-012-9441-8

Venkata Mohan S, Chandrasekhar K (2011a) Self-induced bio-potential and graphite electron accepting conditions enhances petroleum sludge degradation in bio-electrochemical system with simultaneous power generation. Bioresour Technol 102:9532–9541. doi:10.1016/j.biortech.2011.07.038

Venkata Mohan S, Chandrasekhar K (2011b) Solid phase microbial fuel cell (SMFC) for harnessing bioelectricity from composite food waste fermentation: influence of electrode assembly and buffering capacity. Bioresour Technol 102:7077–7085. doi:10.1016/j.biortech.2011.04.039

Vologni V, Kakarla R, Angelidaki I, Min B (2013) Increased power generation from primary sludge by a submersible microbial fuel cell and optimum operational conditions. Bioprocess Biosyst Eng 36:635–642. doi:10.1007/s00449-013-0918-2

Walter XA, Greenman J, Taylor B, Ieropoulos IA (2015) Microbial fuel cells continuously fuelled by untreated fresh algal biomass. Algal Res 11:103–107. doi:10.1016/j.algal.2015.06.003

Wang X, Feng Y, Liu J, Lee H, Li C, Li N, Ren N (2010) Sequestration of $CO_2$ discharged from anode by algal cathode in microbial carbon capture cells (MCCs). Biosens Bioelectron 25:2639–2643. doi:10.1016/j.bios.2010.04.036

Xu C, Poon K, Choi MF, Wang R (2015) Using live algae at the anode of a microbial fuel cell to generate electricity. Environ Sci Pollut Res 22:15621–15635. doi:10.1007/s11356-015-4744-8

Zhang X, Hu Q, Sommerfeld M, Puruhito E, Chen Y (2010) Harvesting algal biomass for biofuels using ultrafiltration membranes. Bioresour Technol 101:5297–5304. doi:10.1016/j.biortech.2010.02.007

Zhu LD, Hiltunen E, Antila E, Zhong JJ, Yuan ZH, Wang ZM (2014) Microalgal biofuels: flexible bioenergies for sustainable development. Renew Sustain Energ Rev 30:1035–1046. doi:10.1016/j.rser.2013.11.003

# Part III
# Future Perspectives

# New Generation DNA Sequencing (NGS): Mining for Genes and the Potential of Extremophiles

**Bhagwan Rekadwad and Juan M. Gonzalez**

**Abstract** NGS technologies have provided unprecedented access to the gene pool of all living beings, in particular the large number and diversity of microorganism inhabitants of any given environment, and these methods allow to do it in a culture-independent manner. Using these technologies, a variety of questions can now be approached to solve the unknowns related to the analysis of nucleic acids and cell functioning such as whole-genome sequencing, target-specific re-sequencing, discovery of transcription factor-binding sites, noncoding RNA expression profiling, whole-genome expression analyses, gene evolution and mobility, and phylogeny and phylogenomics, among many other strategies to study the working of cells and the involved mechanisms. Specifically, how cells live under extreme conditions can provide with increasingly important information on the evolutionary mechanisms of living forms as well as the resistance of biomolecules to extreme conditions and the functioning of extreme microorganisms and their biomolecules. This chapter discusses a number of disclosed extremophilic microorganisms and highlights the potential of some of their genes as an example of the application of NGS to explore the biological features of those microbes living under the extreme conditions. NGS technologies allow high-throughput processing of DNA and RNA sequences which enhanced our understanding of the enormous amount of information encoded in these molecules and accelerated magnitude for an acquisition of sequencing data. NGS has greatly revolutionized the potential to investigate cell structure and functioning through the study of genomics and the molecular biology of extremophilic microorganisms and their genes.

**Keywords** Molecular ecology • New NGS • Mining of unkown genes • DNA • Metagenomics • Whole genome sequencing

---

B. Rekadwad (✉)
School of Life Sciences, Swami Ramanand Teerth Marathwada University, Nanded 431606, Maharashtra, India
e-mail: rekadwad@gmail.com

J.M. Gonzalez
Institute of Natural Resources and Agrobiology, Spanish National Research Council, IRNAS-CSIC, Avda. Reina Mercedes 10, 41012 Sevilla, Spain



V.C. Kalia, P. Kumar (eds.), *Microbial Applications Vol.1*,
DOI 10.1007/978-3-319-52666-9_12

# 1 Introduction

NGS is describing a group of different modern sequencing technologies such as the Illumina (Solexa) sequencing (Goodman et al. 2011; Harris et al. 2012; Jin et al. 2013), Roche 454 sequencing (Somboonna et al. 2011), Ion Torrent: Proton/PGM sequencing (Wang et al. 2014), and SOLiD sequencing (Dai et al. 2010; Picardi et al. 2010). All these technologies and some others under development are providing easy massive high-throughput sequence data generation in a much rapid manner than previously imagined. NGS has transformed today's biology through several ways (Schuster 2008):

1. It provides an ever-cheaper and higher-throughput alternative to DNA sequencing rather than traditional technologies such as the Sanger dideoxy sequencing. Smallest genome can now be sequenced in a day time scale.
2. This high-throughput genome sequencing is fruitful in the discovery of new genes and several regulatory elements, including those that are associated with the production of secondary metabolites such as enzymes (Table 1) (Coughlan et al. 2015), therapeutic proteins, antibiotics, etc.
3. Targeted sequencing is allowed to identify the specific genes responsible for the stability of cell and genome in extreme environments. Besides a series of novel discovery related to all biological fields, NGS allows, for example, the discovery and analysis of novel metabolic pathways and novel genes working under extreme conditions (Grada and Weinbrecht 2013).
4. The RNA-seq is a strong alternative technique which provides quantitative and qualitative information on whole-cell gene expression, i.e., information of the entire transcriptome in a single experiment without requiring previous knowledge of the given unknown genome sequence. RNA-seq is a strong alternative to microarrays in gene expression studies (Yu et al. 2014a, b).

The huge amount of sequences being generated by NGS technologies leads to massive submissions and really high rates of increase of sequence data in DNA repositories (i.e., GenBank, among others) including nucleotide sequences from the genome (from a single organism), metagenome (from a complex community composed by many organisms), and transcriptome (the RNA resulting from the genes expressed in an organism) studies and amino acid sequences corresponding to novel protein-encoding genes. This offers unique novel opportunities for further work on data mining aiming to the search of new genes and functionalities. At present, the analysis of sequencing data or DNA data mining is the major bottleneck on the processing of genomic and sequencing data and consequently the understanding of this huge new information containing all the capabilities of living cells which is coded in the cell's DNA and genome.

In the present chapter, we have described biotechnological applications of newly discovered extremophilic genes using NGS technology in healthcare and medicine, food, agriculture, and use of natural resources through the sustainable way.

**Table 1** Genes for novel industrially important enzymes discovered using functional metagenomics

| Enzyme | References |
|---|---|
| Alkaline serine protease | Pushpam et al. (2011), Biver et al. (2013a) |
| Amidase (11) | Uchiyama and Miyazaki (2010) |
| Carboxylic ester hydrolases (3) | Biver and Vandenbol (2013) |
| Cellulase (showing β-glucosidase activity) | Jiang et al. (2009) |
| Esterase | Kim et al. 2006 |
| Esterase (protease-insensitive feruloyl) | Cheng et al. (2012a) |
| Esterase (thermostable family VII with high stability in organic solvent) | Kang et al. (2011) |
| Esterase (thermostable) | Rhee et al. (2005) |
| Esterases (12), endo-β-1,4-glucanases (9) and cyclodextrinase (1) | Ferrer et al. (2005b) |
| Esterases (2) | Chu et al. (2008) |
| Esterases (2) | Elend et al. (2006) |
| Esterases (2) | Ouyang et al. (2013) |
| Esterases (5) | Ferrer et al. (2005a) |
| Fibrinolytic metalloprotease (Zn-dependent) | Lee et al. (2007) |
| Glycosylhydrolase | Palackal et al. (2007) |
| Lipase (alkaline-stable family IV) | Peng et al. (2014) |
| Lipase (cold-active) | Elend et al. (2007) |
| Lipase (moderately thermostable) | Faoro et al. (2012) |
| Lipolytic activity (37) | Nacke et al. (2011) |
| Lipolytic clones (6) | Lee et al. (2004) |
| Lipolytic enzymes (4) | Lass et al. (2011) |
| Nitrilase (137) | Robertson et al. (2004) |
| Serine proteases (2) | Neveu et al. (2011) |
| Tannase (halotolerant and moderately thermostable) | Yao et al. (2011) |
| UDP glycosyltransferase (2, one is a novel macroside glycosyltransferase) | Rabausch et al. (2013) |
| Xylanase | Cheng et al. (2012b) |
| α-Amylase (low pH, thermostable) | Lim et al. (2007) |
| α-Amylase (periplasmic) | Pooja et al. (2015) |
| β-Agarase | Voget et al. (2003) |
| β-Galactosidase | Wang et al. (2012) |
| β-Galactosidase (cold-active) | Vester et al. (2014) |
| β-Galactosidase (cold-adapted) | Wang et al. (2010) |
| β-Glucanases (3) | Walter et al. (2005) |

## 2 Data Mining Through NGS Technology

The largest number and improved quality of massive DNA sequencing data being incorporated in public DNA repositories and the sequences being generated by researchers worldwide are providing with a massive amount of information which requires further work and intense focus to search for genes and to understand how different sets of information are encoded. Deciphering that enormous amount of information is a major goal for the next decades of scientific work.

Herein, we present some simple examples where sequencing data are analyzed in the search for answers to the functioning and application of microbial features and capabilities. There are many other cases not contemplated in this chapter on data mining studies successfully retrieving important information from genomic data. The authors recommend searching the literature available on your topic of interest for the most recent advancements on genome analyses. Herein, some case studies are presented (Fig. 1) to show just as an example of the huge effort that scientists around the world are carrying out to improve our understanding of cell functioning and their potential applications.

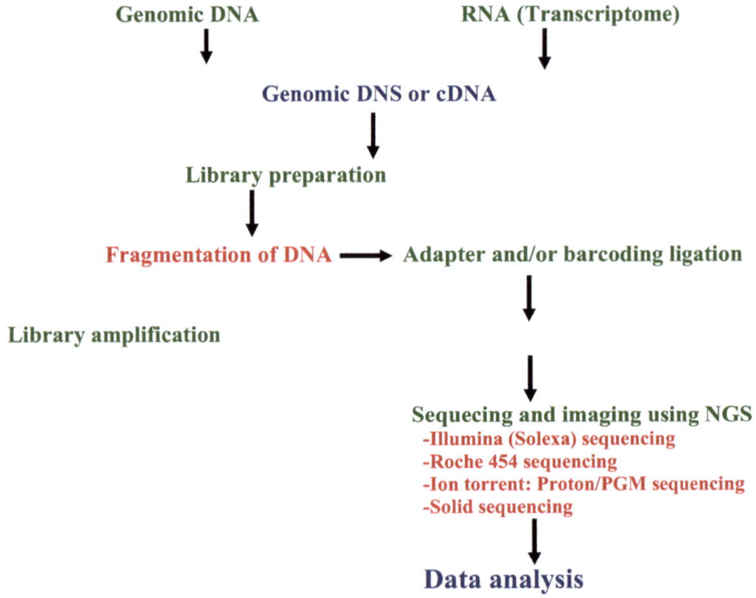


**Fig. 1** Overview of new generation sequencing technology

## 3 Sulfur Oxidation in *Acidithiobacillus thiooxidans* Under Extremes

*A. thiooxidans* is a unique chemo-litho-auto-trophic extremophilic bacterium grows on inorganic molecules that it uses to obtain molecules for biosynthesis of biomass and the energy required to support its growth and life. *A. thiooxidans* obtains energy from oxidation (Oxid$^n$) of inorganic elemental sulfur ($S^o$) and sulfur compounds to carry out various bioprocesses such as bioleaching/biomining (recovery of Copper, Cu) (Quatrini et al. 2009). The exploration of its physiology was under cover because of the lack of published data on *A. thiooxidans* genome (Valdés et al. 2008; Ghosh and Dam 2009; Chen et al. 2012). This problem is solved by NGS by constructing a preliminary model of *A. thiooxidans* whole genome and studied the genes involved in key energy pathways like $S^o$ Oxid$^n$. The *A. thiooxidans* A01 whole genome has been fully sequenced and annotated (Yin et al. 2014). As a consequence, different features have been identified and this helps to understand how these cells work and thrive on inorganic molecules, mainly on $S^o$ and metals under hostile living conditions only because of enzymes involved in cellular processes, e.g., sulfur de-oxygenase (SDO), sulfide quinine reductase (SQR), thiosulfate-quinone oxidoreductase (TQR), tetrathionate hydrolase (TeTH), $S^o$ Oxid$^n$ protein system (Sox—SoxYA and SoxYZ), and electron transport components (ETC) of Sox system (Holmes and Bonnefoy 2007; Valdes et al. 2008). SOR was detected in the draft genome sequence of *A. thiooxidans* A01A, and multiple sequence alignments of related protein showing the similarity among different related genes were investigated (Chen et al. 2005; Urich et al. 2005, 2006). A putative pathway, additionally, explained the catalysis of sulfite ($SO_3^{-2}$) to sulfate ($SO_4^{-2}$) carried out by phospho-adenosine-phospho-phosphate reductase (PAPSR) and adenylyl-sulfate kinase (APS). This conversion of $SO_3^{-2}$ to $SO_4^{-2}$ is different from the pathways that exist in *Acidithiobacillus caldus* which is catalyzed by sulfate-adenylyl transferase (SAT) (Mangold et al. 2011; Bobadilla Fazzini et al. 2013). Moreover, the quantitative analysis through real-time PCR allowed to understand the level of expression of a specific gene. $S^o$ Oxid$^n$ were strongly expressed in medium containing $S^o$ and scarcely expressed in medium containing thiosulfate ($Na_2S_2O_3$) during the mid-log phase of growth. Based on these previous studies, the $S^o$ Oxid$^n$ model was constructed to understand $S^o$ Oxid$^n$ pathways in *A. thiooxidans* A01 which provide information about $S^o$ and $Na_2S_2O_3$ physiology and many other potential functions for key sulfur oxidation present in *A. thiooxidans* genome (Yin et al. 2014).

## 4 NGS for Typing and Detection of Cluster of Antibiotic Resistance Genes Among Bacteria Clones

Antibiotic resistance is a major concern in the medical field because an increasing number of bacterial species are becoming progressively resistant to these drugs and soon we will run out of antibiotics to treat specific infections. Microbes thriving under antibiotic treatments experience a set of changes leading to potential adaptation to these new extreme environmental conditions. The study of these adaptive processes is essential to understand how to maintain sensitivity and so how to be able to use antibiotics to fight infections. NGS has a key role in typing results and detection of antibiotic resistance genes in a quite short time for any viable bacterial clone. Hence, these reliable technologies can be used to track the transmission of antibiotic resistance from parent bacteria to their clones.

Herein, we present an example of the application of NGS technologies to the analysis of antibiotic resistance in a medical scenario. The performance of this stunning technology was evaluated during the outbreak of sequence type 131 (ST131) *Escherichia coli* in hospitals in the Netherlands (Mohammadi et al. 2005; Stamatakis 2006). The BioDetection system was used to perform the NGS assay. Total 56 extended spectrums of β-lactamases (ESBL) *E. coli* clones have been collected in two independent surveys (in the month of March 2013 and May 2013). The *E. coli* typing results were compared with amplified fragment length polymorphism (AFLP). The largest cluster of the 31 genes was identified from distinguishable clones in closely adjacent wards. These results indicate that there was spread of bacteria clones in nursing home environment. Thus, BioDetection assay successfully detected the ESBL gene cluster (Bar-Joseph et al. 2001; Dhanji et al. 2010; Clermont et al. 2013). On other hand, the BioDetection assay detected ESBL genes in all sequence type 131 (ST131) *E. coli* clones with 98% sensitivity. But, it is very difficult to discriminate between ESBL and non-ESBL TEM and SHV beta-lactamases or to specify CTX-M genes by the group. For this, it is necessary to evaluate the performance of BioDetection system for typing of *E. coli* through comparison of obtained results with other reported/experimental AFLP results (Wirth et al. 2006; Doumith et al. 2012; Veenemans et al. 2014; Giovanni et al. 2015).

## 5 Applications of NGS in Molecular Ecology of Non-model Organisms (MENMO)

As we are probably aware, advanced molecular biology techniques have generated giant data on non-model organisms at a very reasonable cost. Today, we are able to perform virtual studies on any organism using bioinformatics approach (go genomic); in turn the impacts of molecular ecology (using NGS) can be studied well in advance. This wonderful technology can be applied to study the ecology of any extreme environment on population genomics and conservation genetic studies as where non-model organisms/species are unavailable. Therefore, it is necessary to have appropriate knowledge of various possibilities (Ekblom and Galindo 2011). Using new NGS technologies, the pitfalls and drawbacks can be highlighted in advance. Using such technologies, we can highlight some of the pitfalls and drawbacks by raising points such as transcriptome characterization (Vera et al.

2008; Wall et al. 2009; Maas et al. 2015), gene expression profiling (Garcia-Reyero et al. 2008; Barakat et al. 2009; Harr and Turner 2010; Sarkar et al. 2015), candidate gene finding (Zou et al. 2008; Kristiansson et al. 2009; Ekblom et al. 2010), whole-genome sequencing (Hillier et al. 2008; Hohenlohe et al. 2010), targeted sequencing (Alvarez et al. 2009; Buée et al. 2009; Andersson et al. 2010; Mamanova et al. 2010), large-scale identification and development of molecular markers (Jakobsson et al. 2008; Novembre et al. 2008; Slate et al. 2009; Santure et al. 2010; Bhushan et al. 2013; Kekre et al. 2015), nucleotide variation profiling (Stinchcombe and Hoekstra 2007; Schwarz et al. 2009; Elmer et al. 2010; Futschik and Schlotterer 2010; Galindo et al. 2010; Lennon et al. 2010; Renaut et al. 2010; Faison et al. 2015), epigenetics (Salmon et al. 2005; Johnson et al. 2006; Barski et al. 2007; Bhinge et al. 2007; Kouzarides 2007; Taylor et al. 2007; Bossdorf et al. 2008; Cokus et al. 2008; Hurd and Nelson 2009; Simon et al. 2009; Pomraning et al. 2009), and data analysis (Huse et al. 2007; Pepke et al. 2009; Trapnell and Salzberg 2009; Wall et al. 2009; Cannon et al. 2010).

The new generation sequencing technology have applications for the study of microorganisms in the environment, both natural and man-made sites, and above all to those thriving under extreme conditions, such us very high or low temperatures, very acidic or basic pH values, and extreme salinity, and the use of unique nutrient sources, thriving under toxic conditions or in the presence of treatments with biocides (Rampelotto 2010), are fields where NGS is about to experience a very rapid and productive development. Future work will start to unveil the large number of aspects to be deciphered that will decisively contribute to the better understanding of ecological and physiological concerns related to these microbes and the potential application of this new knowledge into the biotechnological industry. The ways NGS will be applied and the outcomes of applying these novel technologies to the study of microorganisms remain still fiction and they are far from being defined. Only time will tell us the long listing of success stories to be reported soon in a base to the great advantages that NGS can provide to environmental microbiology and, in general, to any biological field.

## 6 Perspectives

The long list of applications and uses of NGS in biotechnology, medical, and environmental sciences is a singular case supporting a large fraction of today's rapid development of the biological field. Specifically, the high resistance to a number of environmental stressing factors and unusual features of extremophilic microorganisms is an area for growth to be expected in the next decade. Using NGS, scientists will better understand how life is possible in our planet and explain how life can exist in extremes found on Earth and, similarly, perhaps in other planets. Besides, the biotechnological use of these singular properties will certainly influence their industrial use and a rapid development of a green and sustainable economy.

**Acknowledgment** The authors wish to thank Dr. V. C. Kalia, chief scientist of CSIR-Institute of Genomics and Integrative Biology, and professor of Academy of Scientific and Innovative Research, Delhi University campus, Delhi (India), for giving continuous any possible opportunity to write this chapter. BNR is also thankful to the coauthor of this chapter Dr. Juan M Gonzalez, senior scientist, Institute of Natural Resources and Agrobiology, Spanish National Research Council, Seville, Spain, for his talented and well-versed advice, untiring efforts, timely suggestions, rational and constructive criticisms, and constant encouragement right from the beginning till the final shaping of this chapter in the present form.

## References

Alvarez LA, Exton DA, Timmis KN, Suggett DJ, McGenity TJ (2009) Characterization of marine isoprene-degrading communities. Environ Microbiol 11:3280–3291. doi:10.1111/j.1462-2920.2009.02069.x

Andersson AF, Riemann L, Bertilsson S (2010) Pyrosequencing reveals contrasting seasonal dynamics of taxa within Baltic Sea bacterioplankton communities. ISME J 4:171–181. doi:10.1038/ismej.2009.108

Barakat A, DiLoreto DS, Zhang Y, Smith C, Baier K, Powell WA, Wheeler N, Sederoff R, Carlson JE (2009) Comparison of the transcriptomes of American chestnut (*Castanea dentata*) and Chinese chestnut (*Castanea mollissima*) in response to the chestnut blight infection. BMC Plant Biol 9:51. doi:10.1186/1471-2229-9-51

Bar-Joseph Z, Gifford DD, Jaakkola TT (2001) Fast optimal leaf ordering for hierarchical clustering. Bioinformatics 17:S22–S29. doi:10.1093/bioinformatics/17.suppl_1.S22

Barski A, Cuddapah S, Cui K, Roh TY, Schones DE, Wang Z, Wei G, Chepelev I, Zhao K (2007) High-resolution profiling of histone methylations in the human genome. Cell 129:823–837. doi:10.1016/j.cell.2007.05.009

Bhinge AA, Kim J, Euskirchen GM, Snyder M, Iyer VR (2007) Mapping the chromosomal targets of STAT1 by sequence tag analysis of genomic enrichment (STAGE). Genome Res 17:910–916. doi:10.1101/gr.5574907

Bhushan A, Joshi J, Shankar P, Kushwah J, Raju SC, Purohit HJ, Kalia VC (2013) Development of genomic tools for the identification of certain *Pseudomonas* up to species level. Indian J Microbiol 53:253–263. doi:10.1007/s12088-013-0412-1

Biver S, Vandenbol M (2013) Characterization of three new carboxylic ester hydrolases isolated by functional screening of a forest soil metagenomic library. J Ind Microbiol Biotechnol 40:191–200. doi:10.1007/s10295-012-1217-7

Biver S, Portetelle D, Vandenbol M (2013a) Characterization of a new oxidant-stable serine protease isolated by functional metagenomics. Springerplus 2:410. doi:10.1186/2193-1801-2-410

Bobadilla Fazzini RA, Cortés MP, Padilla L, Maturana D, Budinich M, Maass A, Parada P (2013) Stoichiometric modeling of oxidation of reduced inorganic sulfur compounds (Riscs) in *Acidithiobacillus thiooxidans*. Biotechnol Bioeng 110:2242–2251. doi:10.1002/bit.24875

Bossdorf O, Richards CL, Pigliucci M (2008) Epigenetics for ecologists. Ecol Lett 11:106–115. doi:10.1111/j.1461-0248.2007.01130.x

Buée M, Reich M, Murat C, Morin E, Nilsson RH, Uroz S, Martin F (2009) 454 pyrosequencing analyses of forest soils reveal an unexpectedly high fungal diversity. New Phytologist 184:449–456. doi:10.1111/j.1469-8137.2009.03003.x

Cannon CH, Kua C-S, Zhang D, Harting JR (2010) Assembly free comparative genomics of short-read sequence data discovers the needles in the haystack. Mol Ecol 19:147–161. doi:10.1111/j.1365-294X.2009.04484.x

Chen Z, Jiang C, She Q, Liu S, Zhou P (2005) Key role of cysteine residues in catalysis and subcellular localization of sulfur oxygenase-reductase of *Acidianus tengchongensis*. Appl Environ Microbiol 71:621–628. doi:10.1128/AEM.71.2.621-628.2005

Chen L, Ren Y, Lin J, Liu X, Pang X, Lin J (2012) *Acidithiobacillus caldus* sulfur oxidation model based on transcriptome analysis between the wild type and sulfur oxygenase reductase defective mutant. PLoS One 7:e39470. doi:10.1371/journal.pone.0039470

Cheng F, Sheng J, Cai T, Jin J, Liu W, Lin Y, Du Y, Zhang M, Shen L (2012a) A protease-in sensitive feruloyl esterase from China holstein cow rumen metagenomic library: expression, characterization, and utilization in ferulic acid release from wheat straw. J Agric Food Chem 60:2546–2553. doi:10.1021/jf204556u

Cheng FS, Sheng JP, Dong RB, Men YJ, Gan L, Shen L (2012b) Novel xylanase from a holstein cattle rumen metagenomic library and its application in xylooligosaccharide and ferulic acid production from wheat straw. J Agric Food Chem 60:12516–12524. doi:10.1021/jf302337w

Chu XM, He HZ, Guo CQ, Sun BL (2008) Identification of two novel esterases from a marine metagenomic library derived from South China Sea. Appl Environ Microbiol 80:615–625. doi:10.1007/s00253-008-1566-3

Clermont O, Christenson JK, Denamur E, Gordon DM (2013) The Clermont *Escherichia coli* phylo-typing method revisited: improvement of specificity and detection of new phylo-groups. Environ Microbiol Rep 5:5865. doi:10.1111/1758-2229.12019

Cokus SJ, Feng S, Zhang X, Chen Z, Merriman B, Haudenschild CD, Pradhan S, Nelson SF, Pellegrini M, Jacobsen SE (2008) Shotgun bisulfite sequencing of the *Arabidopsis* genome reveals DNA methylation patterning. Nature 452:215–219. doi:10.1038/nature06745

Coughlan LM, Cotter PD, Hill C, Alvarez-Ordóñez A (2015) Biotechnological applications of functional metagenomics in the food and pharmaceutical industries. Front Microbiol 6:672. doi:10.3389/fmicb.2015.00672

Dai M, Thompson RC, Maher C, Contreras-Galindo R, Kaplan MH, Markovitz DM, Omenn G, Meng F (2010) NGSQC: cross-platform quality analysis pipeline for deep sequencing data. BMC Genomics 11:S7. doi:10.1186/1471-2164-11-S4-S7

Dhanji H, Doumith M, Clermont O, Denamur E (2010) Real-time PCR for detection of the O25b-ST131 clone of *Escherichia coli* and its CTX-M- 15-like extended-spectrum beta-lactamases. Int J Antimicrob Agents 36:355–358. doi:10.1016/j.ijantimicag.2010.06.007

Doumith M, Day MJ, Hope R, Waine J, Woodford N (2012) Improved multiplex PCR strategy for rapid assignment of the four major *Escherichia coli* phylogenetic groups. J Clin Microbiol 50:3108–3110. doi:10.1128/JCM.01468-12

Ekblom R, Galindo J (2011) Applications of next generation sequencing in molecular ecology of non-model organisms. Heredity 107:1–15. doi:10.1038/hdy.2010.152

Ekblom R, Balakrishnan CN, Burke T, Slate J (2010) Digital gene expression analysis of the zebra finch genome. BMC Genomics 11:219. doi:10.1186/1471-2164-11-219

Elend C, Schmeisser C, Leggewie C, Babiak P, Carballeira JD, Steele HL, Reymond JL, Jaeger KE, Streit WR (2006) Isolation and biochemical characterization of two novel metagenome-derived esterases. Appl Environ Microbiol 72:3637–3645. doi:10.1128/AEM.72.5.3637-3645. 2006

Elend C, Schmeisser C, Hoebenreich H, Steele HL, Streit WR (2007) Isolation and characterization of a metagenome-derived and cold-active lipase with high stereo specificity for (R)-ibuprofenesters. J Biotechnol 130:370–377. doi:10.1016/j.jbiotec.2007.05.015

Elmer KR, Fan S, Gunter HM, Jones JC, Boekhoff S, Kuraku S, Mayer A (2010) Rapid evolution and selection inferred from the transcriptomes of sympatric Crater Lake Cichlid fishes. Mol Ecol 19:197–211. doi:10.1111/j.1365-294X.2009.04488.x

Faison WJ, Rostovtsev A, Castro-Nallar E, Crandall KA, Chumakov K, Simonyan V, Mazumder R (2015) Whole genome single-nucleotide variation profile-based phylogenetic tree building methods for analysis of viral, bacterial and human genomes. Genomics 104:1–7. doi:10. 1016/j.ygeno.2014.06.001

Faoro H, Glogauer A, Couto GH, de Souza EM, Rigo LU, Cruz LM, Monteiro RA, Pedrosa Fde O (2012) Characterization of a new *Acidobacteria*-derived moderately thermostable lipase from a Brazilian Atlantic forest soil metagenome. FEMS Microbiol Ecol 81:386–394. doi:10.1111/j.1574-6941.2012.01361.x

Ferrer M, Golyshina OV, Chernikova TN, Khachane AN, Martins Dos Santos VA, Yakimov MM, Timmis KN, Golyshin PN (2005a) Microbial enzymes mined from the Urania deep-sea hypersaline anoxic basin. Chem Biol 12:895–904. doi:10.1016/j.chembiol.2005.05.020

Ferrer M, Golyshina OV, Chernikova TN, Khachane AN, Reyes-Duarte D, DosSantos V, Strompl C, Elborough K, Jarvis G, Neef A, Yakimov MM, Timmis KN, Golyshin PN (2005b) Novel hydrolase diversity retrieved from a metagenome library of bovine rumen microflora. Environ Microbiol 7:1996–2010. doi:10.1111/j.1462-2920.2005.00920.x

Futschik A, Schlotterer C (2010) The next generation of molecular markers from massively parallel sequencing of pooled DNA samples. Genetics 186:207–218. doi:10.1534/genetics.110.114397

Galindo J, Grahame JW, Butlin RK (2010) An EST-based genome scan using 454 sequencing in the marine snail *Littorina saxatilis*. J Evol Biol 23:2004–2016. doi:10.1111/j.1420-9101.2010.02071.x

Garcia-Reyero N, Griffitt RJ, Liu L, Kroll KJ, Farmerie WG, Barber DS, Denslow ND (2008) Construction of a robust microarray from a non-model species largemouth bass, *Micropterus salmoides* (Lacepede), using pyrosequencing technology. J Fish Biol 72:2354–2376. doi:10.1111/j.1095-8649.2008.01904.x

Ghosh W, Dam B (2009) Biochemistry and molecular biology of lithotrophic sulfur oxidation by taxonomically and ecologically diverse bacteria and archaea. FEMS Microbiol Rev 33:999–1043. doi:10.1111/j.1574-6976.2009.00187.x

Giovanni G, Creti R, Pompilio A, Bonaventura GD (2015) An overview of various typing methods for clinical epidemiology of the emerging pathogen *Stenotrophomonas maltophilia*. Diagn Microbiol Infect Dis 81:219–226. doi:10.1016/j.diagmicrobio.2014.11.005

Goodman AL, Wu M, Gordon JI (2011) Identifying microbial fitness determinants by insertion sequencing using genome-wide transposon mutant libraries. Nat Protoc 6:1969–1980. doi:10.1038/nprot.2011.417

Grada A, Weinbrecht K (2013) Next-generation sequencing: methodology and application. J Invest Dermatol 133:e11. doi:10.1038/jid.2013.248

Harr B, Turner LM (2010) Genome-wide analysis of alternative splicing evolution among Mus subspecies. Mol Ecol 19:228–239. doi:10.1111/j.1365-294X.2009.04490.x

Harris SR, Clarke IN, Seth-Smith HM, Solomon AW, Cutcliffe LT, Marsh P, Skilton RJ, Holland MJ, Mabey D, Peeling RW, Lewis DA, Spratt BG, Unemo M, Persson K, Bjartling C, Brunham R, de Vries HJ, Morré SA, Speksnijder A, Bébéar CM, Clerc M, de Barbeyrac B, Parkhill J, Thomson NR (2012) Whole-genome analysis of diverse *Chlamydia trachomatis* strains identifies phylogenetic relationships masked by current clinical typing. Nat Genet 44:413–419. doi:10.1038/ng.2214

Hillier LW, Marth GT, Quinlan AR, Dooling D, Fewell G, Barnett D, Fox P, Glasscock JI, Hickenbotham M, Huang W, Magrini VJ, Richt RJ, Sander SN, Stewart DA, Stromberg M, Tsung EF, Wylie T, Schedl T, Wilson RK, Mardis ER (2008) Whole-genome sequencing and variant discovery in *C. elegans*. Nat Methods 5:183–188. doi:10.1038/nmeth.1179

Hohenlohe PA, Bassham S, Etter PD, Stiffler N, Johnson EA, Cresko WA (2010) Population genomics of parallel adaptation in three spine stickleback using sequenced RAD tags. PLoS Genet 6:e1000862. doi:10.1371/journal.pgen.1000862

Holmes DS, Bonnefoy V (2007) Genetic and bioinformatic insights into iron and sulfur oxidation mechanisms of bioleaching organisms. In: Rawlings DE, Johnson DB (eds) Biomining. Springer, Berlin, pp 281–307. doi:10.1007/978-3-540-34911-2_14

Hurd PJ, Nelson CJ (2009) Advantages of next-generation sequencing versus the microarray in epigenetic research. Brief Funct Genomic Proteomic 8:174–183. doi:10.1093/bfgp/elp013

Huse S, Huber J, Morrison H, Sogin M, Welch D (2007) Accuracy and quality of massively parallel DNA pyrosequencing. Genome Biol 8:R143. doi:10.1186/gb-2007-8-7-r143

Jakobsson M, Scholz SW, Scheet P, Gibbs JR, Van Liere JM, Fung H, Szpiech ZA, Degnan JH, Wang K, Guerreiro R, Bras JM, Schymick JC, Hernandez DG, Traynor BJ, Simon-Sanchez J, Matarin M, Britton A, van de Leemput J, Rafferty I, Bucan M, Cann HM, Hardy JA, Rosenberg NA, Singleto AB (2008) Genotype, haplotype and copy-number variation in worldwide human populations. Nature 451:998. doi:10.1038/nature06742

Jiang CJ, Ma GF, Li SX, Hu TT, Che ZQ, Shen PH, Yan B, Wu B (2009) Characterization of a novel beta-glucosidase-like activity from a soil metagenome. J Microbiol 47:542–548. doi:10.1007/s12275-009-0024-y

Jin F, Li Y, Dixon JR, Selvaraj S, Ye Z, Li Y, Yen C, Schimt AD, Espinoza CA, Ren B (2013) A high-resolution map of the three-dimensional chromatin interactome in human cells. Nature 503:290–294. doi:10.1038/nature12644

Johnson SM, Tan FJ, McCullough HL, Riordan DP, Fire AZ (2006) Flexibility and constraint in the nucleosome core landscape of *Caenorhabditis elegans* chromatin. Genome Res 16:1505–1516. doi:10.1101/gr.5560806

Kang CH, Oh KH, Lee MH, Oh TK, Kim BH, Yoon J (2011) A novel family VII esterase with industrial potential from compost metagenomic library. Microb Cell Fact 10:41. doi:10.1186/1475-2859-10-41

Kekre A, Bhushan A, Kumar P, Kalia VC (2015) Genome wide analysis for searching novel markers to rapidly identify *Clostridium* strains. Indian J Microbiol. doi:10.1007/s12088-015-0535-7

Kim YJ, Choi GS, Kim SB, Yoon GS, Kim YS, Ryu YW (2006) Screening and characterization of a novel esterase from a metagenomic library. Protein Expr Purif 45:315–323. doi:10.1016/j.pep.2005.06.008

Kouzarides T (2007) Chromatin modifications and their function. Cell 128:693–705. doi:10.1016/j.cell.2007.02.005

Kristiansson E, Asker N, Förlin L, Larsson DJ (2009) Characterization of the *Zoarces viviparus* liver transcriptome using massively parallel pyrosequencing. BMC Genomics 10:345. doi:10.1186/1471-2164-10-345

Lass A, Zimmermann R, Oberer M, Zechner R (2011) Lipolysis—a highly regulated multi-enzyme complex mediates the catabolism of cellular fat stores. Prog Lipid Res 50:14–27. doi:10.1016/j.plipres.2010.10.004

Lee SW, Won K, Lim HK, Kim JC, Choi GJ, Cho KY (2004) Screening for novel lipolytic enzymes from uncultured soil microorganisms. Appl Environ Microbiol 65:720–726. doi:10.1007/s00253-004-1722-3

Lee DG, Jeon JH, Jang MK, Kim NY, Lee JH, Lee JH, Kim SJ, Kim GD, Lee SH (2007) Screening and characterization of a novel fibrinolytic metalloprotease from a metagenomic library. Biotechnol Lett 29:465–472. doi:10.1007/s10529-006-9263-8

Lennon N, Lintner RE, Anderson S, Alvarez P, Barry A, Brockman W, Daza R, Erlich RL, Giannoukos G, Green L, Hollinger A, Hoover CA, Jaffe DB, Juhn F, McCarthy D, Perrin D, Ponchner K, Powers KL, Rizzolo K, Robbins D, Ryan E, Russ C, Sparrow T, Stalker J, Steelman S, Weiand M, Zimmer A, Henn MR, Nusbaum C, Nicol R (2010) A scalable, fully automated process for construction of sequence-ready barcoded libraries for 454. Genome Biol 11:R15. doi:10.1186/gb-2010-11-2-r15

Lim JK, Hyun SL, Yun JK, Seung SB, Jeong HJ, Sung GK, Jung-Hyun L (2007) Critical Factors to high thermostability of an α-amylase from hyperthermophilic archaeon *Thermococcus onnurineus* NA1. J Microbiol Biotechnol 17:1242–1248

Maas AE, Lawson GL, Tarrant AM (2015) Transcriptome-wide analysis of the response of the thecosome pteropod *Clio pyramidata* to short-term $CO_2$ exposure. Comp Biochem Physiol Part D Genomics Proteomics 16:1–9. doi:10.1016/j.cbd.2015.06.002

Mamanova L, Coffey AJ, Scott CE, Kozarewa I, Turner EH, Kumar A, Howard E, Shendure J, Turner DJ (2010) Target-enrichment strategies for next generation sequencing. Nat Methods 7:111–118. doi:10.1038/nmeth.1419

Mangold S, Valdes J, Holmes DS, Dopson M (2011) Sulfur metabolism in the extreme acidophile *Acidithiobacillus caldus*. Front Microbiol 2:17. doi:10.3389/fmicb.2011.00017

Mohammadi T, Reesink HW, Pietersz RN, Vandenbroucke-Grauls CM, Savelkoul PH (2005) Amplified-fragment length polymorphism analysis of *Propionibacterium* isolates implicated in contamination of blood products. Br J Haematol 131:403–409. doi:10.1111/j.1365-2141.2005.05771.x

Nacke H, Will C, Herzog S, Nowka B, Engelhaupt M, Daniel R (2011) Identification of novel lipolytic genes and gene families by screening of metagenomic libraries derived from soil samples of the German biodiversity exploratories. FEMS Microbiol Ecol 78:188–201. doi:10.1111/j.1574-6941.2011.01088.x

Neveu J, Regeard C, Dubow MS (2011) Isolation and characterization of two serine proteases from metagenomic libraries of the Gobi and Death Valley deserts. Appl Microbiol Biotechnol 91:635–644. doi:10.1007/s00253-011-3256-9

Novembre J, Johnson T, Bryc K, Kutalik Z, Boyko AR, Auton A, Indap A, King KS, Bergmann S, Nelson MR, Stephens M, Bustamant CD (2008) Genes mirror geography within Europe. Nature 456:98. doi:10.1038/nature07331

Ouyang LM, Liu JY, Qiao M, Xu JH (2013) Isolation and biochemical characterization of two novel metagenome derived esterases. Appl Biochem Biotechnol 169:15–28. doi:10.1007/s12010-012-9949-4

Palackal N, Lyon CS, Zaidi S, Luginbuhl P, Dupree P, Goubet F, Macomber JL, Short JM, Hazlewood GP, Robertson DE, Steer BA (2007) A multifunctional hybrid glycosyl hydrolase discovered in an uncultured microbial consortium from ruminant gut. Appl Microbiol Biotechnol 74:113–124. doi:10.1007/s00253-006-0645-6

Peng Q, Wang X, Shang M, Huang J, Guan G, Li Y, Shi B (2014) Isolation of a novel alkaline-stable lipase from a metagenomic library and its specific application for milk fat flavor production. Microb Cell Fact 13:1. doi: 10.1186/1475-2859-13-1

Pepke S, Wold B, Mortazavi A (2009) Computation for ChIPseq and RNA-seq studies. Nat Methods 6:S22–S32. doi:10.1038/nmeth.1371

Picardi E, Horner DS, Chiara M, Schiavon R, Valle G, Pesole G (2010) Large-scale detection and analysis of RNA editing in grape mtDNA by RNA deep-sequencing. Nucleic Acids Res 38:4755–4767. doi:10.1093/nar/gkq202

Pomraning KR, Smith KM, Freitag M (2009) Genome-wide high throughput analysis of DNA methylation in eukaryotes. Methods 47:142–150. doi:10.1016/j.ymeth.2008.09.022

Pooja S, Pushpanathan M, Jayashree S, Gunasekaran P, Rajendhran J (2015) Identification of periplasmic alpha-amylase from cow dung metagenome by product induced gene expression profiling (Pigex). Indian J Microbiol 55:57–65. doi:10.1007/s12088-014-0487-3

Pushpam PL, Rajesh T, Gunasekaran P (2011) Identification and characterization of alkaline serine protease from goat skin surface metagenome. AMB Express 1:3. doi:10.1186/2191-0855-1-3

Quatrini R, Appia-Ayme C, Denis Y, Jedlicki E, Holmes DS, Bonnefoy V (2009) Extending the models for iron and sulfur oxidation in the extreme acidophile *Acidithiobacillus ferrooxidans*. BMC Genomics 10:394. doi:10.1186/1471-2164-10-394

Rabausch U, Juergensen J, Ilmberger N, Bohnke S, Fischer S, Schubach B, Schulte M, Streit WR (2013) Functional screening of metagenome and genome libraries for detection of novel flavonoid-modifying enzymes. Appl Environ Microbiol 79:4551–4563. doi:10.1128/AEM.01077-13

Rampelotto PH (2010) Resistance of microorganisms to extreme environmental conditions and its contribution to astrobiology. Sustainability 2:1602–1623. doi:10.3390/su2061602

Renaut S, Nolte AW, Bernatchez L (2010) Mining transcriptome sequences towards identifying adaptive single nucleotide polymorphisms in lake whitefish species pairs (*Coregonus* spp. *Salmonidae*). Mol Ecol 19:115–131. doi:10.1111/j.1365-294X.2009.04477.x

Rhee JK, Ahn DG, Kim YG, Oh JW (2005) New thermophilic and thermostable esterase with sequence similarity to the hormone-sensitive lipase family, cloned from a metagenomic library. Appl Environ Microbiol 71:817–825. doi:10.1128/AEM.71.2.817-825.2005

Robertson DE, Chaplin JA, Desantis G, Podar M, Madden M, Chi E, Richardson T, Milan A, Miller M, Weiner DP, Wong K, McQuaid J, Farwell B, Preston LA, Tan X, Snead MA, Keller M, Mathur E, Kretz PL, Burk MJ, Short JM (2004) Exploring nitrilase sequence space for enantioselective catalysis. Appl Environ Microbiol 70:2429–2436. doi:10.1128/AEM.70.4.2429-2436.2004

Salmon A, Ainouche ML, Wendel JF (2005) Genetic and epigenetic consequences of recent hybridization and polyploidy in Spartina (Poaceae). Mol Ecol 14:1163–1175. doi:10.1111/j.1365-294X.2005.02488.x

Santure AW, Stapley J, Ball AD, Birkhead TR, Burke T, Slate J (2010) On the use of large marker panels to estimate inbreeding and relatedness: empirical and simulation studies of a pedigreed zebra finch population typed at 771 SNPs. Mol Ecol 19:1439–1451. doi:10.1111/j.1365-294X.2010.04554.x

Sarkar A, Pazhani GP, Dharanidharan R, Ghosh A, Ramamurthy T (2015) Detection of integron-associated gene cassettes and other antimicrobial resistance genes in enterotoxigenic *Bacteroides fragilis*. Anaerobe 33:18–24. doi:10.1016/j.anaerobe.2015.01.008

Schuster SC (2008) Next-generation sequencing transforms today's biology. Nat Methods 5:16–18. doi:10.1038/nmeth1156

Schwarz D, Robertson H, Feder J, Varala K, Hudson M, Ragland G et al (2009) Sympatric ecological speciation meets pyrosequencing: sampling the transcriptome of the apple maggot *Rhagoletis pomonella*. BMC Genomics 10:633. doi:10.1186/1471-2164-10-633

Simon SA, Zhai J, Nandety RS, McCormick KP, Zeng J, Mejia D, Meyers BC (2009) Short-read sequencing technologies for transcriptional analyses. Ann Rev Plant Biol 60:305–333. doi:10.1146/annurev.arplant.043008.092032

Slate J, Gratten J, Beraldi D, Stapley J, Hale M, Pemberton J (2009) Gene mapping in the wild with SNPs: guidelines and future directions. Genetica 136:97–107. doi:10.1007/s10709-008-9317-z

Somboonna N, Wan R, Ojcius DM, Pettengill MA, Joseph SJ, Chang A, Hsu R, Read TD, Dean D (2011) Hypervirulent *Chlamydia trachomatis* clinical strain is a recombinant between lymphogranuloma venereum (L(2)) and D lineages. MBio 2:e00045-11. doi:10.1128/mBio.00045-11

Stamatakis A (2006) RAxML-VI-HPC: maximum likelihood-based phylogenetic analyses with thousands of taxa and mixed models. Bioinformatics 22:2688–2690. doi:10.1093/bioinformatics/btl446

Stinchcombe JR, Hoekstra HE (2007) Combining population genomics and quantitative genetics: finding the genes underlying ecologically important traits. Heredity 100:158–170. doi:10.1038/sj.hdy.6800937

Taylor KH, Kramer RS, Davis JW, Guo J, Duff DJ, Xu D, Caldwell CW, Shi H (2007) Ultradeep bisulfite sequencing analysis of DNA methylation patterns in multiple gene promoters by 454 sequencing. Cancer Res 67:8511–8518. doi:10.1158/0008-5472.CAN-07-1016

Trapnell C, Salzberg SL (2009) How to map billions of short reads onto genomes. Nat Biotech 27:455–457. doi:10.1038/nbt0509-455

Uchiyama T, Miyazaki K (2010) Product-induced gene expression, a product-responsive reporter assay used to screen metagenomic libraries for enzyme-encoding genes. Appl Environ Microbiol 76:7029–7035. doi:10.1128/AEM.00464-10

Urich T, Coelho R, Kletzina A, Frazao C (2005) The sulfur oxygenase reductase from *Acidianus ambivalens* is an icosa-tetramer as shown by crystallization and Patterson analysis. BBA Proteins Proteomics 1747:267–270. doi:10.1016/j.bbapap.2004.11.015

Urich T, Gomes CM, Kletzin A, Frazão C (2006) X-ray structure of a self compartmentalizing sulfur cycle metalloenzyme. Science 311:992–996

Valdes J, Pedroso I, Quatrini R, Dodson RJ, Tettelin H, Blake R, Eisen JA, Holmes DS (2008) *Acidithiobacillus ferrooxidans* metabolism: from genome sequence to industrial applications. BMC Genomics 9:597. doi:10.1186/1471-2164-9-597

Valdés J, Pedroso I, Quatrini R, Holmes DS (2008) Comparative genome analysis of *Acidithiobacillus ferrooxidans. A. thiooxidans* and *A. caldus*: insights into their metabolism and ecophysiology. Hydrometallurgy 94:180–184. doi:10.1016/j.hydromet.2008.05.039

Veenemans J, Overdevest IT, Snelders E, Willemsen I, Hendriks Y, Adesokan A, Doran G, Bruso S, Rolfe A, Pettersson A, Kluytmans JA (2014) Next-generation sequencing for typing and detection of resistance genes: performance of a new commercial method during an outbreak of extended-spectrum-beta-lactamase-producing *Escherichia coli.* J Clin Microbiol 52:2454–2460. doi:10.1128/JCM.00313-14

Vera JC, Wheat CW, Fescemyer HW, Frilander MJ, Crawford DL, Hanski I, Marden JH (2008) Rapid transcriptome characterization for a nonmodel organism using 454 pyrosequencing. Mol Ecol 17:1636–1647. doi:10.1111/j.1365-294X.2008.03666.x

Vester JK, Glaring MA, Stougaard P (2014) Discovery of novel enzymes with industrial potential from a cold and alkaline environment by a combination of functional metagenomics and culturing. Microb Cell Fact 13:72. doi:10.1186/1475-2859-13-72

Voget S, Leggewie C, Uesbeck A, Raasch C, Jaeger KE, Streit WR (2003) Prospecting for novel biocatalysts in a soil metagenome. Appl Environ Microbiol 69:6235–6242. doi:10.1128/AEM.69.10.6235-6242.2003

Wall PK, Leebens-Mack J, Chanderbali AS, Barakat A, Wolcott E, Liang H, Landherr L, Tomsho LP, Hu Y, Carlson JE, Ma H, Schuster SC, Soltis DE, Soltis PS, Altman N, dePamphilis CW (2009) Comparison of next generation sequencing technologies for transcriptome characterization. BMC Genomics 10:347. doi:10.1186/1471-2164-10-347

Walter J, Mangold M, Tannock GW (2005) Construction, analysis, and beta-glucanase screening of a bacterial artificial chromosome library from the large-bowel microbiota of mice. Appl Environ Microbiol 71:2347–2354. doi:10.1128/AEM.71.5.2347-2354.2005

Wang K, Li G, Yu SQ, Zhang CT, Liu YH (2010) A novel metagenome-derived beta-galactosidase: gene cloning, over expression, purification and characterization. Appl Microbiol Biotechnol 88:155–165. doi:10.1007/s00253-010-2744-7

Wang K, Lu Y, Liang WQ, Wang SD, Jiang Y, Huang R, Liu YH (2012) Enzymatic synthesis of galacto-oligo saccharides in an organic-aqueous biphasic system by a novel beta-galactosidase from a metagenomic library. J Agric Food Chem 60:3940–3946. doi:10.1021/jf300890d

Wang Y, Wen Z, Shen J, Cheng W, Li J, Qin X, Ma D, Shi Y (2014) Comparison of the performance of Ion Torrent chips in noninvasive prenatal trisomy detection. J Hum Genet 59:393–396. doi:10.1038/jhg.2014.40

Wirth T, Falush D, Lan R, Colles F, Mensa P, Wieler LH, Karch H, Reeves PR, Maiden MC, Ochman H, Achtman M (2006) Sex and virulence in *Escherichia coli*: an evolutionary perspective. Mol Microbiol 60:1136–1151. doi:10.1111/j.1365-2958.2006.05172.x

Yao J, Fan XJ, Lu Y, Liu YH (2011) Isolation and characterization of a novel tannase from a metagenomic library. J Agric Food Chem 59:3812–3818. doi:10.1021/jf104394m

Yin H, Zhang X, Li X, He Z, Liang Y, Guo X, Hu Q, Xiao Y, Cong J, Ma L, Niu J, Liu X (2014) Whole-genome sequencing reveals novel insights into sulfur oxidation in the extremophile *Acidithiobacillus thiooxidans*. BMC Microbiol 14:179. doi:10.1186/1471-2180-14-179

Yu Y, Zhao C, Zhenqiang S, Wang C, Fuscoe JC, Tong W, Shi L (2014a) Comprehensive RNA-Seq transcriptomic profiling across 11 organs, 4 ages, and 2 sexes of Fischer 344 rats. Sci Data 1:140013. doi:10.1038/sdata.2014.13

Yu Y, Fuscoe JC, Zhao C, Guo C, Jia M, Qing T, Bannon DI, Lancashire L, Bao W, Du T, Luo H, Su Z, Jones WD, Moland CL, Branham WS, Qian F, Ning B, Li Y, Hong H, Guo L, Mei N, Shi T, Wang KY, Wolfinger RD, Nikolsky Y, Walker SJ, Duerksen-Hughes P, Mason CE, Tong W, Thierry-Mieg J, Thierry-Mieg D, Shi L, Wang C (2014b) A rat RNA-Seq transcriptomic BodyMap across 11 organs and 4 developmental stages. Nat Commun 5:3230. doi:10.1038/ncomms4230

Zou Z, Najar F, Wang Y, Roe B, Jiang H (2008) Pyrosequence analysis of expressed sequence tags for *Manduca sexta* hemolymph proteins involved in immune responses. Insect Biochem Mol Biol 38:677–682. doi:10.1016/j.ibmb.2008.03.009

# Impact of Next-Generation Sequencing Technology in Plant–Microbe Interaction Study

**Archana Kumari, Samson Sumer, Bharati Jalan, Pyniarlang Lyngdoh Nongbri, and Mostaque Ahmed Laskar**

**Abstract** Next-generation sequencing (NGS) technologies have revolutionized the biological research during the last few years. Nowadays due to this high-throughput technique, it is quite easy to produce huge amount of sequencing data at low cost. In the past years, plant–microbe interaction study was not an easy task. This review will give a broad idea about the importance of NGS in plant–microbe interaction study specifically for those microorganisms which play a great role in the interaction. Due to difference in sequencing systems, it is quite tough to overcome the problem regarding different types of errors. We are emphasizing on the importance of NGS data in plant–microbe interaction including the analysis of different microbial communities (using amplicon sequencing, Cross linking and sequencing of hybrids etc.). Screened research articles which are based on plant–microbe interaction study were used here to conclude the novel methods of plant–microbe interaction.

**Keywords** Next generation sequencing • Plant microbiota • Amplicon sequencing • Metagenomics • Transcriptomics

## 1 Introduction

Plants survive in adverse environmental conditions due to the association of various microorganisms which are present below ground in the rhizosphere and above the ground in the phyllosphere. Lorenz Hiltner a German Scientist defined the term rhizosphere which means roots surrounded by soil. He discovered many important microbes which play a major role in plant growth and health. From the time of Hiltner to the present day, various researches have been done to learn about the mechanism of plant–microbe interactions (Hartmann et al. 2008; Bulgarelli et al. 2013). These beneficial microorganisms are present as endophytes residing within the plant or as epiphytes residing on plant surface or near the roots. The importance

A. Kumari (✉) • S. Sumer • B. Jalan • P.L. Nongbri • M.A. Laskar
Department of Biotechnology, St Anthony's College, Shillong, Meghalaya 793001, India
e-mail: archana.bioinfo87@gmail.com

 
V.C. Kalia, P. Kumar (eds.), *Microbial Applications Vol.1*,
DOI 10.1007/978-3-319-52666-9_13

of rhizosphere is more than that of phyllosphere because it is not only directly involved in plant nutrition and health but also involved in the good maintenance of microorganisms. These microorganisms are involved in carbon sequestration, ecosystem functioning, and nutrient cycling in terrestrial ecosystems (Berg 2009; Newton et al. 2010).

The microbial communities associated with plants are recognized specifically based on the microenvironment posed by crops and their cultivars. Thus, plants can be categorized based on the distinct microenvironments such as endorhiza, phyllosphere, spermosphere, carposphere, etc. These microenvironments are maintained with the help of various environmental factors for specific natural life. An interesting fact which apprehensions about the microbial populations on plant system is that their survival mechanism like how they are surviving nicely although they are coming from different sources (Berg et al. 2005; Vorholt 2012). Scientists revealed this fact with the help of the dormant pathogenic bacteria which usually get colonized although they act as a good platform for seed microbiome. In contrast plants are always interacting with diverse microbes present in wind or water, and some have capacity to colonize the phyllosphere (Fürnkranz et al. 2012; Bragina et al. 2012).

Insights into the rhizosphere microorganism plant interactions could be obtained due to the advancement in molecular biology techniques as well as in bioinformatics (Hartmann et al. 2009). Using stable isotope probing (SIP) approach, Haichar et al. (2008) revealed the structures of plant–host habitat and various bacterial communities. Other good examples of work that utilized molecular biology and bioinformatics are of Lundberg and Bulgarelli individually in *Arabidopsis thaliana*, in which they identified only specific bacterial communities present in the roots of *Arabidopsis thaliana* the model plant. Two bacterial species are present in the roots of *Arabidopsis*: *Proteobacteria* which are responsible for the regulation of growth-promoting factors and *Actinobacteria* which are responsible for the production of antimicrobial compounds (Bulgarelli et al. 2012; Lundberg et al. 2012). The abovementioned workers proposed that *Arabidopsis* itself is involved in the recruitment of a group of microbes which benefit its basic functions during specific environmental conditions.

Due to the commensal lifestyle of some microbes, neither they do any damage to the plant nor do they involve in plant growth promotion. The mechanism behind plant–microbe interaction is still not completely known. So, various questions arise for plant-related microbiota such as "Who are they?," "How they are surviving their life in particular environment?," "How do they interact with plant system?," "How they are beneficial for each other?," and "How they are affecting the plant growth and development directly or indirectly?" Solving all above questions will help to understand the whole mechanism of plant–microbe interaction and also help to identify those microorganisms which can be used in the near future to increase crop yield. In agriculture, plant microbiome interactions act as a fuel to increase the yield naturally (Berg 2009). Some good examples are stress protection products, biofertilizers, biocontrol, etc. Nowadays, there are vast growing markets for these bioproducts, but they are suffering from some specific problems like unpredictable possessions under field, short shelf life, and risk calculations. Advancement in biotechnology has played a major role for development of advanced bioproducts using "omics" approach. In this area NGS has a great influence on the (a) discovery of new possessions for biocontrol along with plant growth-promoting factors, (b) optimization of different processes, (c) stabilization of outcome under field trial, and (d) risk calculation studies.

To answer the abovementioned questions, it is necessary to mine and annotate the genes involved in the plant–microbe associations from the genomes of both partners. Whole genome study of entire microbial communities, in other words metagenomics studies, will provide insights into the composition of such communities. Information on the physiological aspects of these microbial communities vis-à-vis their association with plants can be obtained from metagenomics studies (Niedringhaus et al. 2011). Reverse genetics approaches can be used to study the metabolic activities and gene regulatory mechanisms of the microbial cells that are in association with plants. In biological research next-generation sequencing (NGS) technologies have great impact because they provide a new platform to answer all those questions which possibly could not be solved before because of financial and technical restrictions. NGS technologies have provided the opportunity for finding answers to crucial questions in plant–microbe interactions with great speed (Schadt et al. 2010).

In this present review, we have discussed the importance of NGS in plant–microbe interaction studies. We have presented an overview of the specific requirement as well as function of different types of sequencing systems including their sources of errors and biases and other important matters. Specific focus is on the advantages of NGS techniques in studying microorganisms associated with plants. This review gives a brief outlook about what the scientist community will probably study in the near future.

## 2 Next-Generation Sequencing (NGS) Platforms

Currently, there are different types of platforms available for NGS. Instruments used for NGS can be classified as second- and third-generation sequencing technologies (Liu et al. 2012). It is quite difficult to categorize these instruments (Pareek et al. 2011); nonetheless we summarized in Table 1 all available methods for second- and third-generation sequencing. Roche 454, Illumina, and Life Technologies instruments come under second-generation sequencing technology. The second-generation sequencing technology is based on SOLiD (the Sequencing by Oligonucleotide Ligation and Detection) and sometimes Ion Torrent sequencers (Schadt et al. 2010). By Pacific Biosciences, the PacBio RS is the only single system which is commercially available for third-generation sequencing.

**Table 1** Different NGS platforms

| Company name | Sequencing principle | Sequencing platforms | Method for library construction | Modifications of nucleotides | Major sequencing error | Signal identification |
|---|---|---|---|---|---|---|
| Illumina | Flexible sequencing by synthesis | Illumina | Bridge PCR amplification | End blocked fluorescent nucleotides | Substitutions | Optical measurement of fluorescent emission |
| Life Technologies (2005–2008) | Sequencing by ligation | SOLiD 4 | PCR on microbeads | Two base encoded fluorescent oligonucleotides | Substitutions | Optical measurement of fluorescent emission |
| Life Technologies (2008–2010) | Semiconductor-based sequencing by synthesis | Ion PGM | PCR on microbeads | None | Indels | Transistor-based detection |
| Pacific Biosciences | Single-molecule, real-time DNA sequencing by synthesis | PacBio RS | Not applied | Phosphor (P)-linked fluorescent nucleotides | Indels | Real-time optical detection of fluorescent dye |
| Roche | Pyrosequencing | 454 FLX titanium | PCR on microbead | None | Indels | Optical detection of light |

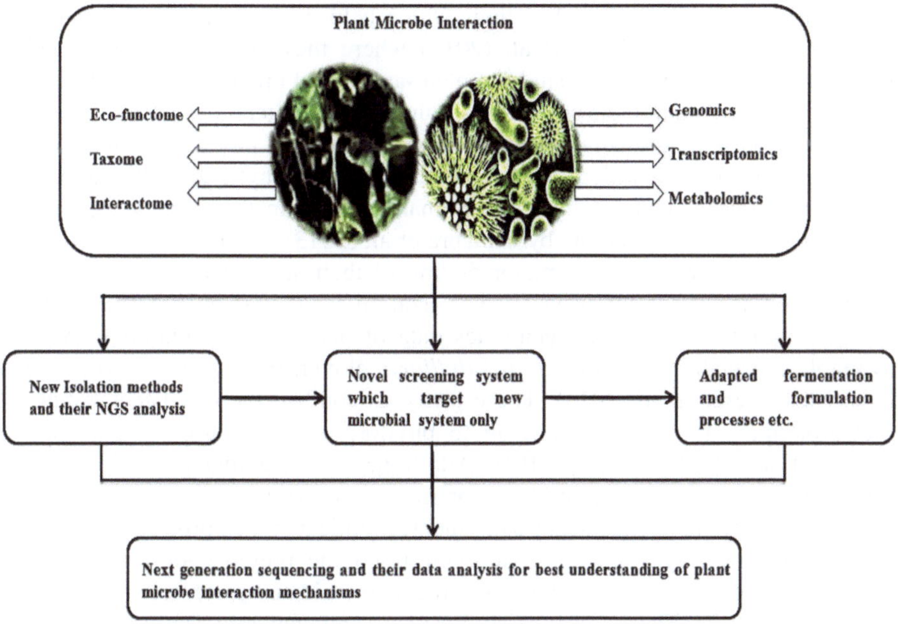


**Fig. 1** Different approaches of studying plant-microbe interaction using NGS

# 3 Plant Microbiota and NGS

Plants' natural habitats are consisting high diversity of microorganisms. Because of good correlation between above ground and below ground, a large number of microbial diversities can be expected (Heijden et al. 1998; Thompson and Milos 2011). In agricultural systems, sometimes under intense environmental conditions, some microbes grow well and also benefited the plants. Conversely, natural ecosystems are especially well managed by mosses, plant growth-promoting bacteria, etc. Therefore, using NGS study of plants with respect to its eco-function, taxome, and interactome including the study of genomics, transcriptomics and metabolomics of microorganisms will ultimately reveal the hidden mechanism behind its interaction and survival (Fig. 1).

## *3.1 Shotgun Sequencing and Metagenomics Study*

Till today, only few studies in metagenomics related to plant-associated microorganisms were completed using shotgun sequencing (Table 3). Currently, Roche 454 sequencing technology is mostly used for these types of studies. Recently, Mendes et al. (2014) found that the epiphytic rhizosphere microbiome is present in soybean, and they also characterized their taxonomic as well as functional composition. Also, beneficial functions which help in plant growth and nutrition were identified by Sessitsch et al. (2012) where they used Sanger sequencing technology for metagenomics study of plant-associated microorganisms (Sessitsch et al. 2012). Unno and their co-workers found that metagenomes present in rhizosphere increased the growth of plant due to the presence of phytic acid. Some of the unique genes were identified that encode enzymes for phytic acid utilization, for example, citrate synthase (Unno and Shinano 2013). In the same year, another metagenomics study was done by Chhabra et al. (2013), where they constructed a fosmid library in *E. coli*. The major finding of their study was they screened an assay which has mineral phosphate solubilization capacity (Chhabra et al. 2013).

From various studies, metagenomics data of microbial communities are now available especially from *Arabidopsis thaliana*, clover, rice, tomato, soybean, and tamarisk (Ottesen et al. 2013). These analyses confirm the consistent nature of metaproteome of bacteria belonging to phyllospheres of various plant species (Knief et al. 2012; Vorholt 2012). Additionally, these metagenomic datasets revealed the presence of microbial community at phylum level. On the other hand, the comparative analysis of data (metagenomic and metaproteomic) between rhizosphere and phyllosphere in rice confirmed the presence of very complex microbiota and a very clear vision about metagenomic and proteomic composition (Knief et al. 2012). The analysis of phyllosphere metagenomic datasets in incorporation with other metagenomic datasets used to monitor some genes which are involved in energy generation from light, i.e., photosynthesis (Atamna-Ismaeel et al. 2012; Vorholt 2012).

Some especial kind of metagenomic project was also completed to obtain a whole sequence of plant pathogens which can't be cultured. *Candidatus Liberibacter asiaticus* which is the main causative agent of citrus huanglongbing is transmitted through phloem feeding insects. Metagenomics study is performed using 454 sequencing technology (Delmotte et al. 2009; Duan et al. 2009). Sequencing resulted as ~38 contigs which were further confirmed by PCR reactions. Complete genome data analysis exposed that there was huge reduction in its genome with respect to highly divergent member of the family *Rhizobiaceae* because of their intracellular lifestyle (Fig. 2).

### 3.2 Study Through Amplicon Sequencing

Nowadays, most popular method to study plant–microbe interaction is amplicon sequencing (Fig. 2). This method is mostly used to distinguish between the rhizosphere and phyllosphere communities. Roche 454 sequencing is mostly used for this purpose, but some researchers also used the Ion PGM platform or the Illumina MiSeq. The amplicon size in phyllosphere studies varied from ~1000 to 10,000 reads per sample (Yergeau et al. 2014a, b), but longer reads can also be obtained with the help of 454 FLX+ instrument (Jiang et al. 2013; Perazzolli et al. 2014). From the last few years, NGS amplicon sequencing was exclusively used for

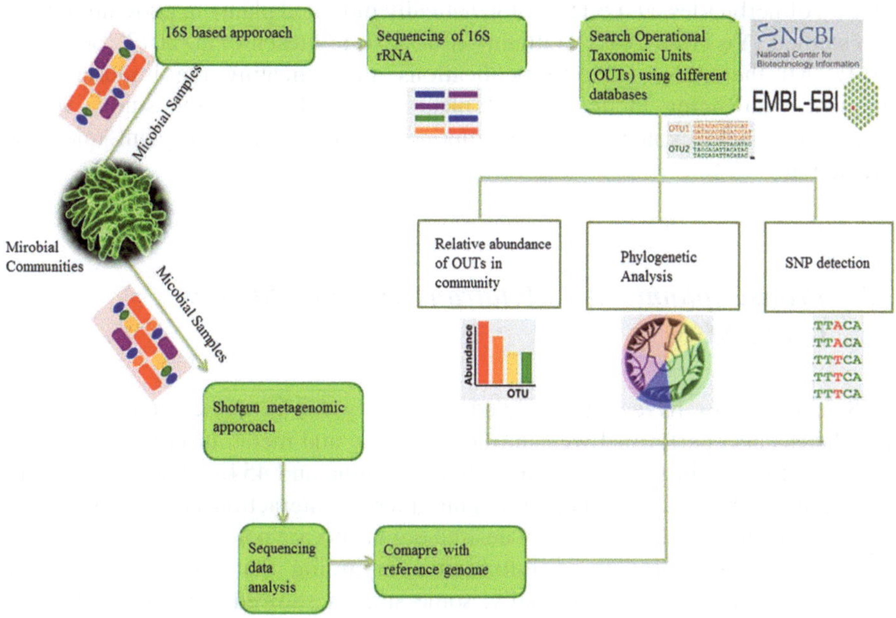


**Fig. 2** Basic pipeline to study metagenomics using shotgun sequencing and amplicon sequencing for microbial communities

bacterial or fungal communities study (Kavamura et al. 2013), where the study of phyllosphere communities of bacteria based on the 16S rRNA gene and fungal communities was based on the ITS region (Bokulich et al. 2014). *chiA* is the only marker gene found by amplicon sequencing in plant microbiota (Cretoiu et al. 2012). The key objectives for this particular study were to find the diversity of this gene in different habitats.

Amplicon sequencing method is used to understand the reason of plant colonization in phyllosphere due to particular type of plant microbiota (Maignien et al. 2014). Also, the amplicon sequencing solved the numerous questions regarding plant microbiota in rhizosphere like biogeographical distribution of various microorganisms (Gottel et al. 2011; Peiffer et al. 2013), factors affecting host–plant interaction (Navarrete et al. 2013), factors affecting plant growth as well as nutrition (Lundberg et al. 2012), different soil types (Zhang et al. 2013), etc. Some scientists focused on the exploration of endomycorrhiza and ectomycorrhizal (Badri et al. 2013a, b). From various studies now it has been clear that plant plays a major role in microbiota selection (Rastogi et al. 2012; Reisberg et al. 2013); also its various mechanisms that affect the whole controlling process have been studied (Badri et al. 2009; Bodenhausen et al. 2013).

However, the importance of colonization of various plant compartments (Barriuso et al. 2010; Redford et al. 2010; Rosenzweig et al. 2012), role of specific treatments during plant cultivation like irrigation (Lumini et al. 2010; Dumbrell et al. 2011; Dohrmann et al. 2013), various aspects of bioremediation, major impacts of herbicides, and effects of genetically modified plants in agriculture were also studied (Yu et al. 2012; Williams et al. 2013; Ottesen et al. 2013; Bell et al. 2014). All the above selected publications are indicating the importance of amplicon sequencing in the field of plant–microbial interaction study. Surely, further study will explore the other important factors related to plant–microbial interaction.

### *3.3 Transcriptomic and Metatranscriptomic Studies Using NGS*

NGS technologies have not restricted itself only for genomics sequencing but nonetheless also performed well in transcriptomic and metatranscriptomic studies in plants (Fig. 3). In present scenario, both Illumina and 454 technologies are the most extensively used technology for plant–microbe interaction study (Thakur et al. 2013). When the whole genomes of desired organisms are not sequenced, then NGS is used to explore the whole information regarding that particular organism (Tremblay et al. 2012). Additionally, some studied showed the parallel analysis of the host and the pathogen interaction using transcriptome data (Weßling et al. 2012). These types of analysis are mostly dependent on the plant–host interaction pattern (Zhuang et al. 2012).

Recently, some metatranscriptomic studies were done using these advanced technologies. Firstly, Chaparro and co-workers studied the metatranscriptome in *Arabidopsis* plant. They studied the role of different microbial communities in rhizosphere at different plant development stages. They also found that these microbial genes were also involved in the regulation of various metabolic pathways (Chaparro et al. 2014). Similar study has been performed by Yergeau and their colleagues where they compared different composition in the rhizosphere of willow with bulk soil (soil was contaminated with organic pollutants). In this study they confirmed that various genes involved in hydrocarbon degradation also genes involved in carbon and amino acid uptake upregulated in the rhizosphere (Fernandez et al. 2012; Yergeau et al. 2014a, b).

## 4 Major Challenges in Sequencing of Pathogen Genomes and Metagenomes

Whole genome sequencing started only after the successful completion of small segment DNA sequencing. These advanced technologies markedly increased the area of research in phytopathogen and also in study of intra-strain diversity of a pathogen species (Mardis 2008). The major challenge lies on the genome assemblies for eukaryotic filamentous phytopathogens because of their large genome size

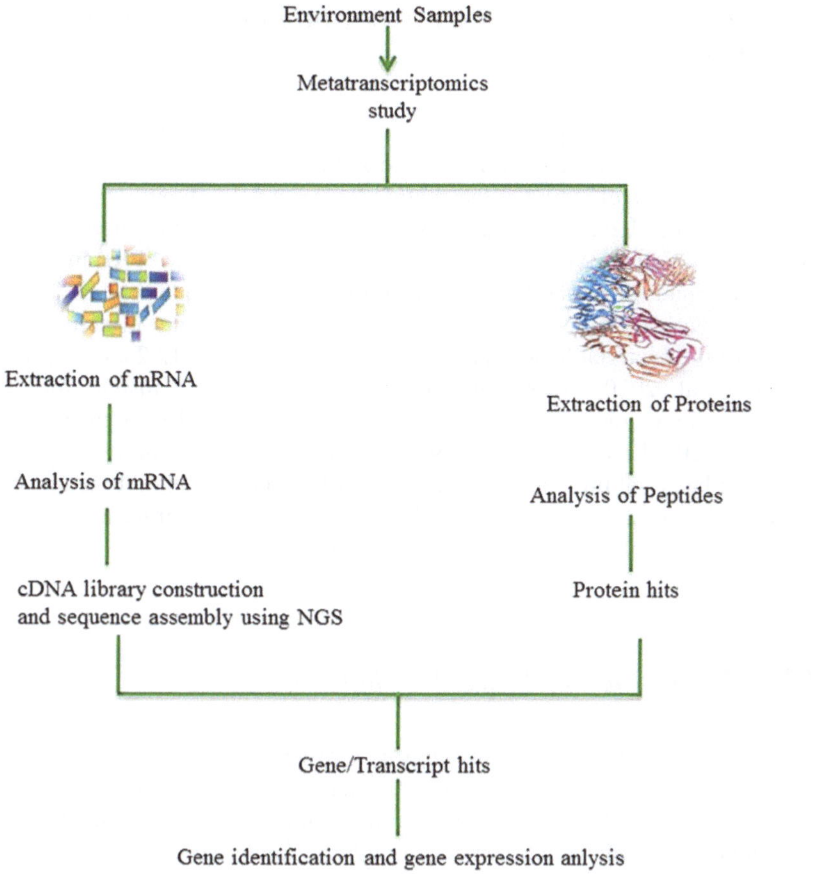


**Fig. 3** Workflow for transcriptomic and metatranscriptomic study of plant-associated microbes for gene identification and expression analysis

approximately 18.7 Mb–180 Mb (Schirawski et al. 2010; Wicker et al. 2013). Their large genome size can be a reason for differences in ploidy levels also due to the presence of transposable elements (Schmidt and Panstruga 2011). Additionally, the presence of these elements causes troubles in the contigs assembly. They also cause difficulties to identify specific difference in karyotypes when they are compared to strains that differ in virulence. In spite of these challenges, scientists revealed some important mechanisms for different filamentous phytopathogens (Schmidt and Panstruga 2011). Currently, the major challenges in the study of plant microbiota are complex rhizosphere and diverse soil communities which prevent the completion of de novo assembly. Therefore, third-generation sequencing technologies such as PacBio can help to overcome these problems (Proctor 2011).

### 4.1 Modifications in Pattern-recognition Receptors and Their Functions

The innate immunity of plant is based on the microbe-associated molecular patterns (MAMPs) by pattern-recognition receptors (PRRs). Identification of novel PRRs is considered as less effective at the protein level because of very low abundance. A novel method called the 5C strategy was used to overcome this problem by Patrick Boyle. First of all the MAMP probe is interlinked with cognate PRR in the presence of UV rays and further both dissolved from the membrane. The chemical analysis and characterization demonstrates that this strategy is quite effective for cross-linking of the bacterial flagellin (MAMP flg22) to its PRR FLS2 and its purification (Albert et al. 2010). This methodology provided a new insight for detecting unknown interacting molecules of proteins of interest and also for the identification of novel PRRs. LRR-receptor kinases FLS2and EFR of *Arabidopsis thaliana* that detect the bacterial proteins flagellin and EF-Tu using a method based on chimeric variants (Albert and Felix 2010; Doehlemann and Hemetsberger 2013) is also one of the best example of pattern recognition receptor.

### 4.2 Modification of Fungi by Plant Viruses

Highly improved tool in plant biology is virus-induced gene silencing (VIGS) to know the function of genes through transient silencing. Peter and Donato confirmed the role of virus-induced gene silencing in filamentous fungi having essential mechanisms to replicate and infect in *Colletotrichum* species. Moreover, genetically modified viruses (GMV) can be used for the expression analysis of foreign genes like green fluorescent protein (GFP). GMV have been observed to be quite helpful in gene silencing experiments. As advanced techniques are very fast and effective so these techniques could be utilized for the manipulation of untransformable fungi such as biotrophic fungi (Lu et al. 2003; Mascia et al. 2014).

### 4.3 Gene Silencing and Epigenetics in Plants

In present scenario it has been found that heritable genetics or gene transfer from one species to another might be influenced by RNAi-mediated gene silencing. The model plant *Arabidopsis* infected with *Pseudomonas syringae* pv. tomato (PstDC3000) showed higher resistance to the pathogens. This mechanism confirmed that hypomethylated SA-dependent genes responsible for downregulation of genes; so indirectly involved to increase resistance in *Arabidopsis* (Luna et al. 2012). Parallel study has been performed in rice plants also. These results are representing the possibility of heritable epigenetics in heritable variation within species (Stroud et al. 2013; Mascia et al. 2014).

## 5 Cultivation-Independent Methods for Plant Microbiota Study

In recent times, cultivation-independent methods in amalgamation through NGS are gaining new insight for analyzing arrangement and utilities of the plant-colonizing microbial communities. The sequencing data of especial genes from different microbial community containing taxonomically information regarding its DNA is known as marker gene analysis or in other words amplicon sequencing. It permits characterization of different microbial communities with respect to their relative abundances and identities. Massive sequence data are produced by marker gene sequencing. These data delivers noticeable facts related to taxonomy of different microbial communities in contrast with other profiling techniques (Nocker et al. 2007). Additionally, the metagenomics approach salvages the information contained within the whole genome of a particular microbial population through shotgun sequencing (Riesenfeld et al. 2004; Delmotte et al. 2009). Additionally, metagenomics study also allows the characterization, function analysis, and metabolic pathway involvement. Bioinformatics analysis helps to determine the possible prospective of microbial communities in corresponding metabolic pathways (Table 2). Not only metagenomics but also metatranscriptomics is playing a great role in the regulation of gene expression under different environmental conditions. Analysis mostly done by reverse transcription along with random shotgun sequencing of isolated RNA from the microbes. Further, these analyses complemented with metaproteomics or metaproteogenomics (Riesenfeld et al. 2004) which regulates the expression of various protein products under specific experimental conditions. For the completion of metaproteogenomics analysis, it needs a reference gene for protein identification (Knief et al. 2012).

## 6 Characterizing the Plant-Associated Microbiota

Even though numerous filamentous phytopathogen bacteria and only some of them acting like mutualists (Knief et al. 2011; Schenk et al. 2012; Knief 2014) have been studied in the research laboratory, the exact good cultivation surroundings for these plant-related bacteria and fungi are still mysterious, which inhibits their extensive study via isolate cultures (Hugenholtz 2002). The asymptomatic plants generally provide different surroundings for the survival of various plant-associated microorganisms. Different studies using NGS of these microbial habitats act as a major breakthrough in the discovery of novel taxa (Xu et al. 2012). Also they influenced

**Table 2** Cultivation-independent methods for plant microbiota study

| Techniques applied | Software used | Advantages | Disadvantages | Applications |
|---|---|---|---|---|
| Marker gene analysis | mothur, QIIME, and amplicon noise | Easy method for characterization of new and rare species | Biases during PCR amplifications | Discovery of novel species including their taxonomic profiling |
| Metagenomics | For assembly: IDBA-UD, Ray Meta, and MetaVelvet<br>For profiling: MLTreeMap, AMPHORA, mOTU, and MetaPhlAn<br>For function analysis: MG-RAST, IMG/M, and CAMERA | Unbiased profiling and it allows genomic studies of uncultured microbial species | Very lower coverage with respect to the marker gene sequencing<br>Very low abundance<br>Anticipated gene functions are not matched to expressed protein content | Discovery of novel species including their taxonomic profiling<br>Also taxonomic binning and genome reconstruction as well as study of functional, metabolic potential, and evolutionary relationships |
| Metatranscriptomics | For de novo assembly: IDBA-MT<br>For mapping: Bowtie2 and BWA-SW<br>For function analysis: MG-RAST and CAMERA | Determination of novel transcripts and sensitive detection method | Presence of rRNA in samples | Study of active function and pathways analysis |
| Metaproteogenomics | Mascot (for protein identification); MG-RAST and CAMERA (for function) | Good estimation of functional activities with proteomics as compare to transcriptomics | Requires reference genes for protein identification | Study of active function and pathways analysis |

Note: References for above discussed method Berg and Smalla (2009), Boisvert et al. (2012), Brady and Salzberg (2009), Caporaso et al. (2010), Glass et al. (2010), Kemler et al. (2013), Koopman and Carstens (2011), Leung et al. (2013), Markowitz et al. (2012), Namiki et al. (2012), Patil et al. (2011), Peng et al. (2012), Quince et al. (2011), Schloss et al. (2009), Segata et al. (2012), Shade et al. (2013), Stark et al. (2010), Sun et al. (2010), Sunagawa et al. (2013), Wu and Scott (2012)

some environmental factors for the discovery of novel effector proteins from microbial communities (Table 3).

**(a) Rhizosphere**

Generally roots evacuate approximately ~11–45% of photosynthetically fixed carbon, which consist of numerous carbon compounds; these carbon compounds are

**Table 3** Revealing facts of plant-associated microbiota using NGS technology

| Sequencing technology | Plant species and type of sequencing as well as sample | Major findings | References |
|---|---|---|---|
| *Rhizosphere* | | | |
| Pyrosequencing (Roche 454) | *A. thaliana* and marker genes, i.e., bacterial 16S rRNA sequencing and fungal ITS sequencing data | Role of ABC transporter mutant in different root exudate compositions from the wild type, with increased excretion of phenolic compounds and reduced sugar excretions, accompanied by higher abundances of OTUs related to beneficial rhizobacteria | Badri et al. (2009) |
| Pyrosequencing (Roche 454) | *A. thaliana* and bacterial 16S rRNA sequencing | Impact of soil bacteria on leaf metabolome | Badri et al. (2013a, b) |
| Pyrosequencing (Roche 454) | Oak and bacterial 16S rRNA sequencing data | Rhizosphere enrichment of proteo-bacteria relative to bulk soil | Uroz et al. (2010) |
| Pyrosequencing (Roche 454) | Sugar beet and PhyloChip (marker gene) | *Gammaproteobacteria* and *Betaproteobacteria* were enriched in soil so that it can suppress *Rhizoctonia solani* infection. On the other hand, *Pseudomonadaceae* strains protected the plant against infection. The protective mechanism of strain was lost in a mutant with a defective non-ribosomal peptide synthase gene | Mendes et al. (2011) |
| Pyrosequencing (Roche 454) | Maize and bacterial 16S rRNA sequencing data | Twenty-seven modern maize inbred strains were studied across five fields. The difference between field and bulk soil versus the rhizosphere accounted for most variation in diversity, and a weak genotype effect was observed within fields. *Burkholderiales*, *Oceanospirillales*, and *Sphingobacteriales* were found to be enriched relative to bulk soil, whereas *Acidobacteria*, *Chloroflexi*, *Planctomycetes*, and *Verrucomicrobia* were depleted | Peiffer et al. (2013) |

(continued)

**Table 3** (continued)

| Sequencing technology | Plant species and type of sequencing as well as sample | Major findings | References |
|---|---|---|---|
| Shotgun sequencing (Roche 454) | Rice and shotgun metagenome and metaproteogenome | Three rice cultivars were studied in one field. *Alphaproteobacteria* (*Rhizobium* spp. and *Methylobacterium* spp.) and *Actinobacteria* (*Microbacterium*) dominated the phyllosphere. Methanol-based methylotrophy linked to *Methylobacterium* dominated the protein repertoire, as well as proteins linked to transport processes and stress response. In the rhizosphere, *Alphaproteobacteria*, *Betaproteobacteria*, and *Deltaproteobacteria* were most abundant, and *Archaea* were present. There was also higher diversity. Proteins linked to methanogenesis and methanotrophy, as well as nitrogen fixation, were found | Knief et al. (2012) |
| 454 GS Flx (Roche 454) | Wheat, oat, pea, and a sad1 oat mutant and metatranscriptome | Analyzed active rhizosphere microbiomes in soil, as well as rhizosphere for three plant species and a sad1 oat mutant that is deficient in producing antifungal avenacins. Interestingly, for the sad1 mutant, the non-fungal eukaryotic rhizosphere community was more strongly altered than the fungal community, suggesting that avenacins in vivo may have effects other than protecting from fungal pathogens | Turner et al. (2013) |
| Roche 454 | *Glycine max r*hizosphere and bulk soil samples taken from mesocosm experiments with soil from soybean fields in Brazil | The rhizosphere community is selected from the bulk soil based on functions related to N, Fe, P, and K metabolism | Mendes et al. (2014) |
| Roche 454 | Rhizosphere samples from greenhouse-grown *Lotus japonicus*; plants of the same age but two different developmental stages grown in presence of phytic acid | Differences in microbial community composition in the rhizosphere of the differently developed plants; identification of genes related to phytic acid utilization | Unno and Shinano (2013) |

(continued)

**Table 3** (continued)

| Sequencing technology | Plant species and type of sequencing as well as sample | Major findings | References |
|---|---|---|---|
| Roche 454 | Barley rhizosphere samples collected from an experimental field in Ireland with 15 years of barley monoculture under low-input mineral management regime | Identification of genes and operons involved in mineral phosphate solubilization in the rhizosphere | Chhabra et al. (2013) |
| | *Phyllosphere* | | |
| Massively parallel sequencing (Roche 454) | *Quercus macrocarpa* and eukaryotic marker gene sequencing (ITS, 18S rRNA or 28S rRNA) | Different fungal communities were studied; they were found to be hyperdiverse and dominated by ascomycetes and *Alternaria. Epicoccum* and *Erysiphe* were the most abundant genera | Jumpponen and Jones (2009) |
| 454 GS Flx (Roche 454) | Fifty-six tree species and bacterial 16S rRNA sequencing | Fifty-six tree species were studied in the same location except for *Pinus ponderosa. Proteobacteria, Bacteroidetes, Actinobacteria*, TM7, and *Firmicutes* were the most abundant | Redford et al. (2010) |
| Pyrosequencing (Roche 454) | *Tamarix* species and bacterial and archaeal 16S rRNA sequencing and eukaryotic marker gene sequencing (ITS, 18S rRNA or 28S rRNA) | Three *Tamarix* species (salt-secreting desert tree) were studied at four different locations. Both location and tree species determined microbial community structure | Finkel et al. (2011) |
| 454 GS Flx (Roche 454) | Six tropical tree species and bacterial 16S rRNA sequencing | Six tree species had largely distinct microbial communities in the same location, with ~3–8% overlap of OTUs between species and ~10–18% overlap of OTUs within species. *Alphaproteobacteria, Gammaproteobacteria*, and *Acidobacteria* were abundant | Kim et al. (2012) |
| 454 GS Flx (Roche 454) | Lettuce and bacterial 16S rRNA sequencing | Core genera of phyllosphere microbiota across 44 fields consisting of *Pseudomonas, Bacillus, Massilia, Arthrobacter*, and *Pantoea* species were studied | Rastogi et al. (2012) |

(continued)

**Table 3** (continued)

| Sequencing technology | Plant species and type of sequencing as well as sample | Major findings | References |
|---|---|---|---|
| Shotgun sequencing (Roche 454) | Clover, soybean, and *A. thaliana* and shotgun metagenome and metaproteogenome | Alphaproteobacterial genera (*Sphingomonas* and *Methylobacterium*) were studied. For *Methylobacterium* spp., multiple proteins related to methanol-based methylotrophy were detected; for *Sphingomonas* spp., many proteins related to carbohydrate uptake were found | Delmotte et al. (2009) |
| Illumina MiSeq | Samples from *Salmonella* enrichment cultures from outdoor-grown tomato (*Solanum lycopersicum*) and tomato leaves and fruits | Differences in metagenomic composition of replicate phyllosphere enrichment cultures; enrichment of *Paenibacillus* on *Salmonella*-selective media | Ottesen et al. (2013) |
| Roche 454 | Leaf samples from field-grown soybean (*G. max*), Switzerland | High consistency in the microbial community composition and their proteomes on different host plants | Delmotte et al. (2009) |
| Roche 454 | Psyllid infected with the endophyte "*Candidatus Liberibacter asiaticus*" | Complete genome sequence of the uncultured plant pathogen and insect symbiont "*Candidatus Liberibacter asiaticus*" | Duan et al. (2009) |
| | | *Endosphere* | |
| Pyrosequencing (Roche 454) | Cottonwood trees and 16S rRNA sequencing and eukaryotic marker gene sequencing (ITS, 18S rRNA, or 28S rRNA) | The rhizosphere was enriched with *Acidobacteria* (~31%) and *Alphaproteobacteria* (~30%) relative to the endosphere. Most endophytes were *Gammaproteobacteria* (~54%) or *Alphaproteobacteria* (~23%). *Pezizomycotina* and *Agarimycotina* were abundant in both the rhizosphere and the endosphere | Gottel et al. (2011) |
| Pyrosequencing (Roche 454) | Pea and eukaryotic marker gene sequencing (ITS, 18S rRNA, or 28S rRNA) | Rhizosphere and endosphere fungal communities of diseased and healthy pea roots were studied across three fields in comparison to bulk soils. Health status and field both had significant effects on fungal community | Xu et al. (2012) |

(continued)

**Table 3** (continued)

| Sequencing technology | Plant species and type of sequencing as well as sample | Major findings | References |
|---|---|---|---|
| | | structure in roots, whereas only field was associated with significant changes of the rhizosphere and soil communities | |

directly or indirectly taking part in primary and secondary metabolism (Badri et al. 2013a, b). These work as energy sources, antimicrobials for soil microorganisms (Berg 2009; Dennis et al. 2010), which leads to the fortification of bacteria in the rhizosphere (Uroz et al. 2010). Likewise, the rhizosphere microbial communities are shaped by plants with the help of growth-promoting factors. These beneficial microorganisms also have biocontrol activity which help to release microorganism-derived antimicrobials; in this manner they indirectly counteracting pathogens (Berendsen et al. 2012; Mendes et al. 2013). For example, disease-suppressive soils, which have the capacity to suppress plant diseases, act as a key microbial beneficial strain, and it is also relevant for antimicrobial gene cluster (Mendes et al. 2011; Hirsch and Mauchline 2012) (Table 3).

In plant rhizosphere the taxonomic profiling studies have endeavored various information regarding plant species, genotype, soil type, different growth stages, and various microbial community structures using NGS. Metagenomics study on rhizosphere and bulk soil microbial species revealed some noticeable consequence on the microbiota of rhizosphere for some plants (Uroz et al. 2010; Peiffer et al. 2013). But in several cases, bonding of microbial diversity of rhizosphere was very strongly related with specific soils (Berendsen et al. 2012). In most of the cases, host genetic makeup was considered as an inconsequential determinant for the rhizosphere microbiome (Lundberg et al. 2012). Therefore, soil bacterial biomes ultimately elucidate the alterations of fields that increase field-specific rhizosphere microorganisms for these plants. Besides all those things, pathogen-specific genes and their secretion were specifically found in root-associated microbiota; this indicates the significance of host innate immune system and plant–microbe interactions (Bulgarelli et al. 2012).

**(b) Phyllosphere**

We can say that phyllosphere is actually subject to harsh environmental conditions. These include high UV radiation, fluctuating temperatures, low water availability, as well as low nutrient availability. Due to cell wall metabolism, plants produce various organic compounds like amino acids, sugars, alcohol, some volatile carbon, etc. (Vorholt 2012). NGS technologies helped a lot to examine the importance of geographical proximity versus species identity of plant phyllosphere microbial communities (Delmotte et al. 2009). Several tree species select its own bacterial communities for excretion of different phytochemicals in the phyllosphere (Whipps et al. 2008). Some metaproteogenomic studies have been performed on few plants like clover, soybean, and *Arabidopsis thaliana*. In terms of similarities, it has been seen that clover, soybean, and *Arabidopsis thaliana* microbial communities in phyllosphere were quite similar (Redford and Fierer 2009; Lopez-Velasco et al. 2011) as compared to the phyllosphere of rice which was more distinct (Rastogi et al. 2012; Schlaeppi et al. 2014) (Table 3).

**(c) Endosphere**
However, there are various endosphere microbial communities whose composition is still conserved and not known by the scientific community. They are trying hard to explore this area. The endosphere acted as an inhabitant for rhizosphere microbial community (Lundberg et al. 2012). In the *Arabidopsis thaliana*, three bacterial families from three different phyla were found that they enriched in the endosphere, even though their interaction patterns with different plants and the significance of these phyla for endosphere communities are still unknown. Model plant *Arabidopsis thaliana* and its close lineages share a taxonomically narrow root microbiota; due to common sharing of root microbiota, they comprise stable community of bacteria (*Actinomycetales*, *Burkholderiales*, and *Flavobacteriales*) (Schlaeppi et al. 2014). The bacterial communities present on root endophytes also initiate the inactivation of lignocellulose (Bulgarelli et al. 2012), which indicates their importance in controlling lignocellulolytic activity in endosphere niche. Other studies showed that bacteria like *Methylobacteriaceae* were present in roots of *Arabidopsis thaliana* (Lundberg et al. 2012). These bacteria act as a facultative methylotrophs so that they can metabolize methanol derived from plant in the phyllosphere microbial community and also to accomplish parallel role in the endosphere. Also, in some other plant species like pea root disease caused by fungal endosphere communities (Xu et al. 2012) (Table 3) showing the usefulness of plant microbiota for the plant survival.

## 7 Application of NGS Technologies in Future Studies Will Increase Understanding in Plant–Microbe Interactions

With the availability of up to third-generation sequencing platforms, a number of limitations have been already overcome. Particularly, libraries preparation and their sequencing become very easier and faster with respect to the costs per base and time. These technologies are providing in detail information related to physiological and chemical potential of microbial communities related to plants. A large number of limitations of NGS, for example, higher sequencing error or low-quality reads, can be reduced by using well-designed sequence data analysis methods. The third-generation sequencing will lead us in a new era where we can obtain even more and longer reads, improve day-by-day technologies, and remove all the errors in NGS analysis.

Some other limitation of NGS studies is a huge amount of sequences that represent unknown genes of known or unknown organisms. Sometimes for those

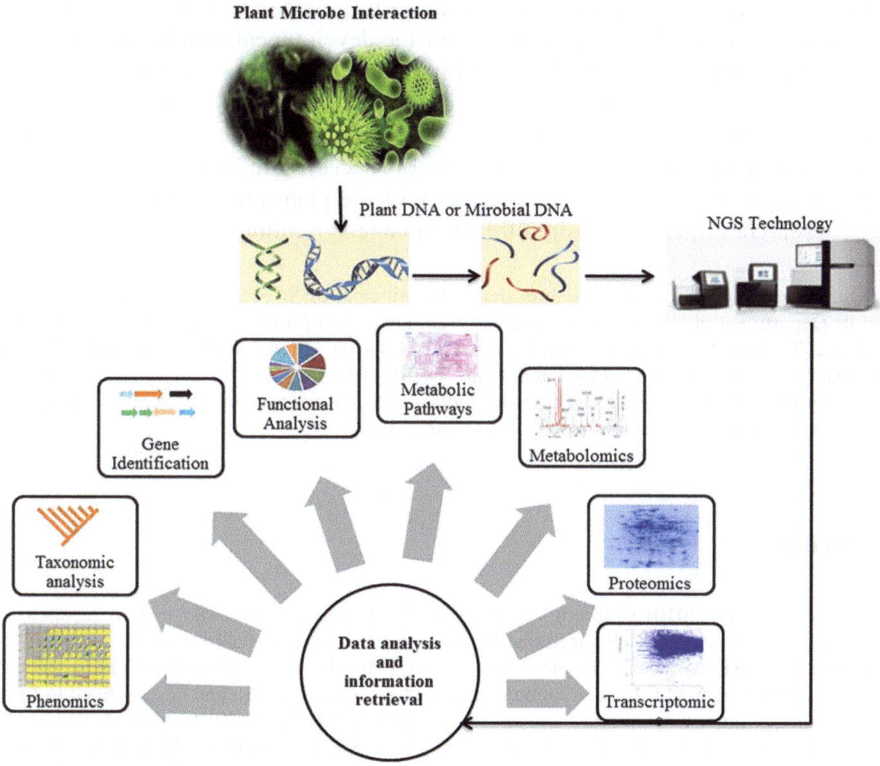


**Fig. 4** Applications of NGS in various fields

sequences, no homolog is found in public databases that also creates a problem for further information retrieval. Linking those unknown genes and finding their function, their phylogeny, and their other properties are still very challenging tasks (Fig. 4). A large number of sequencing projects are going on for gathering the information for different types of microbial strains and their respective ecosystems to overcome these limitations (Brown et al. 2012). All these information play a vital role in the study of plant-associated microorganisms. This information may build a model system so that it can explain or predict microbial interactions in the phyllosphere, endosphere, and rhizosphere under various environmental conditions.

## 8 Conclusions

Novel methods based on NGS techniques will have a great impact on science field. They can help in the detection of new bio-resources and novel plant growth-promoting agents with their high-performance speed and good efficiency. Plant–

microbe interaction may play a vital role to create new perspectives for sustainable agriculture. Plants are the major source for the development of new microbes and various bioactive compounds because huge diversity is present within the plant microbiome. Using NGS techniques we can explore the novel microbes associated with plants which may help to plants for stress resistance during adverse environmental conditions, although a large number of successful information were already reported regarding diversity and specificity of the plant microbiome. Altogether, these researches open new insight for sustainable agriculture.

**Acknowledgment** The authors wish to thank the principal of St. Anthony's College, Shillong, and its Department of Biotechnology is greatly acknowledged for providing necessary funds and facilities. Technical support provided by Dr. V.C. Kalia, Dr. Prasun Kumar, Dr. Gopal Kumar Prajapati, and also Dr. Dev Mani Pandey and Dr. Raju Poddar is fully acknowledged for the execution of above work. DBT, New Delhi, India, is greatly acknowledged for providing Bioinformatics Infrastructure Facility (DBT-BIF) at our institute.

## References

Albert M, Felix G (2010) Chimeric receptors of the *Arabidopsis thaliana* pattern recognition receptors EFR and FLS2. Plant Signal Behav 5:1430–1432. doi:10.4161/psb.5.11.13312

Albert M, Jehle AK, Mueller K, Eisele C, Lipschis M, Felix G (2010) *Arabidopsis thaliana* pattern recognition receptors for bacterial elongation factor Tu and flagellin can be combined to form functional chimeric receptors. J Biol Chem 285:19035–19042. doi:10.1074/jbc.M110.124800

Atamna-Ismaeel N, Finkel O, Glaser F, von Mering C, Vorholt JA, Koblížek M, Belkin S, Béjà O (2012) Bacterial anoxygenic photosynthesis on plant leaf surfaces. Environ Microbiol Rep 4:209–216. doi:10.1111/j.1758-2229.2011.00323.x

Badri DV, Quintana N, El Kassis EG, Kim HK, Choi YH, Sugiyama A, Verpoorte R, Martinoia E, Manter DK, Vivanco JM (2009) An ABC transporter mutation alters root exudation of phytochemicals that provoke an overhaul of natural soil microbiota. Plant Physiol 151:2006–2017. doi:10.1104/pp.109.147462

Badri DV, Chaparro JM, Zhang R, Shen Q, Vivanco JM (2013a) Application of natural blends of phytochemicals derived from the root exudates of *Arabidopsis* to the soil reveal that phenolic-related compounds predominantly modulate the soil microbiome. J Biol Chem 288:4502–4512. doi:10.1074/jbc.M112.433300

Badri DV, Zolla G, Bakker MG, Manter DK, Vivanco JM (2013b) Potential impact of soil microbiomes on the leaf metabolome and on herbivore feeding behavior. New Phytol 198:264–273. doi:10.1111/nph.12124

Barriuso J, Marín S, Mellado RP (2010) Effect of the herbicide glyphosate on glyphosate-tolerant maize rhizobacterial communities: a comparison with pre-emergency applied herbicide consisting of a combination of acetochlor and terbuthylazine. Environ Microbiol 12:1021–1030. doi:10.1111/j.1462-2920.2009.02146.x

Bell TH, Hassan SE-D, Lauron-Moreau A, Al-Otaibi F, Hijri M, Yergeau E, St-Arnaud M (2014) Linkage between bacterial and fungal rhizosphere communities in hydrocarbon-contaminated soils is related to plant phylogeny. ISME J 8:331–343. doi:10.1038/ismej.2013.149

Berendsen RL, Pieterse CM, Bakker PA (2012) The rhizosphere microbiome and plant health. Trends Plant Sci 17:478–486. doi:10.1016/j.tplants.2012.04.001

Berg G (2009) Plant–microbe interactions promoting plant growth and health: perspectives for controlled use of microorganisms in agriculture. J Appl Microbiol Biotechnol 84:11–18. doi:10.1007/s00253-009-2092-7

Berg G, Smalla K (2009) Plant species and soil type cooperatively shape the structure and function of microbial communities in the rhizosphere. FEMS Microbiol Ecol 68:1–13. doi:10.1111/j.1574-6941.2009.00654.x

Berg G, Krechel A, Ditz M, Sikora RA, Ulrich A, Hallmann J (2005) Endophytic and ectophytic potato-associated bacterial communities differ in structure and antagonistic function against plant pathogenic fungi. FEMS Microbiol Ecol 51:215–229. doi:10.1016/j.femsec.2004.08.006

Bodenhausen N, Horton MW, Bergelson J (2013) Bacterial communities associated with the leaves and the roots of *Arabidopsis thaliana*. PLoS One 8:e56329. doi:10.1371/journal.pone.0056329

Boisvert S, Raymond F, Godzaridis É, Laviolette F, Corbeil J (2012) Ray meta: scalable *de novo* metagenome assembly and profiling. Genome Biol 13:R122. doi:10.1186/gb-2012-13-12-r122

Bokulich NA, Thorngate JH, Richardson PM, Mills DA (2014) Microbial biogeography of wine grapes is conditioned by cultivar, vintage, and climate. Proc Natl Acad Sci USA 111:E139–E148. doi:10.1073/pnas.1317377110

Brady A, Salzberg SL (2009) Phymm and PhymmBL: metagenomic phylogenetic classification with interpolated Markov models. Nat Methods 6:673–676. doi:10.1038/nmeth.1358

Bragina A, Berg C, Cardinale M, Shcherbakov A, Chebotar V, Berg G (2012) *Sphagnum mosses* harbour highly specific bacterial diversity during their whole lifecycle. ISME J 6:802–813. doi:10.1038/ismej.2011.15

Brown SD, Utturkar SM, Klingeman DM, Johnson CM, Martin SL, Land ML, Lu T-YS, Schadt CW, Doktycz MJ, Pelletier DA (2012) Twenty-one genome sequences from Pseudomonas species and 19 genome sequences from diverse bacteria isolated from the rhizosphere and endosphere of *Populus deltoides*. J Bacteriol 194:5991–5993. doi:10.1128/JB.01243-12

Bulgarelli D, Rott M, Schlaeppi K, van Themaat EVL, Ahmadinejad N, Assenza F, Rauf P, Huettel B, Reinhardt R, Schmelzer E (2012) Revealing structure and assembly cues for *Arabidopsis* root-inhabiting bacterial microbiota. Nature 488:91–95. doi:10.1038/nature11336

Bulgarelli D, Schlaeppi K, Spaepen S, van Themaat EVL, Schulze-Lefert P (2013) Structure and functions of the bacterial microbiota of plants. Annu Rev Plant Biol 64:807–838. doi:10.1146/annurev-arplant-050312-120106

Caporaso JG, Kuczynski J, Stombaugh J, Bittinger K, Bushman FD, Costello EK, Fierer N, Pena AG, Goodrich JK, Gordon JI (2010) QIIME allows analysis of high-throughput community sequencing data. Nat Methods 7:335–336. doi:10.1038/nmeth.f.303

Chaparro JM, Badri DV, Vivanco JM (2014) Rhizosphere microbiome assemblage is affected by plant development. ISME J 8:790–803. doi:10.1038/ismej.2013.196

Chhabra S, Brazil D, Morrissey J, Burke JI, O'Gara F, N Dowling D (2013) Characterization of mineral phosphate solubilization traits from a barley rhizosphere soil functional metagenome. Microbiol Open 2:717–724. doi:10.1002/mbo3.110

Cretoiu MS, Kielak AM, Al-Soud WA, Sørensen SJ, van Elsas JD (2012) Mining of unexplored habitats for novel chitinases—*chiA* as a helper gene proxy in metagenomics. Appl Microbiol Biotechnol 94:1347–1358. doi:10.1007/s00253-012-4057-5

Delmotte N, Knief C, Chaffron S, Innerebner G, Roschitzki B, Schlapbach R, von Mering C, Vorholt JA (2009) Community proteogenomics reveals insights into the physiology of phyllosphere bacteria. Proc Natl Acad Sci USA 106:16428–16433. doi:10.1073/pnas.0905240106

Dennis PG, Miller AJ, Hirsch PR (2010) Are root exudates more important than other sources of rhizodeposits in structuring rhizosphere bacterial communities? FEMS Microbiol Ecol 72:313–327. doi:10.1111/j.1574-6941.2010.00860.x

Doehlemann G, Hemetsberger C (2013) Apoplastic immunity and its suppression by filamentous plant pathogens. New Phytol 198:1001–1016. doi:10.1111/nph.12277

Dohrmann AB, Küting M, Jünemann S, Jaenicke S, Schlüter A, Tebbe CC (2013) Importance of rare taxa for bacterial diversity in the rhizosphere of Bt-and conventional maize varieties. ISME J 7:37–49. doi:10.1038/ismej.2012.77

Duan Y, Zhou L, Hall DG, Li W, Doddapaneni H, Lin H, Liu L, Vahling CM, Gabriel DW, Williams KP (2009) Complete genome sequence of citrus huanglongbing bacterium, 'Candidatus Liberibacter asiaticus' obtained through metagenomics. Mol Plant Microbe Interact 22:1011–1020. doi:10.1094/MPMI-22-8-1011

Dumbrell AJ, Ashton PD, Aziz N, Feng G, Nelson M, Dytham C, Fitter AH, Helgason T (2011) Distinct seasonal assemblages of arbuscular mycorrhizal fungi revealed by massively parallel pyrosequencing. New Phytol 190:794–804. doi:10.1111/j.1469-8137.2010.03636.x

el Haichar FZ, Marol C, Berge O, Rangel-Castro JI, Prosser JI, Balesdent JM, Heulin T, Achouak W (2008) Plant host habitat and root exudates shape soil bacterial community structure. ISME J 2:1221–1230. doi:10.1038/ismej.2008.80

Fernandez D, Tisserant E, Talhinhas P, Azinheira H, Vieira A, Petitot AS, Loureiro A, Poulain J, Da Silva C, SILVA M (2012) 454-pyrosequencing of *Coffea arabica* leaves infected by the rust fungus *Hemileia vastatrix* reveals in planta-expressed pathogen-secreted proteins and plant functions in a late compatible plant–rust interaction. Mol Plant Pathol 13:17–37. doi:10.1111/j.1364-3703.2011.00723.x

Finkel OM, Burch AY, Lindow SE, Post AF, Belkin S (2011) Geographical location determines the population structure in phyllosphere microbial communities of a salt-excreting desert tree. Appl Environ Microbiol 77:7647–7655. doi:10.1128/AEM.05565-11

Fürnkranz M, Lukesch B, Müller H, Huss H, Grube M, Berg G (2012) Microbial diversity inside pumpkins: microhabitat-specific communities display a high antagonistic potential against phytopathogens. Microb Ecol 63:418–428. doi:10.1007/s00248-011-9942-4

Glass EM, Wilkening J, Wilke A, Antonopoulos D, Meyer F (2010) Using the metagenomics RAST server (MG-RAST) for analyzing shotgun metagenomes. Cold Spring Harbor Protoc, pdb. prot5368. doi:10.1101/pdb.prot5368

Gottel NR, Castro HF, Kerley M, Yang Z, Pelletier DA, Podar M, Karpinets T, Uberbacher E, Tuskan GA, Vilgalys R (2011) Distinct microbial communities within the endosphere and rhizosphere of *Populus deltoides* roots across contrasting soil types. Appl Environ Microbiol 77:5934–5944. doi:10.1128/AEM.05255-11

Hartmann A, Rothballer M, Schmid M (2008) Lorenz Hiltner, a pioneer in rhizosphere microbial ecology and soil bacteriology research. Plant Soil 312(1–2):7–14. doi:10.1007/s11104-007-9514-z

Hartmann A, Schmid M, Van Tuinen D, Berg G (2009) Plant-driven selection of microbes. Plant Soil 321(1–2):235–257. doi:10.1007/s11104-008-9814-y

Hirsch PR, Mauchline TH (2012) Who's who in the plant root microbiome? Nat Biotechnol 30:961–962. doi:10.1038/nbt.2387

Hugenholtz P (2002) Exploring prokaryotic diversity in the genomic era. Genome Biol 3:1–0003.8

Jiang X-T, Peng X, Deng G-H, Sheng H-F, Wang Y, Zhou H-W, Tam NF-Y (2013) Illumina sequencing of 16S rRNA tag revealed spatial variations of bacterial communities in a mangrove wetland. Microb Ecol 66:96–104. doi:10.1007/s00248-013-0238-8

Jumpponen A, Jones K (2009) Massively parallel 454 sequencing indicates hyperdiverse fungal communities in temperate *Quercus macrocarpa* phyllosphere. New Phytol 184:438–448. doi:10.1111/j.1469-8137.2009.02990.x

Kavamura NV, Taketani RG, Lançoni MD, Andreote FD, Mendes R, Soares de Melo I (2013) Water regime influences bulk soil and rhizosphere of *Cereus jamacaru* bacterial communities in the Brazilian Caatinga biome. PLoS One 8:e73606. doi:10.1371/journal.pone.0073606

Kemler M, Garnas J, Wingfield MJ, Gryzenhout M, Pillay K-A, Slippers B (2013) Ion Torrent PGM as tool for fungal community analysis: a case study of endophytes in *Eucalyptus grandis* reveals high taxonomic diversity. PLoS One 8:e81718. doi:10.1371/journal.pone.0081718

Kim M, Singh D, Lai-Hoe A, Go R, Rahim RA, Ainuddin A, Chun J, Adams JM (2012) Distinctive phyllosphere bacterial communities in tropical trees. Microb Ecol 63:674–681. doi:10.1007/s00248-011-9953-1

Knief C (2014) Analysis of plant microbe interactions in the era of next generation sequencing technologies. Front Plant Sci 5. doi:10.3389/fpls.2014.00216

Knief C, Delmotte N, Vorholt JA (2011) Bacterial adaptation to life in association with plants–a proteomic perspective from culture to in situ conditions. Proteomics 11:3086–3105. doi:10.1002/pmic.201000818

Knief C, Delmotte N, Chaffron S, Stark M, Innerebner G, Wassmann R, von Mering C, Vorholt JA (2012) Metaproteogenomic analysis of microbial communities in the phyllosphere and rhizosphere of rice. ISME J 6:1378–1390. doi:10.1038/ismej.2011.192

Koopman MM, Carstens BC (2011) The microbial phyllogeography of the carnivorous plant *Sarracenia alata*. Microb Ecol 61:750–758. doi:10.1007/s00248-011-9832-9

Leung HC, Yiu S-M, Parkinson J, Chin FY (2013) IDBA-MT: de novo assembler for metatranscriptomic data generated from next-generation sequencing technology. J Comput Biol 20:540–550. doi:10.1089/cmb.2013.0042

Liu L, Li Y, Li S, Hu N, He Y, Pong R, Lin D, Lu L, Law M (2012) Comparison of next-generation sequencing systems. J Biomed Biotechnol 2012:11 . doi:10.1155/2012/251364251364

Lopez-Velasco G, Welbaum G, Boyer R, Mane S, Ponder M (2011) Changes in spinach phylloepiphytic bacteria communities following minimal processing and refrigerated storage described using pyrosequencing of 16S rRNA amplicons. J Appl Microbiol 110:1203–1214. doi:10.1111/j.1365-2672.2011.04969.x

Lu R, Martin-Hernandez AM, Peart JR, Malcuit I, Baulcombe DC (2003) Virus-induced gene silencing in plants. Methods 30:296–303. doi:10.1016/S1046-2023(03)00037-9

Lumini E, Orgiazzi A, Borriello R, Bonfante P, Bianciotto V (2010) Disclosing arbuscular mycorrhizal fungal biodiversity in soil through a land-use gradient using a pyrosequencing approach. Environ Microbiol 12:2165–2179. doi:10.1111/j.1462-2920.2009.02099.x

Luna E, Bruce TJ, Roberts MR, Flors V, Ton J (2012) Next-generation systemic acquired resistance. Plant Physiol 158:844–853. doi:10.1104/pp.111.187468

Lundberg DS, Lebeis SL, Paredes SH, Yourstone S, Gehring J, Malfatti S, Tremblay J, Engelbrektson A, Kunin V, Del Rio TG (2012) Defining the core *Arabidopsis thaliana* root microbiome. Nature 488:86–90. doi:10.1038/nature11237

Maignien L, DeForce EA, Chafee ME, Eren AM, Simmons SL (2014) Ecological succession and stochastic variation in the assembly of *Arabidopsis thaliana* phyllosphere communities. MBio 5:e00682–e00613. doi:10.1128/mBio.00682-13

Mardis ER (2008) Next-generation DNA sequencing methods. Annu Rev Genomics Hum Genet 9:387–402. doi:10.1146/annurev.genom.9.081307.164359

Markowitz VM, Chen I-MA, Chu K, Szeto E, Palaniappan K, Grechkin Y, Ratner A, Jacob B, Pati A, Huntemann M (2012) IMG/M: the integrated metagenome data management and comparative analysis system. Nucleic Acids Res 40:D123–D129. doi:10.1093/nar/gkr975

Mascia T, Nigro F, Abdallah A, Ferrara M, De Stradis A, Faedda R, Palukaitis P, Gallitelli D (2014) Gene silencing and gene expression in phytopathogenic fungi using a plant virus vector. Pro Nat Aca Sci USA 111:4291–4296. doi:10.1073/pnas.1315668111

Mendes R, Kruijt M, de Bruijn I, Dekkers E, van der Voort M, Schneider JH, Piceno YM, DeSantis TZ, Andersen GL, Bakker PA (2011) Deciphering the rhizosphere microbiome for disease-suppressive bacteria. Science 332:1097–1100. doi:10.1126/science.1203980

Mendes R, Garbeva P, Raaijmakers JM (2013) The rhizosphere microbiome: significance of plant beneficial, plant pathogenic, and human pathogenic microorganisms. FEMS Microbiol Rev 37:634–663. doi:10.1111/1574-6976.12028

Mendes LW, Kuramae EE, Navarrete AA, van Veen JA, Tsai SM (2014) Taxonomical and functional microbial community selection in soybean rhizosphere. ISME J 8:1577–1587. doi:10.1038/ismej.2014.17

Namiki T, Hachiya T, Tanaka H, Sakakibara Y (2012) MetaVelvet: an extension of Velvet assembler to de novo metagenome assembly from short sequence reads. Nucleic Acids Res 40:e155–e155. doi:10.1093/nar/gks678

Navarrete AA, Kuramae EE, de Hollander M, Pijl AS, van Veen JA, Tsai SM (2013) Acidobacterial community responses to agricultural management of soybean in Amazon forest soils. FEMS Microbiol Eco 83:607–621. doi:10.1111/1574-6941.12018

Newton AC, Fitt BDL, Atkins SD, Walters DR, Daniell TJ (2010) Pathogenesis, parasitism and mutualism in the trophic space of microbe–plant interactions. Trends Microbiol 18(8):365–373. doi:10.1016/j.tim.2010.06.002

Niedringhaus TP, Milanova D, Kerby MB, Snyder MP, Barron AE (2011) Landscape of next-generation sequencing technologies. Anal Chem 83(12):4327–4341. doi:10.1021/ac2010857

Nocker A, Burr M, Camper AK (2007) Genotypic microbial community profiling: a critical technical review. Microb Ecol 54:276–289. doi:10.1007/s00248-006-9199-5

Ottesen AR, Peña AG, White JR, Pettengill JB, Li C, Allard S, Rideout S, Allard M, Hill T, Evans P (2013) Baseline survey of the anatomical microbial ecology of an important food plant: *Solanum lycopersicum* (tomato). BMC Microbiol 13:114. doi:10.1186/1471-2180-13-114

Pareek CS, Smoczynski R, Tretyn A (2011) Sequencing technologies and genome sequencing. J Appl Genet 52:413–435. doi:10.1007/s13353-011-0057-x

Patil KR, Haider P, Pope PB, Turnbaugh PJ, Morrison M, Scheffer T, McHardy AC (2011) Taxonomic metagenome sequence assignment with structured output models. Nat Methods 8:191–192. doi:10.1038/nmeth0311-191

Peiffer JA, Spor A, Koren O, Jin Z, Tringe SG, Dangl JL, Buckler ES, Ley RE (2013) Diversity and heritability of the maize rhizosphere microbiome under field conditions. Proc Natl Aca Sci USA 110:6548–6553. doi:10.1073/pnas.1302837110

Peng Y, Leung HC, Yiu S-M, Chin FY (2012) IDBA-UD: a de novo assembler for single-cell and metagenomic sequencing data with highly uneven depth. Bioinformatics 28:1420–1428. doi:10.1093/bioinformatics/bts174

Perazzolli M, Antonielli L, Storari M, Puopolo G, Pancher M, Giovannini O, Pindo M, Pertot I (2014) Resilience of the natural phyllosphere microbiota of the grapevine to chemical and biological pesticides. Appl Environ Microbiol 80:3585–3596. doi:10.1128/AEM.00415-00411

Proctor LM (2011) The human microbiome project in 2011 and beyond. Cell Host Microbe 10:287–291. doi:10.1016/j.chom.2011.10.001

Quince C, Lanzen A, Davenport RJ, Turnbaugh PJ (2011) Removing noise from pyrosequenced amplicons. BMC Bioinf 12:38. doi:10.1186/1471-2105-12-38

Rastogi G, Sbodio A, Tech JJ, Suslow TV, Coaker GL, Leveau JH (2012) Leaf microbiota in an agroecosystem: spatiotemporal variation in bacterial community composition on field-grown lettuce. ISME J 6:1812–1822. doi:10.1038/ismej.2012.32

Redford AJ, Fierer N (2009) Bacterial succession on the leaf surface: a novel system for studying successional dynamics. Microb Ecol 58:189–198. doi:10.1007/s00248-009-9495-y

Redford AJ, Bowers RM, Knight R, Linhart Y, Fierer N (2010) The ecology of the phyllosphere: geographic and phylogenetic variability in the distribution of bacteria on tree leaves. Environ Microbiol 12:2885–2893. doi:10.1111/j.1462-2920.2010.02258.x

Reisberg EE, Hildebrandt U, Riederer M, Hentschel U (2013) Distinct phyllosphere bacterial communities on *Arabidopsis* wax mutant leaves. PLoS One 8:e78613. doi:10.1371/journal.pone.0078613

Riesenfeld CS, Schloss PD, Handelsman J (2004) Metagenomics: genomic analysis of microbial communities. Annu Rev Genet 38:525–552. doi:10.1146/annurev.genet.38.072902.091216

Rosenzweig N, Tiedje JM, Quensen JF III, Meng Q, Hao JJ (2012) Microbial communities associated with potato common scab-suppressive soil determined by pyrosequencing analyses. Plant Dis 96:718–725. doi:10.1094/PDIS-07-11-0571

Schadt EE, Turner S, Kasarskis A (2010) A window into third-generation sequencing. Hum Mol Genet 19(R2):R227–R240. doi:10.1093/hmg/ddq416

Schenk PM, Carvalhais LC, Kazan K (2012) Unraveling plant–microbe interactions: can multi-species transcriptomics help? Trends Biotechnol 30:177–184. doi:10.1016/j.tibtech.2011.11.002

Schirawski J, Mannhaupt G, Münch K, Brefort T, Schipper K, Doehlemann G, Di Stasio M, Rössel N, Mendoza-Mendoza Pester D, Müller O, Winterberg B, Meyer E, Ghareeb H, Wollenberg T, Münsterkötter M, Wong P, Walter M, Stukenbrock Güldener U, Kahmann R (2010)

Pathogenicity determinants in smut fungi revealed by genome comparison. Science 330 (6010):1546–1548. doi:10.1126/science.1195330

Schlaeppi K, Dombrowski N, Oter RG, van Themaat EVL, Schulze-Lefert P (2014) Quantitative divergence of the bacterial root microbiota in *Arabidopsis thaliana* relatives. Proc Nat Acad Sci USA 111:585–592. doi:10.1073/pnas.1321597111

Schloss PD, Westcott SL, Ryabin T, Hall JR, Hartmann M, Hollister EB, Lesniewski RA, Oakley BB, Parks DH, Robinson CJ (2009) Introducing mothur: open-source, platform-independent, community-supported software for describing and comparing microbial communities. Appl Envir Microbiol 75:7537–7541. doi:10.1128/AEM.01541-09

Schmidt SM, Panstruga R (2011) Pathogenomics of fungal plant parasites: what have we learnt about pathogenesis? Curr Opin Plant Biol 14:392–399. doi:10.1016/j.pbi.2011.03.006

Segata N, Waldron L, Ballarini A, Narasimhan V, Jousson O, Huttenhower C (2012) Metagenomic microbial community profiling using unique clade-specific marker genes. Nat Methods 9:811–814. doi:10.1038/nmeth.2066

Sessitsch A, Hardoim P, Döring J, Weilharter A, Krause A, Woyke T, Mitter B, Hauberg-Lotte L, Friedrich F, Rahalkar M (2012) Functional characteristics of an endophyte community colonizing rice roots as revealed by metagenomic analysis. Mol Plant Microbe Interact 25:28–36. doi:10.1094/MPMI-08-11-0204

Shade A, McManus PS, Handelsman J (2013) Unexpected diversity during community succession in the apple flower microbiome. MBio 4:e00602–e00612. doi:10.1128/mBio.00602-12

Stark M, Berger SA, Stamatakis A, von Mering C (2010) MLTreeMap-accurate maximum likelihood placement of environmental DNA sequences into taxonomic and functional reference phylogenies. BMC Genomics 11:461. doi:10.1186/1471-2164-11-461

Stroud H, Ding B, Simon SA, Feng S, Bellizzi M, Pellegrini M, Wang G-L, Meyers BC, Jacobsen SE (2013) Plants regenerated from tissue culture contain stable epigenome changes in rice. Elife 2:e00354. doi:10.7554/eLife.00354

Sun S, Chen J, Li W, Altinatas I, Lin A, Peltier S, Stocks K, Allen EE, Ellisman M, Grethe J (2010) Community cyber infrastructure for advanced microbial ecology research and analysis: the CAMERA resource. Nucleic Acids Res 39:D546–D551. doi:10.1093/nar/gkq1102

Sunagawa S, Mende DR, Zeller G, Izquierdo-Carrasco F, Berger SA, Kultima JR, Coelho LP, Arumugam M, Tap J, Nielsen HB (2013) Metagenomic species profiling using universal phylogenetic marker genes. Nat Methods 10:1196–1199. doi:10.1038/nmeth.2693

Thakur K, Chawla V, Bhatti S, Swarnkar MK, Kaur J, Shankar R, Jha G (2013) De novo transcriptome sequencing and analysis for *Venturia inaequalis*, the devastating apple scab pathogen. PLoS One 8:e53937. doi:10.1371/journal.pone.0053937

Thompson JF, Milos PM (2011) The properties and applications of single-molecule DNA sequencing. Genome Biol 12:217. doi:10.1186/gb-2011-12-2-217

Tremblay A, Hosseini P, Li S, Alkharouf N, Matthews B (2012) Identification of genes expressed by *Phakopsora pachyrhizi*, the pathogen causing soybean rust, at a late stage of infection of susceptible soybean leaves. Plant Pathol 61:773–786. doi:10.1111/j.1365-3059.2011.02550.x

Turner TR, Ramakrishnan K, Walshaw J, Heavens D, Alston M, Swarbreck D, Osbourn A, Grant A, Poole PS (2013) Comparative metatranscriptomics reveals kingdom level changes in the rhizosphere microbiome of plants. ISME J 7:2248–2258. doi:10.1038/ismej.2013.119

Unno Y, Shinano T (2013) Metagenomic analysis of the rhizosphere soil microbiome with respect to phytic acid utilization. Microbes Environ 28:20–127. doi:10.1264/jsme2.ME12181

Uroz S, Buée M, Murat C, Frey-Klett P, Martin F (2010) Pyrosequencing reveals a contrasted bacterial diversity between oak rhizosphere and surrounding soil. Environ Microbiol Rep 2:281–288. doi:10.1111/j.1758-2229.2009.00117.x

van der Heijden MG, Klironomos JN, Ursic M, Moutoglis P, Streitwolf-Engel R, Boller T, Wiemken A, Sanders IR (1998) Mycorrhizal fungal diversity determines plant biodiversity, ecosystem variability and productivity. Nature 396:69–72. doi:10.1038/23932

Vorholt JA (2012) Microbial life in the phyllosphere. Nat Rev Microbiol 10:828–840. doi:10.1038/nrmicro2910

Weßling R, Schmidt SM, Micali CO, Knaust F, Reinhardt R, Neumann U, van Themaat EVL, Panstruga R (2012) Transcriptome analysis of enriched *Golovinomyces orontii haustoria* by deep 454 pyrosequencing. Fungal Genet Biol 49:470–482. doi:10.1016/j.fgb.2012.04.001

Whipps J, Hand P, Pink D, Bending GD (2008) Phyllosphere microbiology with special reference to diversity and plant genotype. J Appl Microbiol 105:1744–1755. doi:10.1111/j.1365-2672.2008.03906.x

Wicker T, Oberhaensli S, Parlange F, Buchmann JP, Shatalina M, Roffler S, Ben-David R, Doležel J, Šimková H, Schulze-Spanu PD, Bruggmann R, Amselem J, Quesneville H, Themaat EVL, Paape T, Shimizu KK, Keller B (2013) The wheat mildew genome shows the unique evolution of an obligate biotroph. Nat Genet 45:1092–1096. doi:10.1038/ng.2704

Williams TR, Moyne A-L, Harris LJ, Marco ML (2013) Season, irrigation, leaf age, and *Escherichia coli* inoculation influence the bacterial diversity in the lettuce phyllosphere. PLoS One 8:1–14. doi:10.1371/journal.pone.0068642

Wu M, Scott AJ (2012) Phylogenomic analysis of bacterial and archaeal sequences with AMPHORA2. Bioinformatics 28:1033–1034. doi:10.1093/bioinformatics/bts079

Xu L, Ravnskov S, Larsen J, Nicolaisen M (2012) Linking fungal communities in roots, rhizosphere, and soil to the health status of *Pisum sativum*. FEMS Microbiol Eco 82:736–745. doi:10.1111/j.1574-6941.2012.01445.x

Yergeau E, Sanschagrin S, Maynard C, St-Arnaud M, Greer CW (2014a) Microbial expression profiles in the rhizosphere of willows depend on soil contamination. ISME J 8:344–358. doi:10.1038/ismej.2013.163

Yergeau E, Sanschagrin S, Maynard C, St-Arnaud M, Greer CW (2014b) Microbial expression profiles in the rhizosphere of willows depend on soil contamination. ISME J 8:344–358. doi:10.1038/ismej.2013.163

Yu L, Nicolaisen M, Larsen J, Ravnskov S (2012) Succession of root-associated fungi in *Pisum sativum* during a plant growth cycle as examined by 454 pyrosequencing. Plant Soil 358:225–233. doi:10.1007/s11104-012-1188-5

Zhang W, Wu X, Liu G, Chen T, Zhang G, Dong Z, Yang X, Hu P (2013) Pyrosequencing reveals bacterial diversity in the rhizosphere of three *Phragmites australis* ecotypes. Geomicrobiol J 30:593–599. doi:10.1080/01490451.2012.740145

Zhuang X, McPhee KE, Coram TE, Peever TL, Chilvers MI (2012) Rapid transcriptome characterization and parsing of sequences in a non-model host-pathogen interaction; pea-Sclerotinia sclerotiorum. BMC Genomics 13:668. doi:10.1186/1471-2164-13-668

# An Overview of Next-Generation Sequencing (NGS) Technologies to Study the Molecular Diversity of Genome

**C. Mohinudeen, Manoharan Melvin Joe, Abitha Benson, and Sa Tongmin**

**Abstract** Sequencing is a process used for determining the array of biopolymers applicable in identifying microbial gene arrangements, phenotypes, evolutionary biology, metagenomics, potential drug targets, gene cloning, etc. Commercial sequencers are emerging all around the globe due to rapid development of recombinant DNA (rDNA) technology. For sequencing DNA, next-generation sequencing (NGS) methods provide faster, inexpensive, accurate sequencing of polymers than traditional approaches. The concept of NGS is not quite new; it started in the mid-late 1990s with the successful introduction of methods thereafter, namely, Roche 454 pyrosequencing, Illumina sequencing, SOLiD sequencing, etc. In this book chapter, we briefly elaborate on the above-stated methods and its advantages and disadvantages. Furthermore, we will be discussing the sequencing methods that are under development in the biological research.

**Keywords** Next-generation sequencing (NGS) • Molecular diversity • Bioinformatics • Roche 454 pyrosequencing • SOLiD sequencing • True single-molecule sequencing (tSMS) • Single-molecule real-time (SMRT) sequencing • Ion torrent semiconductor sequencing • Nanopore sequencing

C. Mohinudeen
Department of Microbiology, Vels Institute of Science Technology and Advanced Studies (VISTAS), School of Life Science, Vels University, Chennai, India

M.M. Joe (✉)
Department of Microbiology, Vels Institute of Science Technology and Advanced Studies (VISTAS), School of Life Science, Vels University, Chennai, India

Department of Agricultural Chemistry, Chungbuk National University, Cheongju, Chungbuk 361-763, Republic of Korea
e-mail: micromelvin@gmail.com

A. Benson
Department of Biotechnology, Vels Institute of Science Technology and Advanced Studies (VISTAS), School of Life Science, Vels University, Chennai, India

S. Tongmin (✉)
Department of Agricultural Chemistry, Chungbuk National University, Cheongju, Chungbuk 361-763, Republic of Korea
e-mail: tomsa@chungbuk.ac.kr



V.C. Kalia, P. Kumar (eds.), *Microbial Applications Vol.1*,
DOI 10.1007/978-3-319-52666-9_14

# 1 Introduction

Deoxyribonucleic acid (DNA), a fundamental part of the central dogma of molecular biology, defines the complete organism. After the discovery of DNA in 1953 by two eminent scientists, sequencing its genes for studying the genotype and phenotype of cells was necessary. Frederick Sanger and Coulson in 1975 came up with the method of sequencing DNA (Sanger and Coulson 1975), and due to its drawbacks 2 years later, they developed another method based on chain termination of nucleotide strand (Sanger et al. 1977). It is considered to be the first-generation sequencing technology to be commercialized with a wide range of applications in molecular cloning, breeding, and evolution. In spite of its accuracy and applicability, traditional Sanger sequencer did not find scope because of its high cost and low throughput per run (Yadav et al. 2014). Molecular biologists from every nook of the world wanted to develop a faster, high-throughput, cheaper, and accurate method to sequence DNA.

Next-generation sequencing (NGS) technologies fulfilled the needs of emerging molecular biology laboratories. NGS technologies sequence thousands to millions of strands at the same time (Hall 2007). Roche 454 sequencer emerged as the first NGS platform working under the principle of pyrosequencing (Ronaghi et al. 1996). Following Roche, Illumina and ABI SOLiD commercialized their genome analyzer. HeliScope and PacBio RS sequencer developed by Helicos BioSciences and Pacific BioSciences respectively eliminated the PCR amplification step and consumed the total run time (Harris et al. 2007). Although these platforms utilized fluorophores and optical imaging to detect the sequence reads, Ion Personal Genome Machine of Life Technologies use pH detection system to accurately predict the sequences (Niedringhaus et al. 2011). Read length of the NGS technologies increased remarkably as a cause of nanopore detection system with greater accuracy and short run time. Due to high read length, sequencing larger fragments or even the whole genome is now possible with some required modifications. In spite of the remarkable success with biological nanopores, its inherent drawback such as nanopore instability at high temperatures was noted (Kang et al. 2005). To ameliorate the nanopore technology sustainable for upcoming generation, synthetic solid-state nanopores are under progress for commercialization, hoping that these nanopores will devote effective genetic analysis.

In this chapter, we will be discussing in brief on individual sequencing topics commencing with traditional Sanger sequencing, followed by an interconnection with current NGS technologies, and finally concluding with emerging NGS method. Merits and demerits of different DNA sequencing approaches along with their necessary bioinformatics tools are also included.

# 2 Traditional Method

## 2.1 Sanger Sequencing

Sanger and Coulson in 1975 described their first method known as "plus and minus" for sequencing DNA (Sanger and Coulson 1975), but due to its ineffectiveness Sanger and his co-workers in 1977 launched another sequencing method known as the "chain termination method" that uses chain-terminating inhibitors (Sanger et al. 1977). Since then, DNA sequencing has been carried out in the Sanger sequencer using the chain termination approach as a key principle. Prior to sequencing, the fragmented DNA strands are cloned into a suitable bacterial vector (having high copy number) for in vivo amplification. Then, sequencing starts with amplified template denaturation, annealing of primer to the resulting ssDNA, and extension of annealed primer. After incorporation of fluorescently labeled dideoxynucleotides (ddNTPs), each round of primer extension is randomly terminated. In any given fragment, each fluorescent label on the terminating ddNTP corresponds to a particular nucleotide (either A/T/G/C). Fragments are then separated based on their molecular weight by high-resolution electrophoresis separation method carried out in a capillary polymer gel. A detector coupled to the capillary chamber excites the fluorescent labels of the separated fragments by laser (Fig. 1). The resulting four-channel emission spectra provide the readout also known as "trace," and these

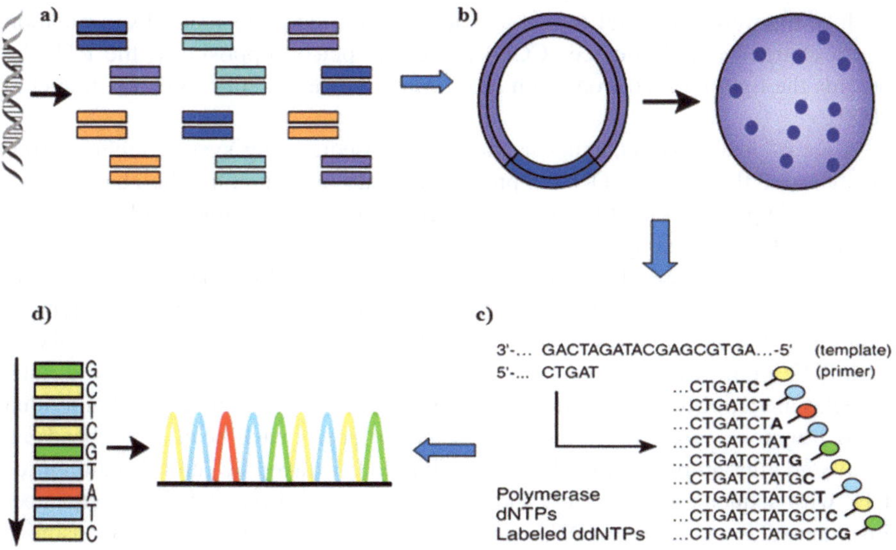


**Fig. 1** Sanger's chain termination sequencing workflow. (**a**) Fragmentation of DNA into small fragments. (**b**) Cloning and amplification of fragments into suitable vector. (**c**) Cycles of DNA sequencing by chain termination and synthesis of ddNTP terminated, labeled products ladder. (**d**) Capillary electrophoresis detection to generate sequencing readout/trace

traces are then translated by base-calling software into DNA sequences (Ewing and Green 1998; Shendure and Ji 2008).

# 3 Current Methods

## *3.1 Roche 454 Pyrosequencing*

Roche 454 sequencing system (http://www.454.com) was the first NGS technology to be commercialized. In 2005, Jonathan Rothberg (the founder of 454 Life Sciences) launched his first commercial NGS platform in the name of GS 20, and later in 2007 Roche Applied Science acquired 454 Life Sciences and marketed the second version of the 454 instrument as GS FLX (Voelkerding et al. 2009). The working principle of this high-throughput sequencer lies in the collaboration of two approaches, namely, single-molecule emulsion PCR (Tawfik and Griffiths 1998) with pyrosequencing technology (developed by Nyren et al. 1993 and refined by Ronaghi et al. 1996, 1998). Pyrosequencing is a DNA "sequencing-by-synthesis" method where synthesis/release of a pyrophosphate takes place during nucleotide incorporation (Ronaghi et al. 1996). A picotiter plate (PTP) is a microfabricated fiber-optic slide in which the pyrosequencing reaction takes place. One PTP measures about 44 μm in diameter and consists of more than a million reaction wells per plate with each well having $3.4 \times 10^6$ picoliter scale holding capacity. The walls of the wells are metal coated to ameliorate the signal-to-noise discrimination, and a charged couple device (CCD) camera is placed opposite to the PTP that records the light emitted from each bead (Margulies et al. 2005; Voelkerding et al. 2009).

A set of reactions occurring in the 454 pyrosequencing system includes fragmentation of the template DNA to prepare a library of several hundred base pairs in length and ligation of the library to 454-specific adaptor oligonucleotides. Next, the library is diluted to single-molecule concentration, followed by their denaturation into single strands and hybridization to Sepharose or Styrofoam beads carrying oligonucleotides complementary to the sequences of the adaptor. The fragment-containing beads (micelles) are emulsified in a "water-in-oil" emulsion mixture with PCR amplification reagents to create individual micelles. Then these isolated micelles are amplified using emulsion PCR (emPCR) to clonally expand individual DNA fragments into 1 million copies on the surface of each bead. Once beads are recovered after amplification, they are individually arrayed by centrifugation into a PTP with sequencing enzymes, and the plate is loaded into GS FLX sequencer. Enzyme ATP sulfurylase, DNA polymerase, luciferase, luciferin, adenosine 5′ phosphosulfate (APS), and apyrase are involved in catalyzing the pyrosequencing reaction (Mardis 2008a, b; Froehlich et al. 2010). During sequencing, the PTP acts as a flow cell that allows the flow of dinucleotide triphosphates (dNTPs) into the wells.

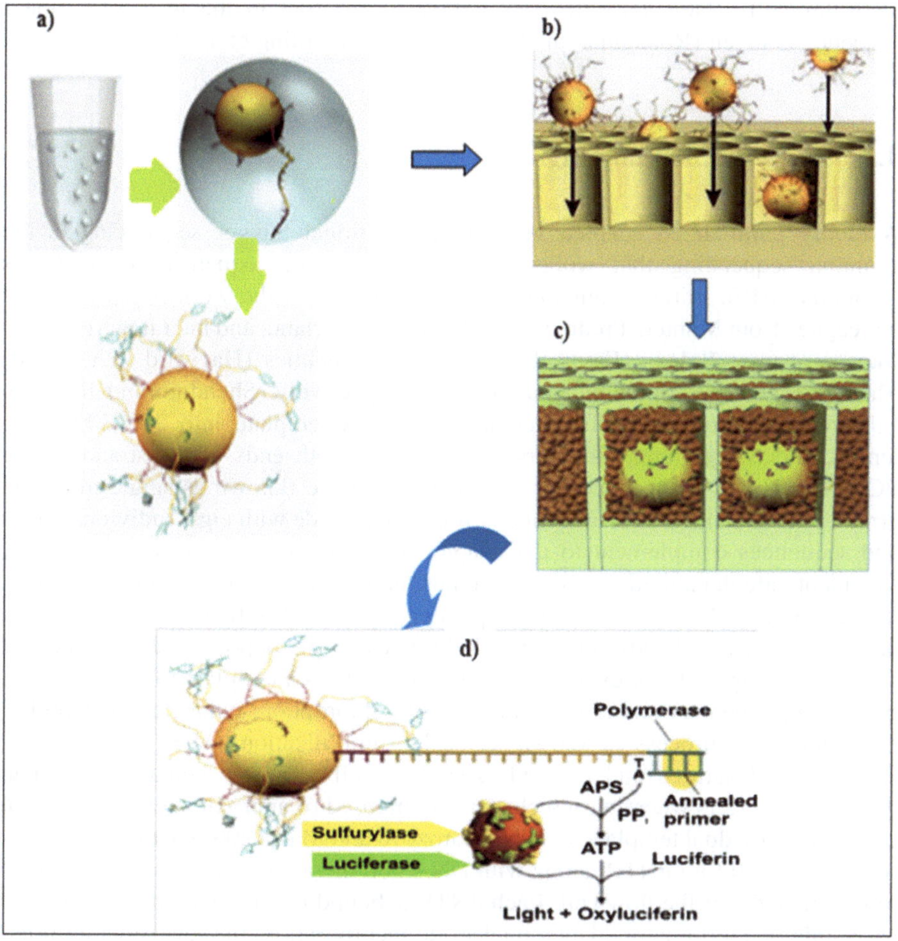


**Fig. 2** 454 Pyrosequencing workflow. (**a**) Emulsification of beads containing DNA fragments ligated to 454-specific adapters, amplification of micelles by emulsion PCR, and emulsion is broken to recover the beads. (**b**) Loading of beads into the picotiter plate (PTP). (**c**) Stacking of beads and pyrosequencing enzymes in the PTP. (**d**) Illustration of pyrosequencing reaction that occurs in a PTP

The enzyme DNA polymerase incorporate these nucleotides as a complementary base to the template strand with the release of inorganic pyrophosphate (PPi). The unmatched bases are degraded by the apyrase enzyme. ATP sulfurylase converts the released PPi into ATP, which then drives the luciferase reporter enzyme to signal luciferin to use the ATP to generate light (Fig. 2). This localized luminescence is transmitted through the PTP and gets recorded on a CCD camera. The amount of light generated is directly proportional to the number of nucleotide incorporated. With the repetitive cycles of pyrosequencing, wells of PTP are subsequently imaged, the signal-to-noise ratio is analyzed and filtered, and algorithmically (Newbler software) a translated output in the form of a linear sequence is given (Ronaghi et al. 1996, 1998; Voelkerding et al. 2009).

## *3.2 Illumina Sequencing*

A highly parallelized adapted version of traditional Sanger sequencing is the Illumina sequencing that works on the principle of sequencing-by-synthesis (Illumina 2010). Initially, this platform originated from the work of Turcatti and colleagues from Manteia Predictive Medicine, Switzerland, and later it merged with four companies Solexa (Essex, UK), Lynx Therapeutics (Hayward, CA, USA), Manteia Predictive Medicine, and Illumina, respectively (Shendure and Ji 2008). Library is constructed by fragmentation of DNA template into 200–300 bp in length, and two different adaptors are ligated to both ends of the strands using PCR. The HiSeq 2500 sequencer uses a solid surface (known as single-molecule array or flow cell) which is an optically transparent slide with eight individual lanes and sequences complement to the adaptor attached on it (Bentley et al. 2008). Fragments are denatured into single strands, and with the help of flanked adaptor, they are annealed to the flow cell complementary sites. The free end adaptor of the array-bound DNA bends down and hybridizes to an opposite adaptor thereby creating a bridge. The bridged DNA strand that acts as a template for the synthesis of its complementary strand is cloned by using fluorophore-unlabeled nucleotides, and hence this process is known as bridge amplification. After amplification, strands are denatured and cleaved. The flow cell will now contain more than 40 million clusters wherein each cluster is composed of approximately 1000 clonal copies of individual template strands (Morozova and Marra 2008). During sequencing reaction, all four dNTPs along with DNA polymerase are added simultaneously to the channels of the flow cell. Each dNTP is bound to a unique fluorescent label reversible dye terminators. Upon nucleotide incorporation, the reversible terminator is cleaved, and the process is repeated until the template strand is made (Fig. 3). A colored fluorescent signal is generated once after the incorporation of each nucleotide, which is then recorded by an imaging device to determine the sequence (Sucher et al. 2012).

## *3.3 SOLiD Sequencing*

Supported oligonucleotide ligation and detection (SOLiD) platform was originated from George Church laboratory in 2005 and later commercialized by Applied Biosystems (Life Technologies) (Shendure et al. 2005). The technology lies in the principle of sequencing by ligation of oligonucleotides to the DNA template. A mixture of short DNA fragments from the library are ligated to oligonucleotide adapters, immobilized on the surface of 1 μm paramagnetic beads, and clonally

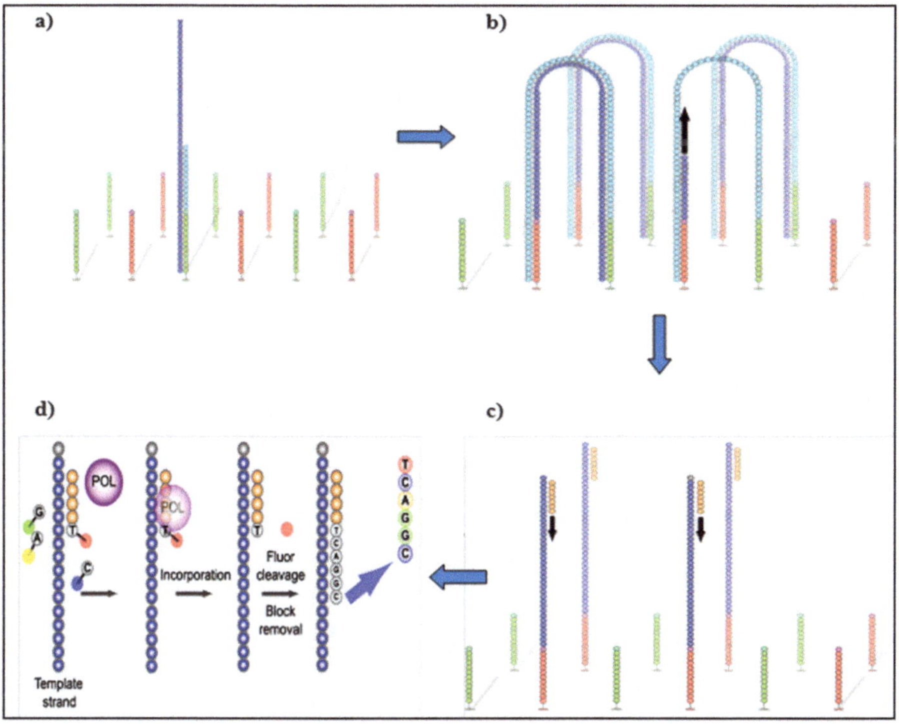


**Fig. 3** Illumina sequencing workflow. (**a**) Hybridization of the adapter-ligated ssDNA to the flow cell. (**b**) Bridge amplification to generate clonally amplified clusters. (**c**) Denatured and cleaved strands are sequenced. (**d**) Polymerase incorporates nucleotides and releases the fluorophore to begin the next synthesis cycle

amplified by emulsion PCR (Dressman et al. 2003). Emulsion is then broken, and beads are recovered and are attached covalently to the surface of specially treated glass slide that is placed into a fluidics cassette within the sequencer to generate dense, disordered array. In this platform, two slides are processed per run, one slide receiving sequencing reactants, and second slide is being imaged (Mardis 2008b). The ligation-based sequencing begins with the annealing of universal primer (*n*) complementary to the array of amplicon-bearing beads that provides a 5′ phosphate substrate for DNA ligase. After annealing, DNA ligase is added along with four semi-degenerate octamer/8-mer fluorescent oligonucleotides in an automated manner within the instrument. The 8-mer fluorescent oligonucleotide is a probe that contains 2 probe-specific bases consisting of one of 16 possible 2-base combinations (TT, GT, and so on), 6 degenerate bases (nnnzzz), a ligation site (on first base), a cleavage site (on fifth base), and 4 different fluorescent dyes (linked to the last base) (Liu et al. 2012). Adjacent to the universal primer (*n*), DNA ligase seals the phosphate backbone when a matching 8-mer probe hybridizes to the DNA template sequence. A fluorescent readout is followed identifying the ligated 8-mer probe, which corresponds to one of the 16 possible combinations. Subsequently, linkage

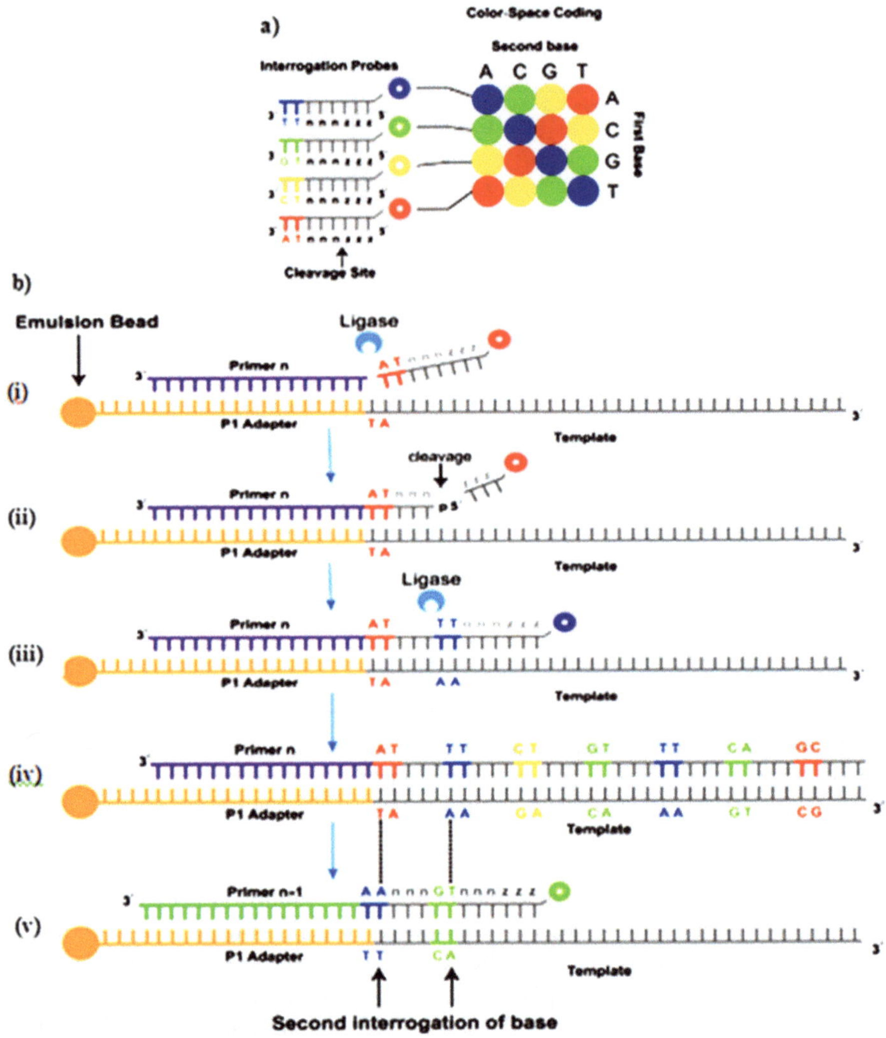


**Fig. 4** SOLiD sequencing workflow. (**a**) Octamer interrogation probes with 2 probe-specific bases on the left and 16 possible 2-base combinations on the right. (**b**) (1) In the first round of sequencing, adapter (P1)-linked template is annealed to the primer (*n*), and the primer is interrogated by a probe having 2 specific bases (e.g., here AT) complementary to the template. (2) Fluorescence emitted during probe annealing and ligation is recorded, and the last three degenerative bases are cleaved. (3) Annealing and ligation of next probe and so on. (4) and (5) The primer (*n*) is fully extended by seven cycles of ligation and the resulting product is denatured from the template for second round of sequencing with a new primer ($n - 1$) and vice versa

between the fifth and sixth base of the 8-mer is chemically cleaved removing the fluorescent group allowing further ligation. Seven ligation cycles (rounds) are performed to extend the first primer (*n*) and then the synthesized strand is denatured from the adapter/template. The second round of sequencing is initiated with hybridization of a new $n - 1$-positioned universal primer (that is offset by 1 base in the adapter sequence) to the synthesized strand and continues the rounds of ligation-mediated sequencing (Fig. 4). Process is repeated each time with a new primer with a successive offset ($n - 2$, $n - 3$, and so on). In this approach, all the templates are sequenced twice (Voelkerding et al. 2009; Shokralla et al. 2012). Finally, 2-base-calling processing software decodes the fluorescence generated from the universal primers into sequence reads.

### 3.4 *True Single-Molecule Sequencing (tSMS)*

True single-molecule DNA sequencing method was first developed by Stephen Quake and colleagues in 2003 and commercialized in 2007 by Helicos BioSciences (Cambridge, MA) as HeliScope sequencer (Braslavsky et al. 2003). It is a unique platform utilizing sequencing-by-synthesis approach without any clonal amplification of the DNA template by PCR (Harris et al. 2007; Pushkarev et al. 2009). Library is prepared by fragmenting template DNA into small, 100–200 bp fragments and polyadenylation of generated DNA fragments with poly-(A) tail at 3′ end, with the final adenosine fluorescently labeled with Cy3 (Shendure and Ji 2008). Denatured polyadenylated template strands hybridized to the poly-(T) oligonucleotides are immobilized on the surface of flow cell at a capture density of up to $100 \times 10^6$ template strands/cm$^2$. The instrument records the position of each fluorescently labeled template on the array prior to sequencing and expects the sequence read in that position. After the positional coordinates are recorded by highly sensitive CCD camera, label is cleaved and washed before sequencing. Sequencing cycle begins by adding DNA polymerase and one of four Cy5-labeled (Cyanine-5) dNTPs to the flow cell resulting in template-dependent extension of DNA strands (Fig. 5). Here sequencing is asynchronous, i.e., during a particular round of sequencing, all the templates will not incorporate a nucleotide (Voelkerding et al. 2009; Anderson and Schrijver 2010). The emitted fluorescence signal is captured, and the images are recorded using CCD camera to determine incorporated nucleotide. After imaging, the label is chemically cleaved and removed by washing, and the cycle is repeated with the next Cy5-labeled dNTP. Therefore this sequencing platform utilizes single DNA molecules as templates rather than clonally amplified clusters and results in a higher sequencing output per run (Deschamps and Campbell 2010).

### 3.5 *Single-Molecule Real-Time (SMRT) Sequencing*

Pacific BioSciences (Menlo Park, CA) commercially introduced PacBio RS sequencer in 2010 for single-molecule real-time (SMRT) sequencing. It is a real-

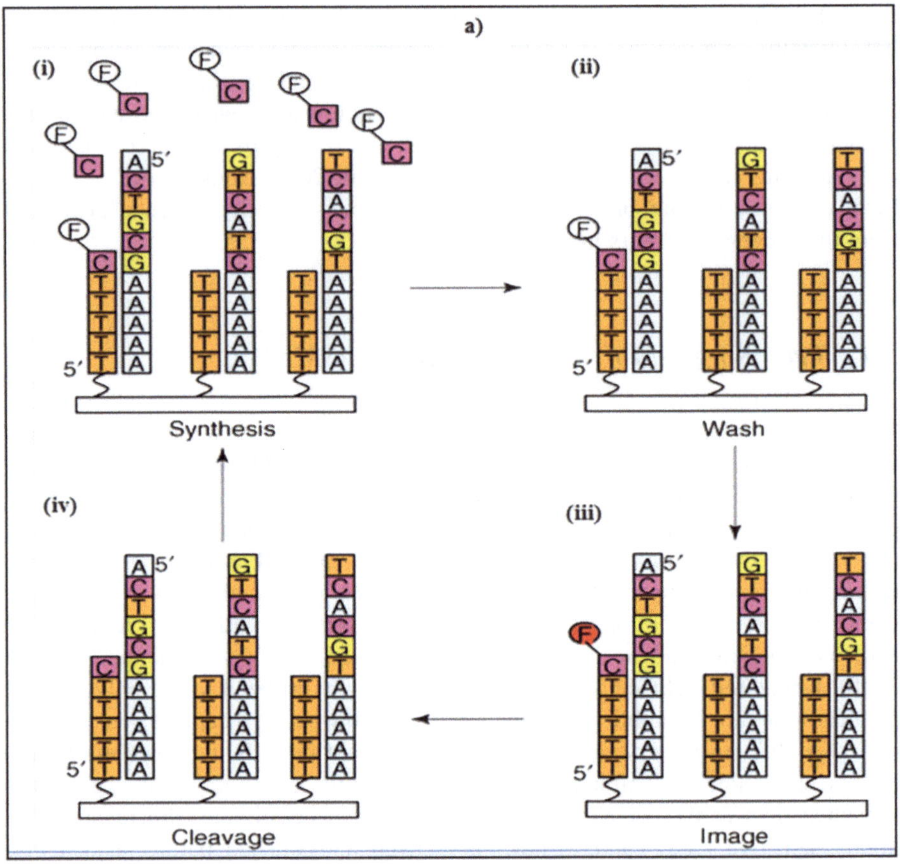


**Fig. 5** True single-molecule sequencing (tSMS) workflow. (**a**) Clockwise illustration of four steps involved in tSMS: (1) Dye-labeled nucleotide is incubated on the flow cell for incorporation. (2) The unincorporated, labeled nucleotide molecules are removed by washing. (3) Then the incorporated Cy-5-labeled nucleotide is imaged by highly sensitive CCD camera. (4) Finally, the dye label is cleaved from the incorporated nucleotide, and the next sequencing cycle follows

time, fluorescent single-molecule sequencing platform that relies on sequencing-by-synthesis approach (Korlach et al. 2010). A dense array nanostructure called as zero-mode waveguide (ZMW) manufactured on chip surface is the chief component of the sequencer. The chip is fabricated by perforating a thin metal film supported by a transparent substrate and contains more than 10–1000 ZMW well, with each well measuring about 10–50 nm in diameter. ZMW is a light-focusing structure that allows for real-time observation of DNA polymerization. Adapters are ligated onto the ends of the template DNA fragments, and a primer complementary to the adapter sequence is annealed. A calculated ratio of primer-annealed DNA templates and DNA polymerase molecules mixture is supplied to the instrument by a diffusion-mediated process. DNA polymerase is engineered to have a decreased rate of polymerization and a unique ability to incorporate fluorescently modified nucleotides (Mardis 2013). Next, fluorescently labeled nucleotides (linked to the phosphate) are added sequentially to the wells, and DNA polymerase incorporates these nucleotides to create DNA strand complementary to the template. The optical system is finely tuned to measure extremely small detection volume ($20 \times 10^{-21}$ L) of fluorescence emitted during reaction (Anderson and Schrijver 2010). Once phosphate diester bond is formed, a fluorescent signal is produced corresponding to the each incorporated nucleotide which is detected by the optical system (Eid et al. 2009). After each incorporation and detection event, fluorophore is cleaved from the growing strand and diffuses out of the wells (Fig. 6). The sequencer incorporates nucleotides at the rate of 10 bases/s, giving rise to a chain of a thousand nucleotides within minutes (Korlach et al. 2008). The

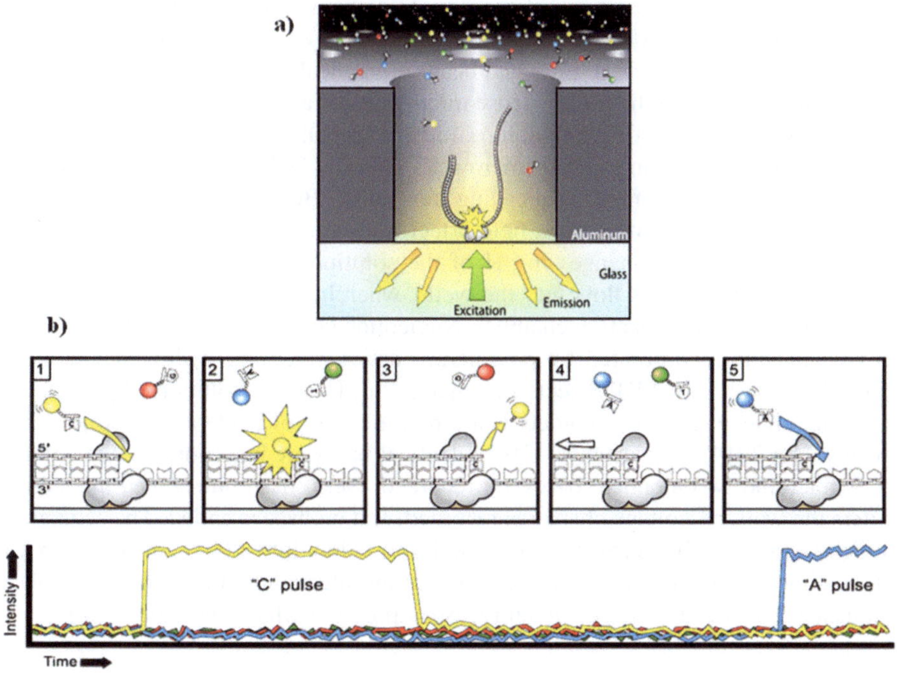


**Fig. 6** Single-molecule real-time (SMRT) sequencing ZMW chip and workflow. (**a**) DNA polymerase bound ssDNA template is immobilized at the bottom of a zero-mode waveguide (ZMW) nanostructure illuminated by laser light at the bottom. Fluorescently labeled nucleotides are added for sequencing the bound template. (**b**) Cyclic event of dinucleotide triphosphate (dNTP) incorporation in a ZMW with detected fluorescence intensity graph. (1) Attachment of phospholinked nucleotide to the template by ZMW bound polymerase. (2) This creates an elevation of fluorescence output on the corresponding color channel. (3) The fluorescence dye linked to the nucleotide pyrophosphate is cleaved after forming the phosphodiester bond with the template, and then it diffuses out of the ZMW ending the fluorescence pulse. (4) and (5) Next nucleotide binds the active site as the polymerase translocates and subsequent pulse is generated

instrument monitors all the wells constantly during the run, performs numerous calculations, condenses the data, and analyzes it to produce sequence reads.

### 3.6 Ion Torrent Semiconductor Sequencing

Ion Personal Genome Machine (PGM) is a fast, simple, massively scalable, versatile, and less costly sequencing technology commercialized in 2010 by Ion Torrent technology, a company that was later acquired by Life Technologies during the same year (Pareek et al. 2011). The sequencer relies on the principle detection of hydrogen ion concentration that is released as a by-product during nucleotide incorporation into the template by DNA polymerase (Rothberg et al. 2011). Library fragments constructed by DNA fragmentation and adapter ligation are hybridized to the bead containing adapter complementary sequences and then amplified by emulsion PCR (emPCR). After successful amplification, emulsion is broken, and the resulting beads are enriched. Enriched beads are annealed to the sequencing primer and are deposited into the highly dense microwells of an Ion Chip along with DNA polymerase by mild centrifugation. Ion Chip is a specialized silicon chip designed to detect pH changes of individual microwells (Mardis 2013). Upper surface of the chip delivers the required sequencing reactants. Beneath the microwells, an ion-sensitive layer followed by a field-effect transistor (FET)/ion sensor sub-layer is aligned to measure changes in pH of the solution (Niedringhaus et al. 2011). Sequentially nucleotides flow into the wells, wherein each well acts an individual DNA polymerization reaction chamber. Nucleotide complements with the template DNA and gets incorporated by polymerase action releasing hydrogen ions ($H^+$) (Fig. 7). As a result, FET detects the change in pH of the solution ($\Delta$pH), and a potential change ($\Delta V$) is recorded as a direct measure of nucleotide incorporation events (Hui 2014). For instance, if two identical bases are incorporated, the voltage is double, and two identical bases are recorded. Next, if an unmatched nucleotide enters the well, no voltage change is recorded, and no base is called. The sequencer directly converts the received chemical information into digital information by processing the signals and by using base-calling algorithms to produce the DNA sequence associated with individual reads (http://www.lifetechnologies.com/).

## 4 Emerging Method

### 4.1 Nanopore Sequencing

Nanopore sequencing technology emerged from the independent work of David Deamer and his colleagues at the University of California (Santa Cruz, CA, USA) in 1996 (Branton et al. 2008). It is a fast-growing real-time sequencing technology, in that sequencing is carried out in a very thin porous membrane with pores ranging in

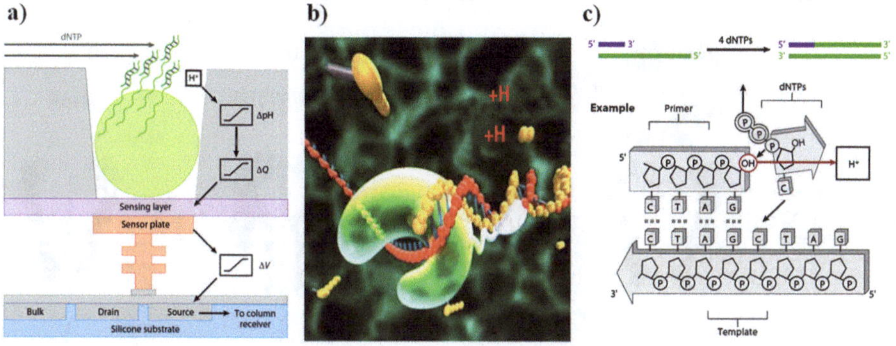


**Fig. 7** Ion Chip and Ion Torrent semiconductor sequencing workflow. (**a**) Ion Torrent and Ion Chip are used for sensing pH changes that occur during sequencing. (**b**) and (**c**) pH-based sequencing reaction in which protons released when nucleotide is incorporated into the template by polymerase

nanoscale size (nanopores). Current is applied across the nanopore, and the negatively charged ssDNA traverses through it toward the positive terminal causing a change in electrical conductivity across the nanopore. The potential difference generated in picoamperes (pA) is measured using a circuit (Gupta 2008). This technology obviates the need to amplify the template fragments by PCR, synchronous washing of the reagent, and synthesis of strand complementary to the template and to reduce time (Ku and Roukos 2013). Based on the kind of nanopore, the technology is categorized into two types, namely, biological nanopore sequencing and solid-state nanopore sequencing.

### 4.1.1 Biological Nanopore Sequencing

In this type of sequencing, α-hemolysin (αHL), a protein nanopore, is held between a phospholipid bilayer separating two chambers filled with KCl solution. Two electrodes are placed on the opposite sides of the bilayer, and an electrical potential is applied during sequencing. The amount of current passing through the nanopore in a given moment varies depending on the shape, size, and length of the nucleotide that blocks the ion flow through the pore (Hui 2014). An exonuclease enzyme is attached to the nanopore surface to cleave individual nucleotide molecules from the DNA as it enters. Negatively charged ssDNA translocates through the membrane as positive potential is applied to the electrode. Translocation velocity varies with different parameters such as electric potential, type of nanopore, and strands of DNA (whether ss/ds) (Steinbock and Radenovic 2015). Ionic current is partially blocked during translocation producing different magnitudes of current disruption that in turn aids in differentiation among the four nucleotides (Fig. 8a).

Specificity of nucleotide detection across the membrane is tenfold higher in current MspA (*Mycobacterium smegmatis* porin A protein) nanopore than classical

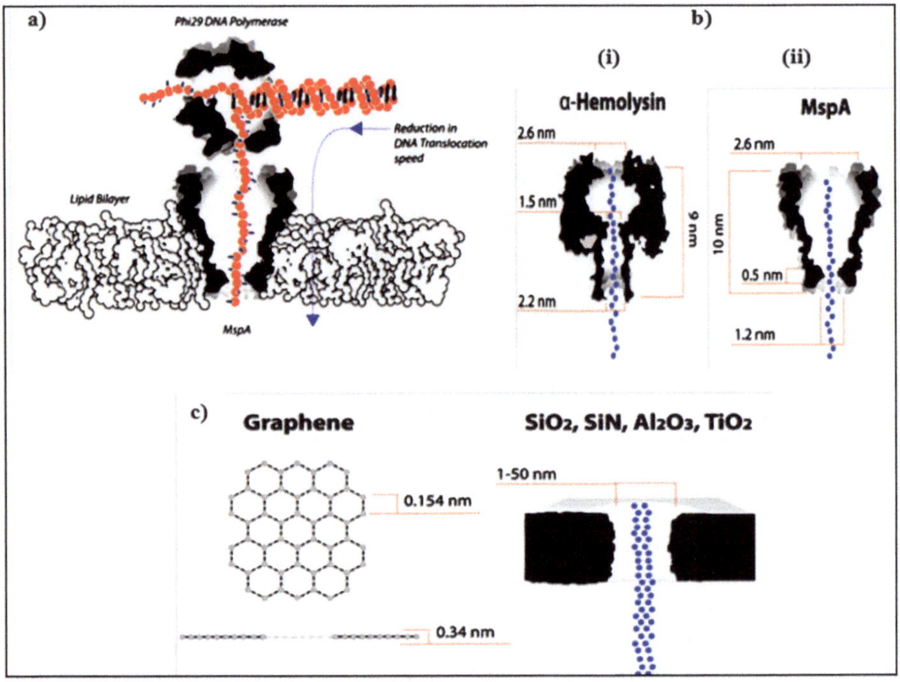


**Fig. 8** Nanopore sequencing technology. (**a**) Nanopore sequencing reaction involves unwinding of dsDNA by φ29 DNA polymerase and controlled translocation of unwounded ssDNA through MspA. (**b**) Biological nanopores: (1) heptameric α-hemolysin (αHL) nanopore and its dimensions and (2) *Mycobacterium smegmatis* porin A (MspA) nanopore and its dimensions. (**c**) Solid-state nanopore

αHL (Manrao et al. 2011) (Fig. 8b). Translocation velocity of DNA can be slowed down for improving detection accuracies by combining MspA with DNA polymerase φ29. Oxford Nanopore Technologies (ONT) is only the leading company that develops, commercializes, and updates these types of bio-nanopore-based sequencing platforms. ONT in collaboration with Professor Daniel Branton, George Church, and Jene Golovchenko of Harvard, David Deamer and Mark Akeson of UCSC, and John Kasianowicz of NIST commercialized their first developer version of the nanopore sequencer "MinION" on 2014 as an early access program (MAP or MinION Access Program) to the researchers and investigators. MinION is a portable device commercialized for the analysis of DNA and RNA in real time. Recently, in May 2015, ONT introduced the second version of the device "MinION" as "MinION MkI" featuring its improvements in performance and easy access. Improvements in library preparation kits and temperature controlling system, an increase in total number of nanopores per flow cell, etc., can be found in MinION MkI. PromethION, a small benchtop platform developed by ONT for high-throughput real-time analysis of the DNA and RNA samples, is currently under the scheme of commercialization (Oxford Nanopore Technologies, UK).

### 4.1.2 Solid-State Nanopore Sequencing

This type of sequencing uses nanopores fabricated mechanically on a solid-state material such as silicon nitrides, silicon oxide or metal oxides, and graphene (Fig. 8c). These artificial nanopores are considered to be more stable than biological nanopores as it eliminates membrane instability and protein positioning drawbacks. IBM in collaboration with 454 Life Sciences is developing artificial nanopores by drilling a 10 nm titanium nitride membrane, which are separated by insulating layers of silica. When DNA strand translocates through the nanopore, the electric field across the layer is flipped resulting in immobilization of the strand and significant measuring of ionic current. Graphene is another new single-atom membrane that is 1–5 nm thick and has pores with a diameter of 5–10 nm (Niedringhaus et al. 2011). Graphene membrane improves sequencing and detection accuracies by allowing the passage of one base at a time (Pennisi 2012). Therefore, these kinds of nanopores are robust in their unique ability and leads researchers toward the next-generation sequencing.

## 5 Bioinformatics Tools

Sequence reads generated from various NGS platforms are analyzed using commercially available bioinformatics tools (Table 1). Each tool has a unique function, categorized as (Lee et al. 2013):

1. Read alignment based on the reference genomes.
2. De novo assembly involves grouping of generated short reads into contigs which are further assembled into scaffolds to reconstruct original genomic DNA of a new species.
3. Identification of single nucleotide genetic variant (i.e., mutation) in the assembled sequence reads.
4. Detection of functional variants altering the protein coding regions of individual genomes.
5. Detection of large alterations such as indels, inversions, and translocation in the DNA (~1 kb–3 Mb).
6. Transcriptome assembly.

## 6 Comparison of Various NGS Platforms

See Table 2.

**Table 1** Bioinformatics tools for analyzing the raw data generated from various NGS platforms

| Category | | Tools | URL | References |
|---|---|---|---|---|
| Reads alignment/assembly tools | Assembly based on reference genomes | BWA | http://maq.sourceforge.net/ | Li and Durbin (2009) |
| | | ELAND | http://www.illumina.com | NA |
| | | GMAP | http://www.gene.com/share/gmap | Wu and Watanabe (2005) |
| | | MAQ | http://maq.sourceforge.net | Li et al. (2008a, b) |
| | | Mosaik | http://bioinformatics.bc.edu/marthlab/Mosaik | Smith et al. (2008a, b) |
| | | RMAP | http://rulai.cshl.edu/rmap | Smith et al. (2008a, b) |
| | | SHRiMP | http://compbio.cs.toronto.edu/shrimp | Rumble et al. (2009) |
| | | SOAP | http://soap.genomics.org.cn | Li et al. (2008a, b) |
| | | SSAHA/SSAHA2 | http://www.sanger.ac.uk/Software/analysis/SSAHA2 | Ning et al. (2001) |
| | De novo assembly | ALLPATHS | http://www.broadinstitute.org/science/programs/genome-biology/crd | Butler et al. (2008) |
| | | Edena | http://www.genomic.ch/edena | Hernandez et al. (2008) |
| | | EULER | http://euler-assembler.ucsd.edu/portal/ | Chaisson and Pevzner (2008) |
| | | Euler-SR | http://euler-assembler.ucsd.edu/portal/?q=team | Chaisson and Pevzner (2008) |
| | | Newbler | http://www.roche-applied-science.com/ | Margulies et al. (2005) |
| | | SHARCGS | http://sharcgs.molgen.mpg.de/ | Dohm et al. (2007) |
| | | SSAKE | http://www.bcgsc.ca/platform/bioinfo/software/ssake | Warren et al. (2007) |
| | | VCAKE | http://sourceforge.net/projects/vcake | Jeck et al. (2007) |
| | | Velvet | http://www.ebi.ac.uk/%7Ezerbino/velvet | Zerbino and Birney (2008) |

(continued)

**Table 1** (continued)

| Category | | Tools | URL | References |
|---|---|---|---|---|
| Variant detection/discovery tool | Single nucleotide variant | ssahaSNP | http://www.sanger.ac.uk/Software/analysis/ssahaSNP | NA |
| | | MAQ | http://maq.sourceforge.net | Li et al. (2008a, b) |
| | Functional variant | B-SIFT | http://research-pub.gene.com/bsift/ | Lee et al. (2009) |
| | | MAPP | http://mendel.stanford.edu/supplementarydata/stone_MAPP_2005 | Stone and Sidow (2005) |
| | Structural variant | BreakDancer | http://genome.wustl.edu/tools/cancer-genomics/ | Chen et al. (2009) |
| | | PEMer | http://sv.gersteinlab.org/pemer/ | Korbel et al. (2009) |
| Transcriptome alignment/assembly | | ERANGE | http://woldlab.caltech.edu/rnaseq/ | Mortazavi et al. (2008) |
| | | NextGENe | http://www.softgenetics.com/NextGENe.html | NA |
| | | Oases | www.ebi.ac.uk/~zerbino/oases/ | Zerbino et al. (2012) |
| | | Trans-AbySS | http://www.bcgsc.ca/platform/bioinfo/software/trans-abyss/ | Robertson et al. (2010) |
| | | Trinity | https://github.com/trinityrnaseq/trinityrnaseq/wiki/ | Grabherr et al. (2011) |

## 7 Conclusion and Perspective

Next-generation sequencing (NGS) technologies contribute rapid, high-throughput, cost-effective probing of genomes of individual organisms. Geneticists have seen a prompt inclination in the sequencing technologies over the past decade. Despite the merits and limitations of each platform, these platforms can be used in tandem for decoding the entire genome of a distinct organism. Modern NGS technologies serve prominent roles in scrutinizing genome-level mutation and gene expression to divulge phenotype of diverse strains. Therefore, these technologies have augmented our minds toward the immense knowledge about the organism and their active systems at molecular level. In the successive years of NGS technologies, progress in state-of-the-art bioinformatics software will be of critical importance to elicit sophisticated amount of generated sequenced reads.

**Table 2** Comparison of various NGS platforms and their advantages and disadvantages

| Company | Platform | Sequencing chemistry | Read length (bp) | Number of reads | Time/run | Throughput/run | Accuracy | Machine cost ($) | Advantages | Disadvantages |
|---|---|---|---|---|---|---|---|---|---|---|
| Applied Biosystems | 3730xl | Chain termination | 400–900 | NA | 20 min–3 h | 1.9–84 kb | 99.999% | 95,000 | High quality and long read length | High cost and low throughput |
| Roche | 454 GS FLX | Pyrosequencing | 700 | 1 million | 23 h | 0.7 Gb | 99.997% | 500,000 | Longer read length improves mapping in repetitive regions and shorter run time | High reagent costs, higher error rate in homopolymer repeats, and low throughput |
| Illumina | HiSeq 2500 | Reversible dye terminators | 36/50/100 | 3 million | 2–11 d | 600 Gb | >99% | 740,000 | Massive throughput and cost-effective | High instrument cost, longer run time, and shorter read length |
| ABI Life Technologies | 5500 SOLiD | Sequencing by ligation | 75 + 35 | 1.4 million | 7 d | 90 Gb | 99.99% | 350,000 | Low reagent costs, high throughput, and inherent error correction by 2-base encoding | Shorter read length and longer run time |
| | Ion Personal Genome Machine (PGM) | pH detection | 35/200/400 | 12 million | 2 h | 2 Gb | >99% | 50,000 | Low instrument cost and shorter run time | High reagent cost and high error rate in homopolymer regions |

| | | | | | | | | | | |
|---|---|---|---|---|---|---|---|---|---|---|
| Helicos BioSciences | Heliscope | Single-molecule sequencing | 25–55 | 600–800 million | 8 d | 37 Gb | 99.99% | 999,000 | No amplification and non-bias representation of templates for sequence-based applications | High error rate, very shorter read length, and high instrument cost |
| Pacific Bioscience | PacBio RS | Real-time single-molecule sequencing | ~3000 | –50 thousand | 2 h | 13 Gb | 84–85% | 700,000 | Low reagent cost, shorter run time, and very long reads with greater potential after 1 kbp | High instrument cost and high error rates |
| Oxford Nanopore Technologies | MinION | Real-time nanopore exonuclease sequencing | 5400–10,000 | NA | Vary according to the experimental need | >10 GB | 96% | 1000 access fee | Very long read length and less costly | 4% error rate |

Ref. Metzker (2010), Pareek et al. (2011), Liu et al. (2012), Lee et al. (2013), Oxford Nanopore Technologies, UK

**Acknowledgements** Authors thank DST-SERB for the Grant No. SB/YS/LS- 79/2013, "Development of endophytic bacterial consortium from selected medicinal plants of Western Ghats of India," and VELS University for their support in this work.

## References

Anderson MW, Schrijver I (2010) Next generation DNA sequencing and the future of genomic medicine. Genes 1:38–69. doi:10.3390/genes1010038

Bentley DR, Balasubramanian S, Swerdlow HP et al (2008) Accurate whole human genome sequencing using reversible terminator chemistry. Nature 456:53–59. doi:10.1038/nature07517

Branton D, Deamer DW, Marziali A et al (2008) The potential and challenges of nanopore sequencing. Nat Biotechnol 26:1146–1153. doi:10.1038/nbt.1495

Braslavsky I, Hebert B, Kartalov E, Quake SR (2003) Sequence information can be obtained from single DNA molecules. Proc Natl Acad Sci U S A 100:3960–3964. doi:10.1073/pnas.0230489100

Butler J, MacCallum I, Kleber M et al (2008) ALLPATHS: de novo assembly of whole-genome shotgun microreads. Genome Res 18:810–820. doi:10.1101/gr.7337908

Chaisson MJ, Pevzner PA (2008) Short read fragment assembly of bacterial genomes. Genome Res 18:324–330. doi:10.1101/gr.7088808

Chen K, Wallis JW, McLellan MD et al (2009) BreakDancer: an algorithm for high-resolution mapping of genomic structural variation. Nat Methods 6:677–681. doi:10.1038/nmeth

Deschamps S, Campbell MA (2010) Utilization of next-generation sequencing platforms in plant genomics and genetic variant discovery. Mol Breed 25:553–570. doi:10.1007/s11032-009-9357-9

Dohm JC, Lottaz C, Borodina T, Himmelbauer H (2007) SHARCGS, a fast and highly accurate shortread assembly algorithm for de novo genomic sequencing. Genome Res 17:1697–1706. doi:10.1101/gr.6435207

Dressman D, Yan H, Traverso G et al (2003) Transforming single DNA molecules into fluorescent magnetic particles for detection and enumeration of genetic variations. Proc Natl Acad Sci 100:8817–8822. doi:10.1073/pnas.1133470100

Eid J, Fehr A, Gray J et al (2009) Real-time DNA sequencing from single polymerase molecules. Science 323:133–138. doi:10.1126/science.1162986

Ewing B, Green P (1998) Base-calling of automated sequencer traces using phred. II. Error probabilities. Genome Res 8:186–194. doi:10.1101/gr.8.3.186

Froehlich T, Heindl D, Roesler A (2010) Miniaturized, high-throughput nucleic acid analysis. U.S. Patent 20100248237 A1. doi:10.1002/jctb.1421

Grabherr MG, Haas BJ, Yassour M et al (2011) Full-length transcriptome assembly from RNA-Seq data without a reference genome. Nat Biotechnol 29:644–652. doi:10.1038/nbt.1883

Gupta PK (2008) Single-molecule DNA sequencing technologies for future genomics research. Trends Biotechnol 26:602–611. doi:10.1016/j.tibtech.2008.07.003

Hall N (2007) Advanced sequencing technologies and their wider impact in microbiology. J Exp Biol 210:1518–1525. doi:10.1242/jeb.001370

Harris T, Buzby P, Jarosz M et al (2007) Optical train and method for tirf single molecule detection and analysis. U.S. Patent 20070070349

Hernandez D, Francois P, Farinelli L et al (2008) De novo bacterial genome sequencing: millions of very short reads assembled on a desktop computer. Genome Res 18:802–809. doi:10.1101/gr.072033

http://www.lifetechnologies.com/

http://www.nanoporetech.com/

http://www.roche-applied-science.com/

Hui P (2014) Next generation sequencing: chemistry, technology and applications. Top Curr Chem 336:1–18. doi:10.1007/128_2012_329

Illumina (2010) Technology spotlight: Illumina® sequencing

Jeck WR, Reinhardt JA, Baltrus DA et al (2007) Extending assembly of short DNA sequences to handle error. Bioinformatics 23:2942–2944. doi:10.1093/bioinformatics/btm451

Kang XF, Gu LQ, Cheley S, Bayley H (2005) Single protein pores containing molecular adapters at high temperatures. Angew Chem Int Ed Engl 44(10):1495–1499. doi:10.1002/anie.200461885

Korbel JO, Abyzov A, Mu XJ et al (2009) PEMer: a computational framework with simulation-based error models for inferring genomic structural variants from massive paired-end sequencing data. Genome Biol 10:R23. doi:10.1186/gb-2009-10-2-r23

Korlach J, Marks PJ, Cicero RL et al (2008) Selective aluminum passivation for targeted immobilization of single DNA polymerase molecules in zero-mode waveguide nanostructures. Proc Natl Acad Sci U S A 105(4):1176–1181. doi:10.1073/pnas.0710982105

Korlach J, Bjornson KP, Chaudhuri BP et al (2010) Real-time DNA sequencing from single polymerase molecules. Methods Enzymol 472:431–455. doi:10.1016/S0076-6879(10)72001-2

Ku CS, Roukos DH (2013) From next-generation sequencing to nanopore sequencing technology: paving the way to personalized genomic medicine. Expert Rev Med Devices 10:1–6. doi:10.1586/erd.12.63

Lee W, Zhang Y, Mukhyala K et al (2009) Bi-directional SIFT predicts a subset of activating mutations. PLoS One 4:e8311. doi:10.1371/journal.pone.0008311

Lee CY, Chiu YC, Wang LB et al (2013) Common applications of next-generation sequencing technologies in genomic research. Transl Cancer Res 2:33–45. doi:10.3978/j.issn.2218-676X.2013.02.09

Li H, Durbin R (2009) Fast and accurate short read alignment with Burrows-Wheeler Transform. Bioinformatics 25:1754–1760. doi:10.1093/bioinformatics/btp324

Li H, Ruan J, Durbin R (2008a) Mapping short DNA sequencing reads and calling variants using mapping quality scores. Genome Res 18:1851–1858. doi:10.1101/gr.078212

Li R et al (2008b) SOAP: short oligonucleotide alignment program. Bioinformatics 24 (5):713–714. doi:10.1093/bioinformatics/btn025

Liu L, Li Y, Li S et al (2012) Comparison of next-generation sequencing systems. J Biomed Biotechnol 2012:251364. doi:10.1155/2012/251364

Manrao E, Derrington I, Pavlenok M et al (2011) Nucleotide discrimination with DNA immobilized in the MspA nanopore. PLoS One 6:e25723. doi:10.1371/journal.pone.0025723

Mardis ER (2008a) Next-generation DNA sequencing methods. Annu Rev Genomics Hum Genet 9:387–402. doi:10.1146/annurev.genom.9.081307.164359

Mardis ER (2008b) The impact of next-generation sequencing technology on genetics. Trends Genet 24:133–141. doi:10.1016/j.tig.2007.12.007

Mardis ER (2013) Next-generation sequencing platforms. Annu Rev Anal Chem 6:287–303. doi:10.1146/annurev-anchem-062012-092628

Margulies M, Egholm M, Altman WE et al (2005) Genome sequencing in open microfabricated high-density picoliter reactors. Nature 437:376–380. doi:10.1038/nature03959

Metzker ML (2010) Sequencing technologies—the next generation. Nat Rev Genet 11:31–46. doi:10.1038/nrg2626

Morozova O, Marra MA (2008) Applications of next-generation sequencing technologies in functional genomics. Genomics 92:255–264. doi:10.1016/j.ygeno.2008.07.001

Mortazavi A, Williams BA, McCue K et al (2008) Mapping and quantifying mammalian transcriptomes by RNA-Seq. Nat Methods 5:621–628. doi:10.1038/nmeth.1226

Niedringhaus TP, Milanova D, Kerby MB et al (2011) Landscape of next-generation sequencing technologies. Anal Chem 83:4327–4341. doi:10.1021/ac2010857

Ning Z, Cox AJ, Mullikin JC (2001) SSAHA: a fast search method for large DNA databases. Genome Res 11(10):1725–1729. doi:10.1101/gr.194201

Nyren P, Pettersson B, Uhlen M (1993) Solid phase DNA mini sequencing by an enzymatic luminometric inorganic pyrophosphate detection assay. Anal Biochem 208:171–175. doi:10.1006/abio.1993.1024

Pareek CS, Smoczynski R, Tretyn A (2011) Sequencing technologies and genome sequencing. J Appl Genet 52:413–435. doi:10.1007/s13353-011-0057-x

Pennisi E (2012) Genome sequencing. Search for pore-fection. Science 336(6081):534–537. doi:10.1126/science.336.6081.534

Pushkarev D, Neff NF, Quake SR (2009) Single-molecule sequencing of an individual human genome. Nat Biotechnol 27:847–850. doi:10.1038/nbt.1561

Robertson G, Schein J, Chiu R et al (2010) De novo assembly and analysis of RNA-seq data. Nat Methods 7:909–912. doi:10.1038/nmeth.1517

Ronaghi M, Karamohamed S, Pettersson B et al (1996) Real-time DNA sequencing using detection of pyrophosphate release. Anal Biochem 242:84–89. doi:10.1006/abio.1996.0432

Ronaghi M, Uhlen M, Nyren P (1998) A sequencing method based on real-time pyrophosphate. Science 281(5375):363–365. doi:10.1126/science.281.5375.363

Rothberg JM, Hinz W, Rearick TM et al (2011) An integrated semiconductor device enabling non-optical genome sequencing. Nature 475:348–352. doi:10.1038/nature10242

Rumble SM, Lacroute P, Dalca AV et al (2009) SHRiMP: accurate mapping of short color-space reads. PLoS Comput Biol 5:e1000386. doi:10.1371/journal.pcbi.1000386

Sanger F, Coulson AR (1975) A rapid method for determining sequences in DNA by primed synthesis with DNA polymerase. J Mol Biol 94:441–448. doi:10.1016/0022-2836(75)90213-2

Sanger F, Nicklen S, Coulson AR (1977) DNA sequencing with chain-terminating inhibitors. Proc Nat Acad Sci U S A 74:5463–5467. doi:10.1073/pnas.74.12.5463

Shendure J, Ji H (2008) Next-generation DNA sequencing. Nat Biotechnol 26:1135–1145. doi:10.1038/nbt1486

Shendure J, Porreca GJ, Reppas NB et al (2005) Accurate multiplex polony sequencing of an evolved bacterial genome. Science 309:1728–1732. doi:10.1126/science.1117389

Shokralla S, Spall JL, Gibson JF, Hajibabaei M (2012) Next-generation sequencing technologies for environmental DNA research. Mol Ecol 21:1794–1805. doi:10.1111/j.1365-294X.2012.05538.x

Smith DR, Quinlan AR, Peckham HE et al (2008a) Rapid whole-genome mutational profiling using next-generation sequencing technologies. Genome Res 18:1638–1642. doi:10.1101/gr.077776

Smith AD, Xuan Z, Zhang MQ (2008b) Using quality scores and longer reads improves accuracy of Solexa read mapping. BMC Bioinf 9:128. doi:10.1186/1471-2105-9-128

Steinbock LJ, Radenovic A (2015) The emergence of nanopores in next generation sequencing. Nanotechnology 26:074003. doi:10.1088/0957-4484/26/7/074003

Stone EA, Sidow A (2005) Physico-chemical constraint violation by missense substitutions mediates impairment of protein function and disease severity. Genome Res 15:978–986. doi:10.1101/gr.3804205

Sucher NJ, Hennell JR, Carles MC (2012) DNA fingerprinting, DNA barcoding, and next generation sequencing technology in plants. Methods Mol Biol 862:13–22. doi:10.1007/978-1-61779-609-8_2

Tawfik DS, Griffiths AD (1998) Man-made cell-like compartments for molecular evolution. Nat Biotechnol 16:652–656. doi:10.1038/nbt0798-652

Voelkerding KV, Dames SA, Durtschi JD (2009) Next-generation sequencing: from basic research to diagnostics. Clin Chem 55:641–658. doi:10.1373/clinchem.2008.112789

Warren RL, Sutton GG, Jones SJ, Holt RA (2007) Assembling millions of short DNA sequences using SSAKE. Bioinformatics 23:500–501. doi:10.1093/bioinformatics/btl629

Wu TD, Watanabe CK (2005) GMAP: a genomic mapping and alignment program for mRNA and EST sequences. Bioinformatics 21:1859–1875. doi:10.1093/bioinformatics/bti310

Yadav NK, Shukla P, Omer A et al (2014) Next generation sequencing: potential and application in drug discovery. Sci World J 2014:802437. doi:10.1155/2014/802437
Zerbino DR, Birney E (2008) Velvet: algorithms for de novo short read assembly using de Bruijn graphs. Genome Res 18:821–829. doi:10.1101/gr.074492.107
Zerbino R, Tobes JM, Bossio ME et al (2012) On the orientation of fibres in structural members fabricated with self-compacting fibre reinforced concrete. Cement Concrete Comp 34(2): 191–200

# Marine Polyextremophiles and Their Biotechnological Applications

**Bhagwan Rekadwad and Chandrahasya Khobragade**

**Abstract** This chapter describes the versatility of marine microorganisms. They have inherent ability to grow and thrive under polyextremes. The bioactive compounds such as hydrolases, unique pigments, alkaloids, peptides, colored antibiotics, exopolysaccharides, siderophores, ectoine, and proteins produced and released under stressful conditions have potential biotechnological applications especially in agriculture, food, health care, and medicine. We have also discussed the possible applications of polyextremophiles in the treatment of cancer and neurodegenerative diseases.

**Keywords** Extremophiles • Extremolytes • Microbial diversity • Secondary metabolites

## 1 Introduction

Marine extremophiles are the organisms that can thrive and reproduce at extremes of salt concentrations (salinity >1.0 M NaCl), pH (>8.0, <5.0), temperature (1–15 °C, >45 °C), and pressure (average 380 atmosphere, >500–1200 atmosphere and beyond), in the presence of high radiations, recalcitrant compounds, heavy metals, and inhibitors. Extremophiles belonging to the *Eubacteria*, *Archaea*, and eukaryotic kingdoms produce extremophilic biomass in ecological niches such as oceans, salt marshes, solar salterns, hypersaline lakes, hot springs, marine hydrothermal vents, and soda lakes. These marine polyextremophiles have great importance and contributed a lot in biotechnological industries. The bioactive compounds such as extremozymes, proteins, and extremolytes are exploited in various bioprocesses and industries. But, it remains to uncover their potential biotechnological applications in health care, food, and agriculture. Very few research groups worldwide are working on molecular mechanisms underlying the potential of such applications (Table 1). The present chapter highlights the applications of marine polyextremophiles.

B. Rekadwad (✉) • C. Khobragade
School of Life Sciences, Swami Ramanand Teerth Marathwada University, Nanded 431606, Maharashtra, India
e-mail: rekadwad@gmail.com



V.C. Kalia, P. Kumar (eds.), *Microbial Applications Vol.1*,
DOI 10.1007/978-3-319-52666-9_15

**Table 1** Polyextremophiles/extremophiles: habitats, survival and defensive strategies, and bioactive compounds and their potential applications in biotechnology

| Extremophiles | Habitat/source/requirement for growth | Survival and defensive strategies | Extremolytes/bioactive compounds | Representative applications | References |
|---|---|---|---|---|---|
| Hyperthermophiles | (a) Deep-sea hydrothermal vents<br>(b) Optimum growth at temperatures above 80 °C | Stabilization of enzymes from stress and freeze drying; protection of oxidative protein damage; reduction of VLS in immunotoxin therapy | Superoxide dismutase, DHAP-dependent aldolases | Dismutation or disproportionation of superoxide free radical anions<br>Synthesis of rare sugars and carbohydrates<br>Generation of potable alcohol, solvents, and acetic acid<br>Catalyze lignin degradation in methylotrophs | Valls and de Lorenzo (2002), Mergeay et al. (2003), Gomes and Steiner (2004), Irwin and Baird (2004), Davilla et al. (2008), Radianingtyas and Wright (2003), Zhu et al. (2013), Falcicchio et al. (2014), Niefar et al. (2015), Dalmaso et al. (2015) |
| Thermophiles | (a) Hot springs, sun-heated soil<br>(b) Artificially heated places such as compost piles, heaps, etc.<br>(c) Grows at temperatures between 45 and 80 °C | | Amylases, cellulase, endoglucanase, xylanase, chitinase, glycosidase, mannanase, proteases, lipases, esterase, DNA polymerases, phytase, phosphatase, lichenase<br>Whole microorganism | Starch processing, oligosaccharide synthesis, paper bleaching, food processing, detergents, genetic engineering<br>Ethanol production, bioremediation, and biomineralization | |
| Halophiles | (a) Oceans, salt marshes, salterns, hypersaline lakes<br>(b) Requires at least 1 M salt for growth | Protection of skin immune cells from UV radiation; enzyme stabilization against heating, freezing, and drying; protection of the skin barrier against water loss and drying out; block of UVA-induced ceramide release in human keratinocytes | Proteases, dehydrogenases, laccases, siderophores, compatible solutes | Peptide synthesis, biocatalysis in organic media, saline water treatment, pharmaceuticals, cosmetic additives | |

| | | | | | |
|---|---|---|---|---|---|
| Acidophiles | (a) Mines, mine drainage, deep-sea hydrothermal vents<br>(b) Optimum pH for growth—below 3 | Maintaining a circumneutral intracellular pH; constant pumping of protons in and out of cytoplasm; acidic polymers of the cell membrane; passive regulation of the cytoplasmic pools of polyamines, and low membrane permeability | Amylases, glucoamylases, proteases, cellulase, oxidases | Starch processing, feed component, desulfurization of coal | |
| Alkaliphiles | (a) Coastal regions, soda lakes<br>(b) Optimum pH for growth—above 8.0 | | Proteases, cellulase, amylases, lipases, cyclodextrinases | Polymer-degrading agents in detergents, food additives | |
| Piezophiles (barophiles) | (a) Oceans<br>(b) Grows under high pressure 380 atmosphere (38 MPa) and above | Homeoviscous adaptation, tight packing of their lipid membranes, and increased levels of unsaturated fatty acids; polyunsaturated fatty acids maintain the membrane fluidity; robust DNA repair systems; highly conserved pressure-regulated operons; presence of heat shock proteins | Whole organisms | Food processing and antibiotic production | |
| Radiophiles (radioresistant) | Tolerance to high doses of radiation | It's extremely efficient DNA repair mechanism enables it to reassemble the fragmented DNA in a relatively short time | Whole microorganism | Bioremediation of radionuclide-contaminated sites | |

(continued)

**Table 1** (continued)

| Extremophiles | Habitat/source/requirement for growth | Survival and defensive strategies | Extremolytes/bioactive compounds | Representative applications | References |
|---|---|---|---|---|---|
| Metalophiles (metallotolerant) | Tolerance to high levels of heavy metals | Transform certain metal species through oxidation, reduction, methylation, and alkylation. Apart from the enzymatic transformations that lead to metal precipitation and immobilization, other biological reactions that generate less poisonous metal species have been applied to bioremediation | Whole microorganism | Ore bleaching, bioremediation, and biomineralization | |
| Eurypsychrophiles (psychrotolerant) | (a) Sea and artificial low temperature places<br>(b) Grows at temperatures above 25 °C but also grow below 15 °C | Translation of cold-evolved enzymes; increased flexibility in the portions of protein structure; presence of cold shock proteins and nucleic acid-binding proteins; reduction in the packing of acyl chains in the cell membranes | Proteases, amylases, lipases, dehydrogenases | Polymer-degrading agents in detergents, biosensors | |
| Stenopsychrophiles (psychrophiles) | (a) Antarctica, deep-sea trenches, and artificial low temperature places<br>(b) Grows at temperatures between 10 and 20 °C | Compared to proteins from mesophiles, psychrophilic proteins show decreased ionic interactions and hydrogen bonds and possess less hydrophobic groups and more charged groups on their | | | |

| | | | | | |
|---|---|---|---|---|---|
| | | surface and longer surface loops. Due to these modifications, at low temperatures, psychrophilic proteins lose their rigidity and gain increased structural flexibility for enhanced catalytic function. As the psychrophilic membranes contain a higher proportion of unsaturated fatty acids, their fluidity and ability to transport nutrients are maintained under very cold conditions. Moreover, the ability to synthesize cold shock or antifreeze proteins as the temperature drops, the more efficient enzyme activity due to alterations in enzyme kinetics, and the stabilization of microtubules enable the psychrophiles to continue their activities | | | |

(continued)

Table 1 (continued)

| Extremophiles | Habitat/source/ requirement for growth | Survival and defensive strategies | Extremolytes/bioactive compounds | Representative applications | References |
|---|---|---|---|---|---|
| Endolithic | Grows inside rocks | Moisture present within the crusts form saturated NaCl solutions. This could be used by the endolithic microorganisms for primary productivity and growth | ? | | |
| Hipolith | Grows on rocks and cold deserts | ? | ? | ? | |
| Oligotroph | Able to grow in environments of scarce nutrients | ? | ? | ? | |
| Geophiles | Soil | Mucoidal layer enveloping cell colonies; biofilm formation as stress response to extreme environmental conditions | | | |
| Toxitolerant | Polluted sites | Tolerates high concentrations of toxic agents (e.g., organic solvents) | Whole microorganism | Removal and detoxification of organic solvent polluted sites such as industries | |
| Xerophiles | (a) Desserts<br>(b) Grows in low water availability, resistant to desiccation | Thick layer of cell wall | Whole microorganism | Made available water to the plants. Help in increased crop productivity. Genes can be transferred to other higher taxa | |

## 2 Hydrolases from Marine Microorganisms

The marine polyextremophiles were investigated for the production of hydrolases. These include amylases, cellulases, peptidases, xylanases, chitinases, pullulanases, beta-xylosidase, lipases, and phytases produced by hyperthermophiles, psychrophiles, halophiles, and piezophiles. These marine extremozymes are stable and function in harsh physicochemical conditions. These are useful in food, fodder, biofuel production, medicine, and pharmaceutical and fine chemical industries (Gomes and Steiner 2004; Dalmaso et al. 2015).

## 3 Bioactive Compound from Marine Microorganisms

Marine microorganisms are always attractive to science. They are capable of producing unique color pigments with broad-ranging pharmacological activities. These have industrial and commercial applications. Microorganisms that produced biologically active and unique compounds include marine *Bacillus*, *Pseudomonas*, *Pseudoalteromonas*, *Streptomyces*, *Vibrio*, and *Cytophaga* isolated from seawater and sediments from sea and coastal region and bacteria associated with marine algae (*Sargassum* and *Codium*). They have produced biotechnologically important products such as alkaloids (prodiginines and tambjamines), indole derivatives (quinines and violacein), macrolides, terpenoids, polyenes, and peptides (Soliev et al. 2011; Soria-Mercado et al. 2012).

Several red, violet, yellow, and red to pink pigments were isolated from marine bacteria. *Serratia marcescens* have produced prodiginines (red-pigmented prodigiosin compounds) as a secondary metabolite. The polyunsaturated hydrocarbon containing 40 carbon molecules is called as carotenes. It exhibits red to pink coloration due to the presence of a wide variety of isoprenoid compounds ($\beta$-carotene, lycopene, phytofluene, and phytoene). These carotenoid or carotenoid-like compounds were produced by marine microorganisms related to the *Cytophaga-Flavobacterium-Bacteroides* group. Similarly, *Salinibacter* has contributed a lot in the production of carotenes in salterns. The marine bacterium *Agrobacterium aurantiacum* have carotenoid biosynthesis gene cluster, which has a role in the production of pigment astaxanthin. *Paracoccus haeundaensis* is another astaxanthin producer isolated from the marine environment. Like *Paracoccus haeundaensis*, *Chromobacterium* has the ability to produce the violet pigment indole derivative – violacein. The pigments (prodiginines, carotenes, violacein, phenazine compounds, quinines, glycosylated and pigmented anthracycline antibiotics (fridamycin D, himalomycin A and B), tambjamines, melanins, and other pigmented compounds) produced by marine microorganisms are biologically active compounds. On the other hand, various deep-sea fungi such as *Acremonium*, *Alternaria*, *Aspergillus*, *Chaetomium*, *Cladosporium*, *Exophiala*, *Engyodontium*, *Fusarium*, *Phoma*, *Penicillium*, *Hormonema*, *Rhodosporidium*, *Rhodotorula*,

*Schizophyllum*, *Tilletiopsis*, *Tritirachium*, and *Sistotrema* produced polyketide compounds, steroid derivatives, indole derivatives, sesquiterpenoids, alkaloid compounds, aromatic compounds, pyrone analogues, sorbicillin derivative, breviane derivative, compounds containing amino acid structure, novel cyclopentenone, trichoderone, prenylxanthones, depsidone-based analogues, citromycetin analogue, diketopiperazine derivatives, hydroxyphenylacetic acid, and other compounds showing inhibitory activities. These are useful in health care, medicine, pharmaceuticals, and cosmetics as antibacterial, antiviral, antimalarial, antiplasmodial, antiprotozoal antibiotic, algicidal, immunosuppressant, anticancer, anti-inflammatory, antiproliferative, antioxidation, cytotoxic, and protecting agents from UV irradiation (Shieh et al. 2003; Yi et al. 2003; Matz et al. 2004; Lee et al. 2004; Zhang et al. 2005; Nakashima et al. 2005; Kim et al. 2007; Williamson et al. 2007; Feher et al. 2008; Yada et al. 2008; Becker et al. 2009; Mayer et al. 2010; Ahmad et al. 2013; Wang et al. 2015; Simon-Colin et al. 2015).

## 4 Exopolysaccharides from Marine Bacteria

The diversity of marine microorganisms producing exopolysaccharides (EPSs) is an unexplored area. Its detailed study may lead to the discovery of new molecules and biocatalysts useful in food products, human therapeutics, and pharmaceuticals. The marine EPS producers secrete capsular polymers in their surrounding environment. These secreted EPS polymers remain attached to the cell membrane through the lipopolysaccharides (LPSs) and give a slimy texture to the colonies of producers. This produced slimy LPS may be slowly dispersed into the environment. EPS produced by marine microorganisms combines with other bacterial polysaccharides such as alginate and chitosan which generate resistance to diseases in host. Additionally, it also increases adhesion to the surfaces, exhibits cell integrity, traps nutrients, and protects the host cells from the impact of toxic compounds and adverse freezing-like conditions (Nicolaus et al. 2010; Freitas et al. 2011; Donot et al. 2012; Mehta et al. 2014; Delbarre-ladrat et al. 2014; Finore et al. 2014).

## 5 Natural Bioactive Products from Marine Hydrothermal Vent Environments

Piezo-acido-hyperthermophiles and piezo-halo-psychrophiles, such as *Streptomyces*, *Micromonospora*, *Rhodococcus marinononascens*, *Bathymodiolus septemdierum*, *Thermococcus* S 557, *Methanococcus jannaschii*, *Bathymodiolus septemdierum*, *Halomonas* LOB-5, *Calyptogena soyoae*, *Thermovibrio ammonificans*, etc., are capable to produce thousands of biologically active neutral compounds. These polyextremophiles have produced microbial metabolites such as

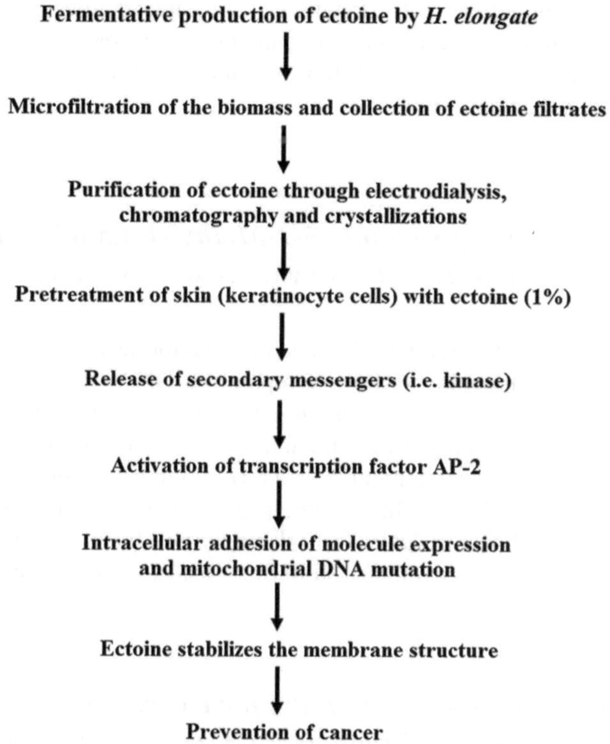


**Fig. 1** Proposed hypothetical mechanism mediated by extremolytes in *H. elongata* (adapted and modified from Copeland et al. 2013)

archaeal glycerol ethers, sterols, loihichelins (A–F amphiphilic peptidic siderophores), ammonificins, and amphiphilic siderophores. These produced industrially important microbial metabolites that are useful in the treatment of cancer (Fig. 1), Alzheimer's, Parkinson's, dementia, and other human diseases (Thornburg et al. 2015; Corinaldesi 2015).

## 6 Future Perspectives

It is vital that marine extremophiles cope and withstand under extreme harsh environmental conditions. They have developed defensive mechanisms to survive in extremes, and their metabolisms play key roles in survival processes. The adaptability of extremophiles arrives from their altered genes and protein, which enables marine extremophiles to produce extremolytes having potential biotechnological applications in the treatment of cancer and degenerative diseases (Alzheimer's, Parkinson's, and dementia) (Calderon et al. 2004; Kanapathipillai et al. 2005; Graf et al. 2008; Kuhlmann et al. 2011; Babu et al. 2015). The produced extremolytes help them to survive and function under harsh physicochemical conditions. Currently, the research is focused on and aiming the polyextremophiles, extremonelles (cell organelles of extremophiles such as mitochondria), and extremolytes' functions in damaging environments. The hypothetical survival mechanisms explain better to understand the survival mechanism of marine extremophiles (Fig. 1).

## 6.1 Use of Ectoine (5-2-Methyl-1,4,5,6-Tetra-Hydro-Pyridine-4-Carobylic Acid) in Cancer Treatment

Exposure to high level/dosage of radiation leads to alteration of DNA structure. If cellular machinery did not repair the DNA, it will produce cancer. Halophilic bacterium *Halobacter elongate* (*H. elongate*) has a mechanism of ectoine biosynthesis, which neutralizes the impact of high UV radiation exposure/dose. This has a role in cancer treatment (Fig. 1). *H. elongate* produce ectoine from aspartate semialdehyde (ASA). The immune-protective effects of ectoine which treat Langerhans cells and protect DNA from damage (i.e., from cancer) are explained using three-step processes summarized in Sect. 15.6.2 of this chapter.

## 6.2 Hypothetical Model for Development of Therapeutic Proteins/Products for Treatment of Neurodegenerative Diseases Using Extremophiles/Extremonelles/Extremolytes

The neurodegenerative diseases (Alzheimer's, Parkinson's, and dementia) are the causative for cell death. The cell death occurred due to oxidative stress (as a result, apoptosis and necrosis occur in healthy cells), which leads to the formation of deadly mitochondrial diseases (http://www.projectsmagazine.eu.com/randd_projects/mitochondrial_mechanisms_of_disease_lessons_from_extremophiles). In these types of mitochondrial diseases, the cell walls were ruptured. These lysed products of the cells will be studied for understanding mechanisms of cell death. On the other hand, investigations are in progress on extremophiles, extremonelles, and their stable extremolytes functioning under harsh environmental conditions. Further research is planned to study the ecology and physiology of extremophiles to understand the surviving properties of extremophiles. It is necessary to identify the macromolecules that make the mitochondria and other cellular organs tenacious to damages. The identification will be carried out for critical proteins and enzymes that avoid the cell death. Also, we have planned to focus and perform studies on synthesis of proteins and enzymes, bioassay, and sequencing of mRNA of protein of interest. This will allow developing of drugs that reduce the efficacy of cell death inducer proteins/enzymes/macromolecules. The developed drug will be studied for its efficacy, nontoxicity, and stability using cell lines. After successful trials on cell lines, the experiments will be planned to carry out on experimental animals.

Thus, the ectoine-mediated neutralization and developments of new drugs/macromolecules may reduce or prevent dehydration of skin and skin aging and may be used in treatments of neurodegenerative diseases such as Alzheimer's, Parkinson's, dementia, and Machado-Joseph disease.

**Conflicts of Interests** Author(s) declares there is no conflict of interests.

**Acknowledgment** The authors wish to thank **Dr. M. V. Deshpande**, Emeritus Scientist, Biochemical Division, CSIR-National Chemical Laboratory, Pune (India), for his valuable guidance. The words are insufficient to thank the **Dr. V. C. Kalia** (Chief Scientist, CSIR-Institute of Genomics and Integrative Biology; Professor, Academy of Scientific and Innovative Research, Delhi University Campus, Delhi (India), for his continuous support. BNR is thankful to the University Grants Commission for the financial support in the form of the postdoctoral fellowship (No. F. PDFSS-2013-14-ST-MAH-4350 – Website).

## References

Ahmad A, Usup G, Mohamad S, Yahya AN, Jafarzade M (2013) Isolation and characterization of pigmented bacteria showing antimicrobial activity from Malaysian marine environment. Malays J Microbiol 9:152–160 https://doaj.org/article/6a0cf251c31c4b0281a8d6500d07fee6

Babu P, Chandel AK, Singh PV (2015) Survival mechanisms of extremophiles, Chap. 2. In: Babu P (ed) Extremophiles and their applications in medical processes. Extremophilic bacteria. Springer International Publishing, Cham, pp 9–23. doi:10.1007/978-3-319-12808-5

Becker MH, Brucker RM, Schwantes CR, Harris RN, Minbiole KPC (2009) The bacterially produced metabolite violacein is associated with survival of amphibians infected with a lethal fungus. Appl Environ Microbiol 71:6635–6638. doi:10.1128/AEM.01294-09

Calderon MI, Vargas C, Rojo F, Iglesias-Guerra F, Csonka LN, Ventosa A, Nieto JJ (2004) Complex regulation of the synthesis of the compatible solute ectoine in the halophilic bacterium *Chromohalobacter salexigens* DSM 3043 T. Microbiology 150:3051–3063. doi:10.1099/mic.0.27122-0

Copeland WE, Wolke D, Angold A, MRCPsych, Costello EJ (2013) Adult psychiatric and suicide outcomes of bullying and being bullied by peers in childhood and adolescence. JAMA Psychiatry 70:419–426. doi:10.1001/jamapsychiatry.2013.504

Corinaldesi C (2015) New perspectives in benthic deep-sea microbial ecology. Front Mar Sci 2:17. doi:10.3389/fmars.2015.00017

Dalmaso GZL, Ferreira D, Vermelho AB (2015) Marine extremophiles: a source of hydrolases for biotechnological applications. Mar Drugs 13:1925–1965. doi:10.3390/md13041925

Davila AF, Gómez-Silva B, Asunción de los Rios, Ascaso C, Olivares H, McKay CP, Wierzchos J (2008) Facilitation of endolithic microbial survival in the hyperarid core of the Atacama Desert by mineral deliquescence. J Geophy Res 113:G01028. doi:10.1029/2007JG000561

Delbarre-ladrat C, Sinquin C, Lebellenger L, Zykwinska A, Colliec-jouault S (2014) Exopolysaccharides produced by marine bacteria and their applications as glycosaminoglycan-likemolecules. Front Chem 2:85. doi:10.3389/fchem.2014.00085

Donot F, Fontana A, Baccou JC, Schorr-Galindo S (2012) Microbial exopolysaccharides: main examples of synthesis, excretion, genetics and extraction. Carbohydr Polym 87:951–962. doi:10.1016/j.carbpol.2011.08.083

Falcicchio P, Wolterink Van-Loo S, Franseen MCR, van deer Oost J (2014) DHAP-dependent aldolases from (hyper) thermophiles: biochemistry and applications. Extremophiles 18:1–13. doi:10.1007/s00792-013-0593-x

Feher D, Barlow RS, Lorenzo PS, Hemscheidt TK (2008) A 2-substituted prodiginine, 2-(*p*-hydroxybenzyl) prodigiosin, from *Pseudoalteromonas rubra*. J Nat Pro 71:1970–1972. doi:10.1021/np800493p

Finore I, Didonato P, Mastascusa V, Nicolaus B, Poli A (2014) Fermentation technologies for the optimization of marine microbial exopolysaccharide production. Mar Drugs 12:3005–3024. doi:10.3390/md12053005

Freitas F, Alves VD, Reis MAM (2011) Advances in bacterial exopolysaccharides: from production to biotechnological applications. Trends Biotechnol 29:388–398. doi:10.1016/j.tibtech.2011.03.008

Gomes J, Steiner W (2004) The biocatalytic potential of extremophiles and extremozymes. Food Technol Biotechnol 42:223–235 http://www.ftb.com.hr/images/pdfarticles/2004/October-December/42-223.pdf

Graf R, Anzali S, Buenger J, Pfluecker F, Driller H (2008) The multifunctional role of ectoine as a natural cell protectant. Clin Dermatol 26:326–333. doi:10.1016/j.clindermatol.2008.01.002

Irwin JA, Baird AW (2004) Extremophiles and their application to veterinary medicine. Ir Vet J 57:348–354. doi:10.1186/2046-0481-57-6-348

Kanapathipillai M, Lentzen G, Sierks M, Park CB (2005) Ectoine and hydroxyectoine inhibit aggregation and neurotoxicity of Alzheimer's b-amyloid. FEBS Lett 579:4775–4780. doi:10.1016/j.febslet.2005.07.057

Kim D, Lee JS, Park YK et al (2007) Biosynthesis of antibiotic prodiginines in the marine bacterium *Hahella chejuensis* KCTC 2396. J Appl Microbiol 102:937–944. doi:10.1111/j.1365-2672.2006.03172.x

Kuhlmann AU, Hoffmann T, Bursy J, Jebbar M, Bremer E (2011) Ectoine and hydroxyectoine as protectants against osmotic and cold stress: uptake through the SigB-controlled Betaine-Choline-Carnitine transporter-type carrier EctT from *Virgibacillus pantothenticus*. J Bacteriol 193:4699–4708. doi:10.1128/JB.05270-11

Lee JH, Kim YS, Choi TJ, Lee WJ, Kim YT (2004) *Paracoccus haeundaensis* sp. nov. a gram-negative, halophilic, astaxanthin-producing bacterium. Int J Syst Evol Microbiol 54:1699–1702. doi:10.1099/ijs.0.63146-0

Matz C, Deines P, Boenigk J, Arndt H, Eberl L, Kjelleberg S, Jürgens K (2004) Impact of violacein producing bacteria on survival and feeding of bacterivorous nanoflagellates. Appl Environ Microbiol 70:1593–1599. doi:10.1128/AEM.70.3.1593-1599.2004

Mayer AM, Glaser KB, Cuevas C, Jacobs RS, Kem W, Little RD, McIntosh JM, Newman DJ, Potts BC, Shuster DE (2010) The odyssey of marine pharmaceuticals: a current pipeline perspective. Trends Pharmacol Sci 31:255–265. doi:10.1016/j.tips.2010.02.005

Mehta A, Sidhu C, Pinnaka AK, Roy Choudhury A (2014) Extracellular polysaccharide production by a novel osmotolerant marine strain of *Alteromonas macleodii* and its application towards biomineralization of silver. PLoS One 9:e98798. doi:10.1371/journal.pone.0098798

Mergeay M, Monchya S, Vallaeys T, Auquier A, Benotman A, Bertin P, Taghavi S, Dunn J, van der Lelie D, Wattiez R (2003) *Ralstonia metallidurans*, a bacterium specifically adapted to toxic metals: towards a catalogue of metal-responsive genes. FEMS Microbiol Rev 27:385–410. doi:10.1016/S0168-6445(03)00045-7

Nakashima T, Kurachi M, Kato Y, Yamaguchi K, Oda T (2005) Characterization of bacterium isolated from the sediment at coastal area of Omura bay in Japan and several biological activities of pigment produced by this isolate. Microbiol Immunol 49:407–415. doi:10.1111/j.1348-0421.2005.tb03744.x

Neifar M, Maktouf S, Ghorbel RE, Jaouani A, Cherif A (2015) Extremophiles as source of bioactive compounds with industrial applications, Chap. 10. In: Gupta VK, Tuohy MG, Lohani M, O'Donovan A (eds) Biotechnology and bioactive compounds: sources and applications, 1st edn. Wiley, Hoboken, NJ. doi:10.1002/9781118733103.ch10

Nicolaus B, Kambourova M, Oner ET (2010) Exopolysaccharides from extremophiles: from fundamentals to biotechnology. Environ Technol 31:1145–1158. doi:10.1080/09593330903552094

Radianingtyas H, Wright FC (2003) Alcohol dehydrogenases from thermophilic and hyperthermophilic archaea and bacteria. FEMS Microbiol Rev 27:593–616. doi:10.1016/S0168-6445(03)00068-8

Shieh WY, Chen YW, Chaw SM, Chiu HH (2003) *Vibrio ruber* sp. nov. a red, facultatively anaerobic, marine bacterium isolated from sea water. Int J Syst Evol Microbiol 53:479–484. doi:10.1099/ijs.0.02307-0

Simon-Colin C, Gueguen Y, Bachere E, Kouzayha A, Saulnier D, Gayet N, Guezennec J (2015) Use of natural antimicrobial peptides and bacterial biopolymers for cultured pearl production. Mar Drugs 13:3732–3744. doi:10.3390/md13063732

Soliev A, Hosokawa K, Enomoto K (2011) Bioactive pigments from marine bacteria: applications and physiological roles. Evid Based Complement Alternat Med 670349:17. doi: 10.1155/2011/670349

Soria-Mercado IE, Villarreal-Gómez LJ, Rivas GG, Sánchez NEA (2012) Bioactive compounds from bacteria associated to marine algae, Chap. 3. In: Sammour RH (ed) Biotechnology—molecular studies and novel applications for improved quality of human life. InTech, Rijeka, ISBN: 978-953-51-0151-2. doi:10.5772/27842

Thornburg CC, Zabriskie TM, McPhail KL (2015) Deep-sea hydrothermal vents: potential hot spots for natural products discovery? J Nat Prod 73(3):489–499. doi:10.1021/np900662k

Valls M, de Lorenzo V (2002) Exploiting the genetic and biochemical capacities of bacteria for the remediation of heavy metal pollution. FEMS Microbiol Rev 26:327–338. doi:10.1016/S0168-6445(02)00114-6

Wang YT, Xue YR, Liu CH (2015) A brief review of bioactive metabolites derived from deep-sea fungi. Mar Drugs 13:4594–4616. doi:10.3390/md13084594

Williamson NR, Fineran PC, Gristwood T, Chawrai SR, Leeper FJ, Salmond GPC (2007) Anticancer and immunosuppressive properties of bacterial prodiginines. Future Microbiol 2:605–618. doi:10.2217/17460913.2.6.605

Yada S, Wang Y, Zou Y, Nagasaki K, Hosokawa K, Osaka I, Arakawa R, Enomoto K (2008) Isolation and characterization of two groups of novel marine bacteria producing violacein. Mar Biotechnol 10:128–132. doi:10.1007/s10126-007-9046-9

Yi H, Chang YH, Oh HW, Bae KS, Chun J (2003) *Zooshikella ganghwensis* gen. nov. sp. nov. isolated from tidal flat sediments. Int J Syst Evol Microbiol 53:1013–1018. doi:10.1099/ijs.0.02521-0

Zhang L, An R, Wang J, Sun N, Zhang S, Hu J, Kuai J (2005) Exploring novel bioactive compounds from marine microbes. Curr Opin Microbiol 8:276–281. doi:10.1016/j.mib.2005.04.008

Zhu H, Liu J, Qu J, Gao X, Pan T, Cui Z, Zhao X, Lu JR (2013) Stress fermentation strategies for production of hyperthermostable superoxide dismutase from *Thermus thermophilus* HB27: effects of ions. Extremophiles 17:995–1002. doi:10.1007/s00972-013-0581-1

MIX
Papier aus verantwortungsvollen Quellen
Paper from responsible sources
FSC® C105338

If you have any concerns about our products,
you can contact us on
**ProductSafety@springernature.com**

In case Publisher is established outside the EU,
the EU authorized representative is:
**Springer Nature Customer Service Center GmbH**
**Europaplatz 3, 69115 Heidelberg, Germany**

Printed by Libri Plureos GmbH
in Hamburg, Germany